# GLOBAL POSITIONING SYSTEM

*Papers*

*published in*

**NAVIGATION**

VOLUME III



*reprinted by*
THE INSTITUTE OF NAVIGATION
Washington, D.C.

*International Standard Book Number:* 0-936406-02-1

Published in the United States of America
The Institute of Navigation
Suite 832
815 Fifteenth Street, N.W.
Washington, D.C. 20005

# PREFACE

It is indeed a rare experience to have been a participant in the formative stages of a development program such as the NAVSTAR Global Positioning System; to have been intimately involved in the struggle to keep a program such as this alive through two administrations; to watch the fruits of these labors capture the imagination of the user community; and now to participate as Editor of *Navigation,* Journal of The Institute of Navigation in publication of this, the third volume of papers devoted entirely to GPS.

Throughout this same period, The Institute of Navigation, through its Technical and Annual Meetings has continued to represent the principal forum for exchange of information on the progress of the GPS program and enlightened dialog among and between the civilian and military user communities, industry and other government agencies.

Volume I of this series was first published in 1980, containing papers from the Summer 1978 issue of *Navigation* (Vol. 25, No. 2), which was dedicated to NAVSTAR GPS and several additional papers published in subsequent editions of the Journal. Now in its third printing, Volume I was supplemented in 1983 by Volume II, which contained a collection of papers published between 1980 and 1983.

As originally conceived, Volume III was to ve devoted entirely to Differential GPS, with specific emphasisi on civil use of the system. Dr. Rudolph M. Kalafus, Chairman of the Radio Technical Commission for Maritime Service (RTCM) Special Committee 104 on Differential GPS and a long-time proponent of widespread civil use of NAVSTAR GPS was chosen by The Institute as Guest Editor for this Volume. As we worked together to develop the theme, it became clear that the scope of Volume III would have to be expanded beyond Differential GPS to include many important papers published in *Navigation* since 1984 related to GPS receiver design, integration with other sensors and up-to-date information on overall GPS operation.

It is hoped that the expanded work will become a ready reference and companion to Volumes I and II.

Stephen W. Gilbert
Editor, *Navigation*

## CONTENTS

# Overview

RUDOLPH M. KALAFUS

U.S. Department of Transportation, TSC, Cambridge, MA

This is the third volume of papers published in *Navigation* on the subject of the Global Positioning System. Since the first volume was published there has been a great deal of effort devoted to exploring new civil applications of the GPS, developing differential techniques, and pushing the system capability to the limit in a number of different ways by combining the satellite measurements with other sensor inputs. Civil agencies are now making serious plans to utilize GPS to generate considerable commercial civil receiver activity. There are a number of companies now preparing receiver designs for what promises eventually to be a large market for a variety of land, marine and aeronautical applications.

This widespread interest has resulted in numerous papers on GPS in the Proceedings of Meetings of the Institute and in *Navigation*. Some of these papers are presented here, divided into the following topics:

- Receiver Design Enhancements
- Differential Operation
- Combined Sensor Operation
- System Operation and Applications

*Receiver Design Enhancements*

Receiver designs aimed at civil markets are being explored in depth to obtain high performance and flexibility in specific applications, and to reduce production costs. Carrier phase measurements are being used in a number of novel ways to improve the performance of both stand-alone receivers, and in relative positioning with both static and dynamic platforms. Methods are being developed to contend with short-term signal fades and bad geometries that can occur in practice. Highly sophisticated military user equipment has been developed for a wide range of applications. The first group of papers addresses these issues.

*Differential Operation*

Differential operation enables users of the Standard Positioning Service to achieve very precise position and navigation measurements with GPS. Standards are being developed for the broadcasts of differential corrections. The pseudolite embodiment of differential GPS holds the promise of simplifying the design of the data link required and providing additional lines of position which can significantly improve service reliability. The codeless version of differential GPS has the attractive feature of obtaining two-frequency measurements with the precision of the Precise Positioning System. Test programs have supported the claims of differential operation. For surveying applications, carrier phase processing is now being used to obtain relative positioning accuracies measured in centimeters. It is anticipated that techniques will eventually be developed to obtain reliable, real-time accuracies of less than a meter in three dimensions for dynamic platforms.

*Combined Sensor Operation*

Combining other navigation sensor information with GPS measurements solves a number of operational problems for some applications. For the next few years, until a full constellation has been deployed, coverage will be intermittent. During the buildup phase, there will be coverage gaps. Even when the full constellation becomes operational, there will be a number of conditions that can occur which result in service outages for unaided GPS receivers. The third group of papers discusses some promising techniques.

*System Operation and Applications*

The last three papers provide information on other important topics. One describes the performance of the GPS control and satellite segments, and documents the high satellite clock stabilities. Another gives a design for a national or continental monitoring system to provide immediate integrity information for aeronautical users. The final paper describes the results of some tests of a GPS receiver in an urban environment.

It is evident that civil user equipment has reached a new stage of maturity, one where the concern for solving the intricacies of the receiver design to get the set to work, is being replaced by the search for methods of enhancing performance and tailoring the designs to particular applications. This trend can be expected to continue and, as engineers strive for even higher accuracies and reliabilities, effects which were previously negligible, such as multipath, will pose many challenging problems in the next few years.

# A Comparison of Multichannel, Sequential and Multiplex GPS Receivers for Air Navigation

ROBERT A. MAHER

Texas Instruments Incorporated, Lewisville, Texas
*Received January 1984*

## ABSTRACT

As the NAVSTAR Global Positioning System (GPS) begins moving beyond the initial development stage, potential air navigation users are going to have a considerable range of choices for set designs. This paper describes three of the main receiver architecture approaches: multichannel, sequential, and multiplex (MUX). These major options are compared according to functional capabilities, effects of dynamics on navigation accuracy, and relative cost. Oscillator *g*-loading effects are addressed. The most appropriate choice for a given military or civil air navigation application likely involves a judicious combination of the various basic set architectures. A general familiarity with NAVSTAR/GPS is assumed.

## INTRODUCTION

The NAVSTAR GPS has been under development for over 10 years. The performance capabilities of the system have been demonstrated with the initial space vehicle (SV) constellation. The mid-1980s will experience a rapid development of various types of GPS sets. Because the system will be relatively new from the perspective of the growing community of potential users, there can naturally be some confusion over the appropriateness and relative cost effectiveness of the different types of GPS sets that will become available.

This paper examines three of the rather traditional approaches to configuring GPS receivers: multichannel, sequential, and multiplex. The three types are described and compared from certain performance and relative cost perspectives. Example configurations are identified for both civil and military air navigation applications. As set-building technique develops and improves, variants of these three basic set types are appearing. It is hoped that the comparisons in this paper will help in assessing both presently available and future receiver types.

## BASIC GPS RECEIVER ARCHITECTURES

For most GPS receiver applications, it is necessary to track or make measurements on signals from at least four SVs.* This four-SV requirement is

*Exceptions to this general requirement for four SVs include known-altitude situations, time transfer situations with known location, interferometric use, and determination of satellite orbits that are substantially inertial.

particularly true for most air navigation applications. Some of the receiver hardware for tracking multiple SVs is naturally common to all SVs. During and after code correlation and Doppler removal on behalf of a particular SV, however, the hardware becomes specialized, at any given time, to only one SV. This SV-specific hardware, referred to as "channel hardware," is shown in Figure 1.

One way to deal with these multi-SV requirements is simply to replicate this channel hardware as many times as the number of SVs that are desired to track. Figure 2 shows which hardware elements are typically common, and which hardware elements must be replicated per receiver channel in a multichannel GPS set. Four such channels can produce a complete navigation solution. Usually, a fifth channel is provided to enable acquisition of a new SV without disturbing the original four or to allow simultaneous measurements on $L_2$.

An alternative to replicating the hardware is to somehow time-share this hardware, rather than to physically replicate it. A time-sharing approach is, of course, attractive in that cost, weight, volume, and power consumption can all be reduced. There are two sharply distinct approaches to this time sharing: sequencing and multiplexing. Each of these two approaches to saving hardware carries its own compromises with respect to full multichannel operation. These performance compromises are discussed in more detail after the sequencing and multiplexing concepts are first described.

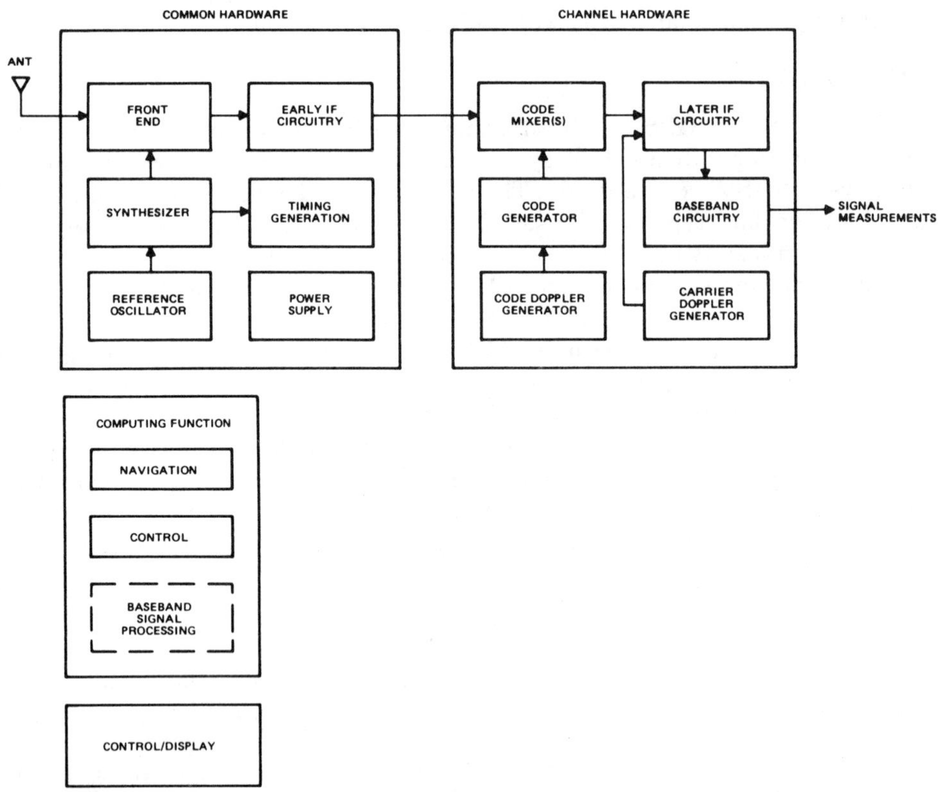


*Fig. 1—Typical GPS hardware functions.*

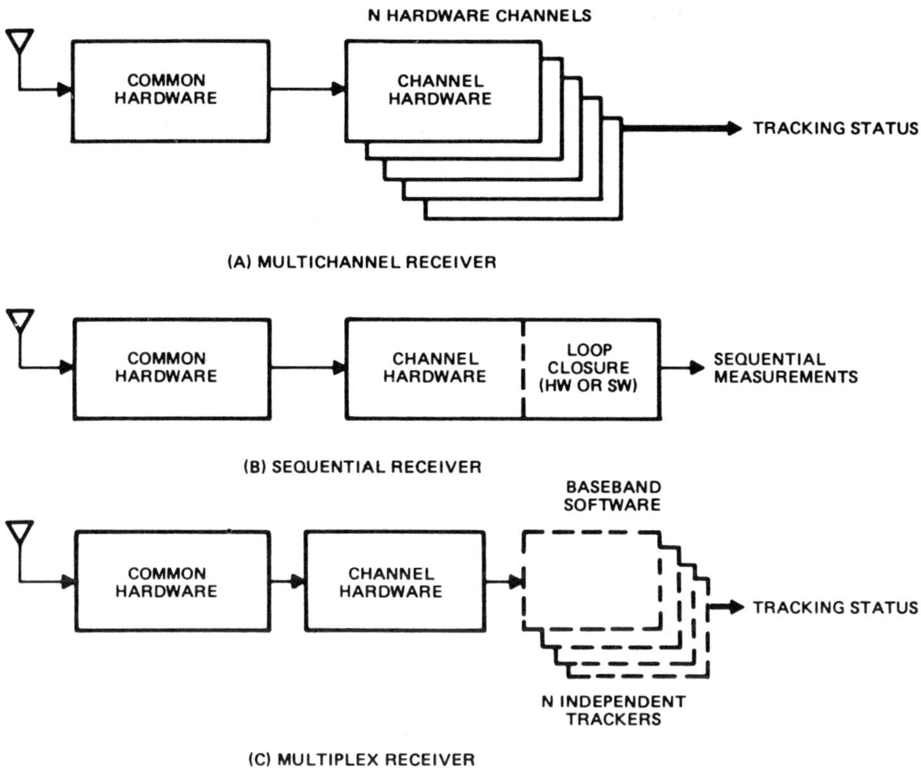


*Fig. 2—GPS receiver types.*

A typical sequencing receiver moves from SV to SV with a dwell on a given SV lasting about 1 second. Hence, four SVs would be cycled through in about 4 seconds. With a typical single-channel sequencing receiver that allows for acquiring a fifth SV, each SV is sampled, or revisited, once every 5 seconds (Figure 3).

A multiplexing receiver cycles through the SVs at a far higher rate, as shown in Figure 4. A typical cycle rate, or sample rate per SV, is 50 Hz, tied to the SV data rate. This sample rate is about 200 times faster than is the case for a single-channel sequencing receiver. This large difference in sample rate lies at the heart of the difference between the sequencing and multiplexing receivers.

To understand the significance of this difference, it is instructive to review a basic principle of sampled-data control systems. (The code and carrier tracking loops in most GPS receivers are feedback control systems.) Stable operation of a feedback control system can be achieved as long as the sample rate is sufficiently greater than the bandwidth of the control loop. As a practical matter, the sample rate must be at least three times the loop bandwidth. A 15-Hz bandwidth third-order phase-lock loop can track the GPS $L_1$ carrier for the dynamics of a maneuvering high-performance aircraft; hence, a sample rate of at least 45 Hz would be necessary in that case. A second-order code-tracking loop not aided by a carrier loop would require a bandwidth of 3 to 6 Hz; with rate-aiding from a carrier loop, the code loop bandwidth can be less than 1 Hz, thus reducing pseudorange measurement noise.

Therefore, it can be seen that sampled data continuous tracking is possible

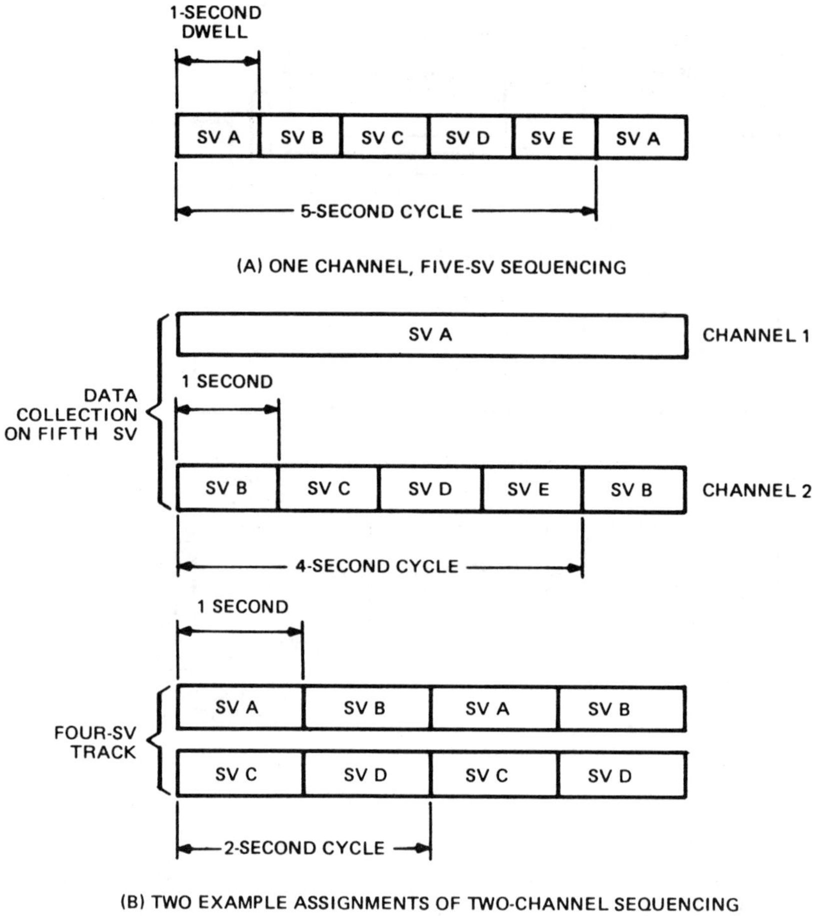


*Fig. 3—Typical timelines for sequential receivers.*

with a multiplex receiver for both carrier and code, and that continuous tracking of higher dynamics is not possible for either carrier or P-code for an unaided sequencing receiver. Rather than effecting sampled data continuous tracking, a sequencing receiver takes a different tack: toward the beginning of a new 1-second dwell on an SV, the receiver does a small search and acquisition of the code phase error and the carrier frequency/phase errors that had built up since the last dwell on this SV a few seconds ago. After these loops settle out, code and carrier measurements are made. The output of a sequencing receiver is, thus, a position and velocity measurement once every 4 or 5 seconds for each SV for a single-channel sequential/receiver (every 2 to 2.5 seconds for a two-channel sequencing receiver). Sequencing receivers with longer cycle times have been implemented. Carrier phase or Doppler count obviously cannot be maintained between dwells.

One interesting variant of sequencing receivers dwells only about 0.2 second on an SV, cycles through several SVs in a couple of seconds, and uses more robust but less accurate automatic frequency control (AFC) rather than phase lock for demodulating SV data.

While an SV is being tracked during a sequential receiver's dwell, the full

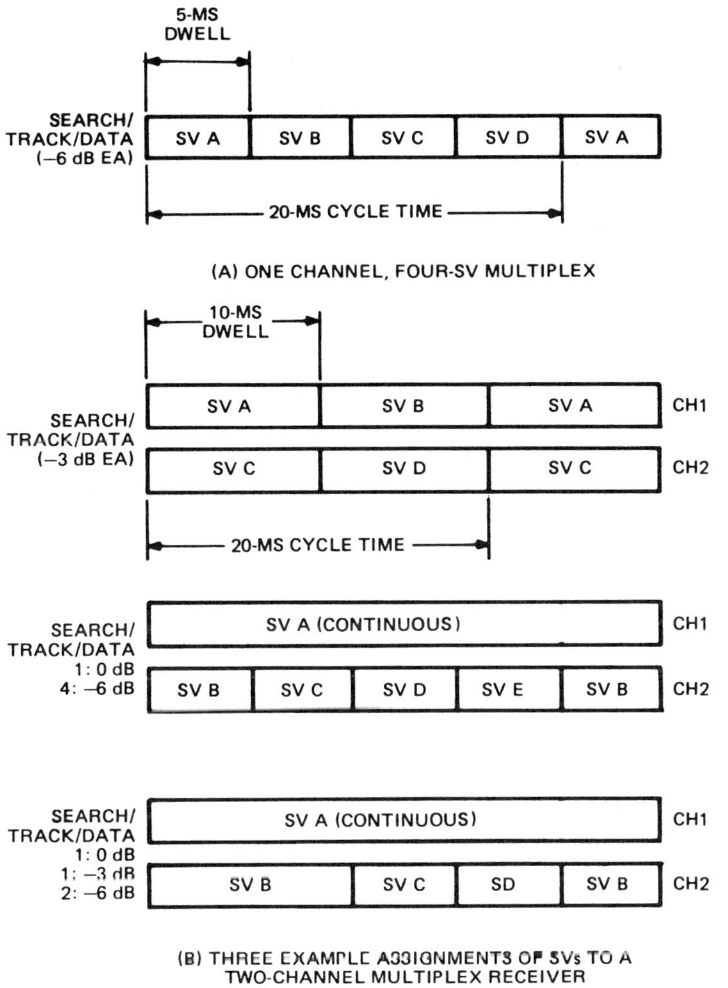


*Fig. 4—Typical timelines for multiplex receivers.*

signal-to-noise ratio (SNR) that would be available to a multi-channel set is available to the sequential receiver's loops. However, during dynamic tracking, the sequencing still suffers an SNR disadvantage relative to a multichannel continuous set. This disadvantage is approximately 2 to 8 dB for a single-channel sequencing set and approximately 1 to 4 dB for a two-channel sequencing set. The exact value is somewhat dependent on the set design, and is very highly dependent on the dynamics to be tracked. The mechanism for this degradation is subtle and concerns the need to reacquire the signal from dwell to dwell. Acquisition always requires more incident SNR than tracking. Whereas narrowing the acquisition bandwidth can reduce search measurement uncertainties, these narrower bandwidths lengthen the acquisition time necessary. Thus, the sequencing timeline is lengthened, the vehicle dynamics are more under-sampled, and the whole tracking situation starts diverging instead of converging. So, the acquisition bandwidths must be kept wider, with the attendant degradation of a few decibels of performance threshold for modest dynamics and many decibels for high dynamics.

During data collection, the sequencing operation of a single-channel sequencing set is interrupted to stay full time on the SV from which data are being collected (e.g., 6 seconds at a time). Because repeated reacquisition of this SV is not necessary during this data dwell, there is no SNR loss for data collection. This data collection, however, must be done during periods of low dynamics, or there will be significant disruptions of the tracking and navigation operations.

A multiplex receiver does maintain continuous sampled data tracking of carrier and code for each SV. Carrier cycle count and carrier phase are maintained indefinitely (unless loss of phase lock), just as in a multichannel set. Because not all the available signal is used during this continuous tracking process, a well-defined penalty in SNR is incurred relative to a multichannel set. This penalty is nominally 3 dB for two-SV multiplex, 6 dB for four-SV multiplex, and 7 dB for five-SV multiplex. The link margin available from the SVs allows this type of operation, and indeed, much successful testing has been done with four-SV multiplex.

A more complete description of multiplex receivers is given in references 1 and 2. Some of the more salient points are mentioned here:

- Code and carrier tracking loops are maintained for all SVs being tracked.
- The SV tracking status (line-of-sight position, velocity, acceleration) is maintained in software.
- At the start of each dwell on an SV, code position (integer and fractional chips) is calculated in software and loaded in the hardware along with code frequency.
- The desired code state is realized in the hardware in less than 400 microseconds. Mechanization of this function is one of the key challenges of multiplex receiver design, particularly for P-code.
- Carrier phase and frequency are also loaded into the hardware at the beginning of each dwell on an SV.
- Local multiplex timing is established and maintained in accordance with data timing of all the SVs. Choice of this timing, along with choice of permutation of SV assignments during the cycle time (20 milliseconds), is made so that data transitions never occur during the dwell on an SV.
- The multiplex timing can be adjusted from time to time, and the SV assignment rearranged, if necessary, without loss of phase tracking.
- A multiplex receiver is feasible only in the context of digital/software loop implementation, because of the required close control of the functions.
- Relative to a nonmultiplex multichannel digital receiver, a multiplex receiver needs 20 to 25 percent more processor throughput.
- For an authorized user, multiplex operation is possible on P-code even with selective availability in effect.

## VARIANTS OF MULTIPLEX

Some variants of the basic multiplex concept can be used to create different receiver types. One technique is $L_1/L_2$ multiplex. Here, the front end of the receiver is switched between $L_1$ and $L_2$ at a submultiple of the multiplex cycle rate. Rather than maintaining separate tracking loops for $L_1$ and $L_2$, the first of two loops is used to track vehicle dynamics using measurements taken on both $L_1$ and $L_2$. This mechanism is called the sum loop. The second loop is a narrower bandwidth difference loop that tracks the $L_1/L_2$ differential only. Sum and difference loops are maintained for both code and carrier independently. Since the code correlation circuitry is common to $L_1$ and $L_2$ and since the $L_1/L_2$ switching rate is high, very accurate measurements of $L_1/L_2$ differentials for code and carrier are possible.

Another variant of multiplex is used for certain inertially-aided, high-jamming applications where data stripping is employed. Data stripping entails recording the digital data during low jamming and using the fact that most of the SV message is repeated and, hence, predictable until the next SV data change. The stored data can then be used to remove the data modulation from the signal, allowing narrower predetection bandwidths, resulting in less squaring loss in a high-jamming situation.

With the data transition problem thus circumvented, a flexible range of multiplex dwell times can be chosen to implement both coherent and incoherent aided tracking operation. Dwell times in excess of 250 milliseconds have been implemented and tested. Various missiles and other unmanned vehicles are typical applications of this high-antijam technology.

Finally, the multiplex concept is a good technique for implementing GPS receivers for attitude measurement. The receiver can rapidly multiplex between several closely spaced antennas using carrier interferometry; widely spaced antennas can use P-code measurements. In any event, the multiplex approach allows high inter-element sample rate, together with the elimination of inter-channel biases that require careful calibration in a multichannel set.

## OSCILLATOR *g*-STABILITY

Oscillator stability is critical to GPS receivers because of the narrow phase-lock bandwidth on signals at L-band and because of the high level of navigation accuracy that is often desired to be obtained from GPS. The oscillator error mechanism that is addressed in this paper is the sensitivity to acceleration. Quartz crystal oscillators change frequency slightly when subjected to *g*-forces. For military sets, it is assumed here that the oscillator *g*-sensitivity is $5 \times 10^{10}/g$. This is 5 parts per 10 billion per *g* and is a very good oscillator. Somewhat better oscillators are available, but $5 \times 10^{10}$ would seem a good number for production military oscillators. For civil applications, the assumed stability is $10^8/g$.

The "pseudo" in GPS pseudorange is removed only to the extent that the oscillator phase (time) and frequency effects are comprehended. With a multi-track (i.e., multiple channel or multiplex) GPS set working with four SVs, the oscillator time and frequency can be estimated accurately, hence calibrating the user clock. If one SV is blocked (e.g., because of banking of the aircraft), then the receiver navigation function has the option of using this recently calibrated clock during this outage period. However, *g*-induced disturbances to the oscillator will introduce random-like short-term errors to the position and velocity measurements.

With a sequencing set, there is always dependence on the oscillator to link sequential GPS measurements typically spread out over a few seconds. Even with four SVs available, the navigation function is, to some extent, always vulnerable to oscillator *g* effects that can add process noise to the Kalman filter during maneuvering.

## COMPARISONS OF FUNCTIONAL CAPABILITIES

The three basic receiver types can be compared from the viewpoint of a few functional capabilities. These capabilities include data collection, bandwidth limitations, navigation filter input data rate, and the availability of integrated Doppler. These functional capability comparisons are summarized in Tables 1 and 2.

For SV data collection, a typical single-channel sequential set must stop its normal sequencing operation and dwell on one satellite for at least 6 seconds to collect one frame of SV data. The collection of further frames of data on this SV can be done in 6-second dwells distributed in time, with measurements on other SVs interspersed to maintain a navigation solution. But the sample rate per SV has been further reduced. The sequential set will be collecting these data one SV at a time with full link margin.

A single-channel four-SV multiplex set can maintain track on all four SVs while collecting data on all of them. The data collection will be with 6 dB less link margin than with a sequential or multichannel set. (The multiplex set can always be designed to operate in a sequential or quasisequential mode for data collection, if desired.)

For applications requiring very high precision, both sequential and multiplex receivers share the advantage of sharing receiver channel hardware. Since a multichannel set has separate code correlation hardware for each SV, special calibration features are usually necessary to reduce interchannel bias effects.

If properly designed, a multichannel set can be regarded as having redundant hardware so that fallback modes of operation can be instituted in the event of a failure in one of the four or five receiver channels.

A curious way to compare the three basic set types is in limits on certain bandwidths. These bandwidths are predetection bandwidth, carrier loop bandwidth, and code loop bandwidth. Predetection bandwidth affects what is called squaring loss and can be significant in jamming situations. Note that in a multiplexing set the multiplex loss that applies to tracking threshold does not apply to the error formation process.

A multiplex receiver, as traditionally designed, has an upper limit of about 15 Hz on the carrier tracking loop bandwidth, because of the 50-Hz sample rate. However, it is possible to double the multiplex sample rate to 100 Hz, thus permitting carrier loop bandwidths up to about 30 Hz.

In multichannel and multiplex receivers, the usual practice is to have the carrier tracking loop rate-aid the code loop. This aiding allows one to substantially reduce the bandwidth of the code loop in spite of high incident (antenna) dynamics. The reduction of code loop bandwidth in turn gives smaller range measurement noise errors. A sequencing receiver without inertial aiding cannot reduce the code loop bandwidth below 5 to 10 Hz because the settling time for the loop starts taking a substantial part of the dwell time.

Table 1—Comparison of functional capabilities

| Receiver Type | Sample Interval Available to Navigation Filter (seconds) | Relative Dynamic Tracking SNR Margin (dB) | Effective Overall Data Rate (bps) | Data Collection Relative SNR (dB) | Regular Navigation Inputs Maintained? | Integrated Doppler Available On All SVs? |
|---|---|---|---|---|---|---|
| ≥Four-channel continuous | Continuous | 0, reference | ≥200 | 0, reference | Yes | Yes |
| Two-channel sequencing | 2 | −1 to −4* | 50 | 0 | Yes | No |
| One-channel sequencing | 5 | −2 to −8* | ≪50, average | 0 | No | No |
| Two-channel MUX | Continuous | −3 | 50<br>200 | 0<br>−3 | Yes | Yes |
| One-channel MUX | Continuous | −6 | 50<br>200 | 0<br>−6 | No<br>Yes | No<br>Yes |

*Highly dependent on dynamics; P-code assumed.

Table 2—Bandwidth Limitations

| Set Type | Minimum Predetection Bandwidth | Minimum Code Loop Bandwidth |
|---|---|---|
| Multichannel | 50 Hz* | No limit |
| Sequencing | 50 Hz* | 5–10 Hz |
| Multiplex | 200 Hz* (four-SV)<br>100 Hz* (two-SV) | No limit |

*Resulting from data, assuming no stripping of data.

## MULTIPLE TRACK

Multichannel and multiplex receivers share the ability to continuously track the code and carrier of multiple SVs. Hence, these two types of receivers can be collectively termed "multiple-track" receivers. Generally speaking, these two receiver types are equivalent except for the multiplex SNR factor of 3 dB for two-SV multiplex, 6 dB for four-SV multiplex, and 7 dB for five-SV multiplex. Both can collect data on multiple SVs in parallel, and both provide true integrated Doppler. Both can provide essentially continuous measurement data to a navigation filter. Both can track substantial dynamics (say 100 $m/s^2$, 50 $m/s^3$).

## ERRORS AS A RESULT OF DYNAMICS

Accuracy comparisons between multitrack and sequential receivers mainly involve errors resulting from dynamics; the accuracy performance of sequential sets is, in fact, significantly dynamics-dependent.

Several factors enter into the effects of dynamics on ultimate navigation accuracy: satellite geometric dilution of precision (GDOP), measurement sample interval, oscillator stability, navigation filter lag, etc., all in addition to the nature of the dynamics themselves.

For the purpose of comparing receivers here, several simplifying assumptions are made. First, certain sinusoidal dynamic cases are used. Next, worst-case peak errors on a given SV are calculated. A constellation GDOP of about 4 would be reasonable, but the peak measurement error on a worst-case SV is multiplied here by an "effective GDOP" of only 2, to account for the fact that in most instances more current measurements on other SVs will contribute to a Kalman filter solution in a way to reduce the peak errors to some extent. This "effective GDOP = 2" number is admittedly arbitrary, but is useful for relative comparisons of different sets. In actuality, different sets will likely have significantly different navigation filter constructions, so these calculations will necessarily be approximate.

Finally, many of the other error sources for GPS are not dealt with because they apply to all sets, do not contribute to dynamic-induced errors, and hence, to first order, do not contribute to performance distinctions between the basic receiver types being discussed here. These other error sources include SV ephemeris errors, ionospheric and tropospheric errors, denial of accuracy (DOA) errors, and receiver noise and quantization errors.

Separate dynamic examples are given for military and for civil applications. There are two military cases: medium-high dynamics associated with heavily stressed helicopter flying, and high dynamics associated with fighter aircraft. Oscillator $g$-sensitivity is $5 \times 10^{-10}/g$.

A benign civil case could be associated with standard enroute maneuvering. A moderate civil case could be associated with a difficult landing approach, and a severe civil case with strong maneuvers. Oscillator $g$-sensitivity is $10^{-8}/g$.

In all cases, a simple sinusoid is used for calculation purposes. For each case, a period $T$ and a peak acceleration are assumed:

$$X = D \cdot \sin \omega t$$

where

$$\begin{aligned} D &= \text{peak displacement} = X_{\text{peak}} \\ \omega &= 2\pi/T \\ T &= \text{assumed period} \\ \dot{X}_{\text{peak}} &= \omega D \\ \ddot{X}_{\text{peak}} &= \omega^2 D, \text{ assumed} \\ \dddot{X}_{\text{peak}} &= \omega^3 D. \end{aligned}$$

The resultant calculations are summarized in Tables 3 and 4. It can be readily seen that the errors associated with multitrack receivers are extremely small. These errors, of course, are only those receiver errors associated with the dynamics (and not with ephemeris, for example), and do not account for any lag in the navigation filter. Also, it is assumed that phase-locked tracking is in effect, with the carrier loop aiding the code loop. It would be highly optimistic to assume that all four SVs are always in track, especially with steep bank angles associated with high-$g$ turns where antenna masking could occur. So, for the multiple-track receivers, a three-SV clock hold case is also given. Here the measurement errors result from oscillator $g$ loading. For the sequential sets, the sampling lag errors swamp out oscillator $g$ effects. (For an inertially aided set, though, the oscillator errors could become a significant relative fraction of the process noise into the Kalman filter.)

## RELATIVE COSTS

Representative relative costs of various GPS receiver systems are given in Table 5.

A sequencing single-channel receiver with dual-frequency ($L_1$, $L_2$) capability is chosen as the cost reference. Although the military $L_1/L_2/P$ case is assumed, civil $L_1$-only, C/A-only sets follow similar ratios. The hardware is broken down into three major categories. The first is the RF/synthesizer/power supply (common hardware). This hardware is the same, except for the extra hardware to do simultaneous $L_1/L_2$ in the case of the five-channel continuous set. The second category is the channel hardware, which is replicated according to the number of receiver channels. The third category of hardware is computing. The costs vary here according to the amount of baseband signal processing and according to the amount of control necessary for each of the configurations.

This paper assumes a more or less traditional approach to receiver design. This assumption means that A/D conversion, if done, is done after code correlation. The GPS sets can and have been built with A/D conversion before code correlation. There will be a tendency in this direction as digital technology

Table 3—Peak dynamics—induced navigation errors (military case)
$X = D \sin \omega t$

| | Helicopter Dynamics | | Fighter Dynamics | |
|---|---|---|---|---|
| | $T = 20$ s ($\omega = 0.3$)<br>$\ddot{X} = 20$ m/s$^2$ $= 2\,g$ peak<br>$D = 200$ m<br>$\dot{X} = 20$ m/s, peak<br>$\dddot{X} = 6$ m/s$^3$, peak | | $T = 6$ s ($\omega = 1.05$)<br>$\ddot{X} = 50$ m/s$^2$ $= 5\,g$ peak<br>$D = 45$ m<br>$\dot{X}_{peak} = 48$ m/s, peak<br>$\dddot{X} = 52$ m/s$^3$, peak | |
| Receiver Type; Number of SVs in Track | Peak Position Error | Peak Velocity Error | Peak Position Error | Peak Velocity Error |
| Multiple track (continuous or MUX) | | | | |
| Four SVs | 0.04 m (0.13 ft) | 0.04 m (delta-range) | 0.04 m (0.13 ft) | 0.04 m (delta-range) |
| Three SVs + oscillator | 0.4 m (1.3 ft) | 0.6 m/s (1.2 kn) | 1 m (3.3 ft) | 1.5 m/s (3 kn) |
| Two-channel sequential ($T_s = 2$ s) | | | | |
| Four SVs | 76 m (250 ft) | 24 m/s (48 kn) | 90 m (295 ft) | 96 m/s (190 kn) |
| One-channel sequential ($T_s = 5$ s) (no data collection) | | | | |
| Four SVs | 200 m (656 ft) | 40 m/s (80 kn) | 90 m (295 ft) | 96 m/s (190 kn) |

Assumptions

- Errors are peak; caused by dynamics effects only
- No ephemeris, etc., errors included
- Effective GDOP for peaks = 2 (constellation GDOP = 4)
- SVs in phase lock
- Velocity state in navigation filter
- No navigation filter lag effects included
- Oscillator $g$-sensitivity $= 5 \times 10^{-10}/g$.

Table 4—Dynamics-induced navigation errors (civil case)
$X = D \sin \omega t$

| Receiver Type; Number of SVs in Track | Enroute Maneuvers<br>$T = 60$ s ($\omega = 0.1$)<br>$\ddot{X} = 1$ m/s$^2$ = 0.1 $g$<br>$D = 100$ m<br>$\dot{X} = 10$ m/s, peak<br>$\dddot{X} = 0.1$ m/s$^3$, peak | | Difficult Approach<br>$T = 30$ s ($\omega = 0.2$)<br>$\ddot{X} = 2$ m/s$^2$ = 0.2 $g$<br>$D = 50$ m<br>$\dot{X} = 10$ m/s, peak<br>$\dddot{X} = 0.4$ m/s$^3$, peak | | Strong Maneuvers<br>$T = 20$ s ($\omega = 0.3$)<br>$\ddot{X} = 5$ m/s$^2$ = 0.5 $g$<br>$D = 50$ m<br>$\dot{X} = 16$ m/s, peak<br>$\dddot{X} = 1.5$ m/s$^3$, peak | |
|---|---|---|---|---|---|---|
| | Peak Position Error | Peak Velocity Error | Peak Position Error | Peak Velocity Error | Peak Position Error | Peak Velocity Error |
| Multiple track (multiple channel or MUX) | | | | | | |
| Four SVs | 0.04 m (0.13 ft) | 0.04 m (delta-range) | 0.04 m (0.13 ft) | 0.04 m (delta-range) | 0.04 m (0.13 ft) | 0.04 m (delta-range) |
| Three SVs + oscillator | 12 m (39 ft) | 0.6 m/s (1.2 kn) | 12 m (39 ft) | 1.2 m/s (2.4 kn) | 20 m (65 ft) | 3 m/s (6 kn) |
| Two-channel sequential ($T_s = 2$s) | | | | | | |
| Four SVs | 4 m (13 ft) | 4 m/s (8 kn) | 9 m (29 ft) | 7.6 m/s (15 kn) | 18 m (59 ft) | 18 m/s (36 kn) |
| One-channel sequential ($T_s = 5$s) (no data collection) | | | | | | |
| Four SVs | 24 m (79 ft) | 10 m/s (20 kn) | 50 m (164 ft) | 12 m/s (34 kn) | 100 m (382 ft) | 34 m/s (68 kn) |

Assumptions

- Errors are peak; caused by dynamics effects only
- No ephemeris, etc., errors included
- Effective GDOP for peaks = 2 (constellation GDOP = 4)
- SVs in phase lock
- Velocity state in navigation filter
- No navigation filter lag effects included
- Oscillator $g$-sensitivity = $10^{-8}/g$.

Table 5—Relative cost ratios of GPS receiver types (for military sets; ratios are very similar for civil sets.)

| One-Channel Sequential Receiver (Reference) | Two-Channel Sequential Receiver | One-Channel Multiplex Receiver | Two-Channel Multiplex Receiver | Five-Channel Continuous Receiver |
|---|---|---|---|---|
| Portion *A*: RF, synthesizer, power supply (common hardware) $A = 0.4\ T$ | $1.0\ \mathrm{A} = 0.40\ T$ | $1.0\ A = 0.40\ T$ | $1.0\ A = 0.40\ T$ | $1.2\ A = 0.48\ T$ (parallel $L_1$ and $L_2$) |
| Portion *B*: IF, code correlation (channel hardware) $\mathrm{B} = 0.25\ T$ | $2.0\ \mathrm{B} = 0.50\ T$ | $B = 0.25\ T$ | $2.0\ B = 0.50\ T$ | $5.0\ B = 1.25\ T$ |
| Portion *C*: computing hardware $\mathrm{C} = 0.35\ T$ | $1.3\ \mathrm{C} = 0.46\ T$ | $1.6\ C = 0.56\ T$ | $1.7\ C = 0.60\ T$ | $1.5\ C = 0.53\ T$ |
| $A + B + C = T$. $T$ = Total Cost of Reference Set | $1.36\ T$ | $1.21\ T$ | $1.50\ T$ | $2.26\ T$ |

advances. To the extent that this digital approach becomes practical, some of the hardware distinctions between set types described herein will tend to blur. The easiest set to implement with precorrelation digitization is a C/A-only sequencing set. Making a multichannel digital receiver, on the other hand, still entails the functional equivalent of multiple simultaneous code generation with individual code dopplers for each SV. The multichannel problem becomes significantly more difficult to implement for a P-code set because of the $\times 10$ higher clock rate, because of the increased complexity of the P-code, and because of the need to provide for the selective availability aspects of the P-code.

## SUMMARY

When the utmost in simplicity is required, a single-channel sequencing GPS receiver can be used to provide a global air navigation function in applications in which high accuracy under dynamics is not important. Conversely, a multichannel set provides the utmost in performance in terms of maximum accuracy under dynamics, availability of integrated doppler, maximum antijam performance, and maximum data collection capability.

A single-channel multiplex set can provide virtually all the benefits of a multichannel set, except at about 6 dB less link margin.

Two-channel sequencing and multiplex sets both provide interesting cost-performance tradeoffs. For a sequencing set, the second channel reduces dynamic lag errors and obviates the need to interrupt navigation to perform data collection. A second multiplex channel provides a set of time slot resources (for example, eight 5-millisecond time slots) that can be adaptively assigned on the basis of SNR needs among several SVs. Some of the salient characteristics of these various receiver types are summarized in Table 6.

It is hoped that the information in this paper can at least serve as a starting point for assessing the appropriateness of a given set type for a given air navigation application.

## ACKNOWLEDGEMENTS

My thanks are given to Kumar Babu, Troy Fuchser, Dennis Henson, Jerry Holmes, Gibb Matlock, Steve Roemerman, and Phil Ward for their helpful suggestions in writing this paper.

## REFERENCES

1. Ward, P. W., "An Advanced NAVSTAR GPS Multiplex Receiver," presented at 1980 Position Location and Navigation Symposium, December 9, 1980.
2. Ward, P. W., "An Advanced NAVSTAR GPS Multiplex Receiver/Navigator for Geophysical Exploration Applications," presented at 52nd Annual International Meeting, October 20, 1982.

Table 6—GPS Set Comparison Summary

| Set Type | Relative Cost | Relative Dynamic Tracking SNR Margin (dB) | Integrated Doppler Available? | Data Collection (bps) | Peak Errors Because of Dynamics | | |
|---|---|---|---|---|---|---|---|
| | | | | | Civil Enroute | Helicopter | Fighter |
| One-channel sequential | 1.00 (reference) | −2 to −8* | No | ⩽50, average; 0 dB; interrupts navigation | 24 m; 10 m/s | 200 m; 40 m/s | 90 m; 96 m/s |
| Two-channel sequential | 1.36 | −1 to −4* | No | 50; 0 dB; no interruption | 4 m; 4 m/s | 76 m; 24 m/s | 90 m; 96 m/s |
| One-channel multiplex | 1.21 | −6 | Yes | 200; −6 dB; no interruption | 0.04 m; 0.04 m delta range | 0.04 m; 0.04 m delta range | 0.04 m; 0.04 m delta range |
| Two-channel multiplex | 1.50 | −3 | Yes | 50/0 dB or 200/−3 dB; no interruption | 0.04 m; 0.04 m delta range | 0.04 m; 0.04 m delta range | 0.04 m; 0.04 m delta range |
| Five-channel set | 2.25 | 0 (reference) | Yes | ≤200; 0 dB, reference; no interruption | 0.04 m; 0.04 m delta range | 0.04 m; 0.04 m delta range | 0.04 m; 0.04 m delta range |

*Highly dependent on dynamics; P-code assumed.

# A Flexible GPS Software Development System and Timing Analyzer for Present and Future Microprocessors

JAMES W. SENNOTT

Bradley University, Peoria, Ill.
*Received January 1984*

## ABSTRACT

Discussed is a timing emulation and operating system which predicts loading for microprocessors and enables efficient generation of software found in such applications as the Global Positioning System. The timing emulator is applied to two classes of user equipment, a conventional fast-multiplexed design and a fully integrated processor featuring continuous state feedback into the spread-spectrum demodulation and tracking loops. Timing data are developed for five representative microprocessors ranging from 8/8 to 32/32 register/bus widths. The advanced integrated processor requires augmentation of the 32/32 processor with a math co-processor. The emulator is used to develop speed requirements on the co-processor. Lastly, the software development system is described. This contains a multi-tasking operating system and matrix library written in the language "C" which may be easily adapted to a range of radio-navigation problems.

## I. INTRODUCTION

This paper describes a system for the efficient generation and testing of microprocessor code for radio navigation receivers, in particular GPS receivers. The goal in developing the system was to reduce manufacturer product development cost, increase performance and shorten the product development cycle. Such a development tool becomes a necessity in a design environment in which more powerful microprocessors are announced monthly. Earlier navigation systems such as LORAN-C and Transit required less demanding and complex software than high performance GPS receivers, and it is becoming more expensive to implement software on new processors. The decision to move to a more cost effective processor requires a careful study of instruction sets and loading, across a range of GPS receiver architectures and end-user performance specifications. The development system and timing analyzer to be described reduces the development time-gap and increases confidence during the microprocessor selection process.

The key to the development system is a portable and efficient GPS operating system, with support routines, written in the language "C". Recently, this language has received much acclaim for its speed on, and portability to, a wide range of microprocessors. The bulk of the operating system, support library and GPS signal processing and navigation tasks may be readily moved from one processor to another, allowing for a much more rapid development cycle than with a conventional assembly language approach. And, unlike other high-level languages, the compiled code runs nearly as fast as assembly language.

A crucial part of the development cycle begins before any code is written. The designer must weigh the benefits of a particular demodulation and filtering scheme against the cost of signal processing. A scheme that today appears prohibitive may be very reasonable in the near future. The development system to be described allows the designer to estimate loading given a rather general description of the microprocessor. Once the selection has been made much of the required code can be generated using a "C" language compiler.

We will first describe the structure of the timing emulator portion of the development system. In the following two sections of the paper we will apply the emulator to two classes of GPS equipment. Representative microprocessors ranging from 8/8 to the new 32/32 machines will be studied. Along the way we will develop the computational load for a high performance integrated processor which provides the maximum possible performance for spread-spectrum radio-navigation user equipment. Lastly, the software development portion of the development system will be described.

## II. TIMING EMULATOR

For the purposes of estimating processor load our emulation must accurately reflect the cycling of processor tasks. Unlike a receiver simulation, in which most of the user equipment operations are actually performed, loading emulation need only reflect the number and type of machine language instructions required of the candidate microprocessor configuration.

For each task execution time must be estimated. For example, satellite state update requires that a sequence of matrix/vector operations be performed in 32-bit signed integer and floating point arithmetic. The resulting processor load could be found by actually programming the sequence in assembly code for each candidate processor configuration, a very tedious task. Another approach would be to program as much of the code as possible in a machine-independent language such as FORTRAN, PASCAL or C. Using available cross-compilers, these source codes could then be converted into machine code. A support library of machine-dependent utilities for floating-point and other multi-precision operations would then be linked to the compiler object code. This is an appealing method in that actual user-equipment software is generated for a variety of target processor configurations. We shall return to this method later in the paper. Unfortunately, compilers are unavailable for many microprocessor configurations of potential interest. And many of the available compilers are inefficient in their generation of machine language code.

The method adopted in the timing emulator is illustrated in Figure 1. Upon operator entry of user equipment specifications and microprocessor configuration/timing data, the GPS operating system model cycles through tasks at the appropriate rates, producing a sequence of "subroutine" calls which are reduced by the subroutine model into primitive machine independent register/register, memory/register, I/O, integer arithmetic, and floating-point subroutine requests. The microprocessor model then maps these into other register, memory, and special arithmetic instructions available on the specified microprocessor hard-

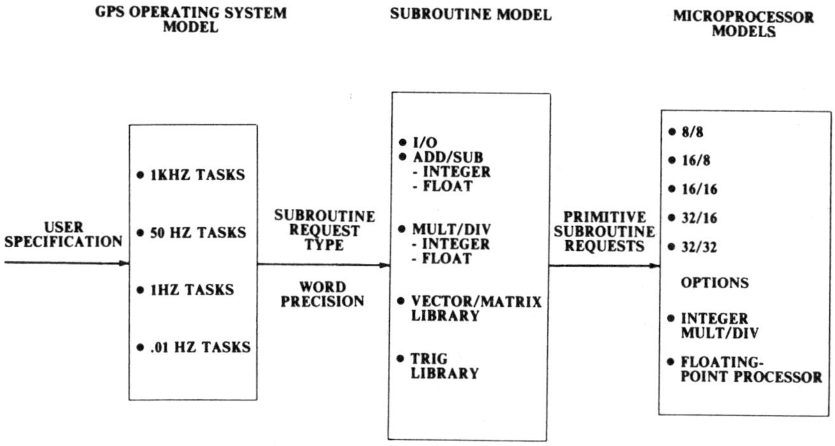


*Fig. 1—Loading estimate generation.*

ware. The operations performed within each of the major emulator blocks will now be discussed.

The GPS system model represents each task as a sequence of calls to the emulator subroutine portion of the emulator. Each request is tagged with a data precision parameter, and with a processor service I.D. which can be used to direct the task to a particular "processor" in a multi-processor configuration. A complete cycle through the GPS processor model represents 100 seconds of real time.

At emulation run time the user is free to define a number of different equipment configurations. Those implemented at present are aided and unaided versions aimed at aviation and land/marine applications. For each the operator can specify the maximum number of satellites to be simultaneously tracked, the number of filter states propagated per navigation dimension, and the filter type.

The subroutine block provides the connection between the high-level GPS processor tasks and the machine-level model. The subroutines that can be serviced range from simple single-byte I/O to floating-point matrix operations. All are converted here into requests for microprocessor service. Simple register/register and register/memory requests pass directly to the microprocessor model portion of the emulator. Other requests are first translated into groups of primitive requests. Matrix subroutines are typical.

Upon receiving a request for a particular matrix operation, the calling parameters are examined. These include the data word length, which may be set to 8, 16, 32 or FP, and the array sizes to be processed. Working with these parameters the number of integer and floating-point add and multiply operations needed to complete the matrix operation is evaluated. The next step is to convert the multiply and add operations into more basic machine independent operations. This is done separately for 8, 16, 32 and FP data formats. The FP format is treated as a special case. If a floating-point processor is available, only those primitive instructions needed to pass data to and from the floating-point processor are tallied.

Upon completion of operations within the subroutine portion of the emulator, all GPS processor requests have been reduced to the following primitive subroutine requests:

Memory-mapped I/O: 8, 16 bits
Register/register: 8, 16, 32 bits
Memory/register: 8, 16, 32 bits
Integer add, sub: 8, 16, 32 bits
Integer multiply, divide: 8, 16, 32 bits
FPP request

The conversion of these into machine loading statistics is the final function performed by the emulator. One or more processors, together with a floating-point processor, may be specified upon emulator initialization. Service requests originating in the GPS processor block may be assigned to a specific processor. Thus a multi-procesor loading estimate can be established. For each processor the following attributes are specified by the operator:

Width of data registers
Width of data path to memory
Average execution time, including instruction fetch, for register/register instructions
Average execution time, including instruction fetch, for memory/register instruction of data register width
Average execution time for integer multiply and divide, for register-length result, if available
Average speed of math co-processor in 32-bit mantissa multiply and divide, if processor available.

In the highly competitive microprocessor marketplace upgraded specifications are announced nearly weekly. Whole microprocessor families appear in the trade literature monthly. Using the timing emulation approach estimates of performance in the GPS application can be quickly established. For illustrative purposes five processors have been selected for investigation. Their basic specifications are given in Table 1.

Each processor is first identified by register and data bus configuration. For example the 6502 processor is in the 8/8 class, with a maximum register width of 8 bits and a data bus of 8 bits. The 68020 is an upgraded 68000 having both register and data bus widths of 32 bits. These basic parameters, together with instruction speed, largely determine the rate at which the multi-precision

Table 1—Microprocessor Specifications

| Processor Class | Processor Model | MS micro sec | PS micro sec | MPY micro sec | DIV micro sec | Clock M Hz. |
|---|---|---|---|---|---|---|
| 8/8 | 6502 | 2 | .5 | — | — | 2 |
| 16/8 | 8085A-2 | 2.6 | .8 | — | — | 5 |
| 16/8 | 68b09 | 2.5 | .5 | 5.5 | — | 2 |
| 32/16 | 68000 | 1.1 | .34 | 5.8 | 13.1 | 12 |
| 32/32 | 68020 | .87 | .27 | 4.6 | 10.5 | 15 |

MS = time for extended address memory to register transfer, word of buss width
PS = time for register to register instruction
MPY = time for register/register multiply
DIV = time for register/register divide

arithmetic tasks within the machine-model portion of the emulator may be performed.

The instructions which underly arithmetic operations are of three types, register/register, register/memory, and integer multiply/divide (if available). For example, the 6502 processor requires 2 microseconds to execute a register/memory data operation of data-bus width (8 bits) while the 68000 transfers 16 bits in 1.1 microseconds. These times include instruction fetch and are for extended direct addressing. Typical register operations require .5 microseconds (8 bits) for the 6502 and .34 microseconds (32 bits) for the 68000. Starting with these specifications, execution times for the complete set of primitive multi-precision subroutines are estimated for each processor class, accounting for the availability of multiply and divide instructions in the instruction set.

## III. FAST MULTIPLEXED PROCESSOR

We first consider a fast-multiplexed architecture. A single code generator and synthesizer are employed in a rapid multiplexing scheme to gather data on up to six satellites every 20 milliseconds. This multiplexing concept has been reported elsewhere[1,2] and has been extensively tested in Phase 1 of the military GPS program. While such an approach cannot yield the same performance as a more costly parallel-channel receiver it responds more rapidly to changes in signal and dynamical conditions than a single-channel "settle-and-dwell" demodulator structure.[3]

Time slotting is flexible with up to twenty distinct slot assignments initialized at the beginning of each 20-millisecond window. For each assigned satellite the correlator generates in parallel six different test statistics. In this way sufficient data is provided to the 50-Hz loop filter tasks for the usual delay-lock and Costas/AFC tracking modes.

A listing of major system tasks is shown in Table 2. Within the slowest two groups are tasks that perform navigation filtering and satellite ephemeris computations.These are largely independent of the hardware architecture, and

Table 2—Fast-multiplexed task summary

| | |
|---|---|
| 1 kHz RATE TASKS<br>—A/D reads<br>—Code generator reset and start<br>—Synthesize setup<br><br>50 Hz RATE TASKS<br>—Test statistic processing<br>—GPS data unpack<br>—Loop filter updates<br>—Amplitude estimates, lock monitor<br>—Delay/frequency/phase estimate update<br>—Computation of next complete set code/ synthesizer offsets<br>—Keyboard service/display requests<br>—Flight director service | 1 Hz RATE TASKS<br>—Satellite state updates<br>—Prediction of pseudoranges<br>—Navigation state updates<br>—Satellite slot assignments checking<br>—Demodulator prepositioning servicing<br>—Computation of RNAV along/ cross-track corrections<br><br>.01 Hz RATE TASKS<br>—Full ephemeris computation update<br>—Orbit linearization<br>—Hardware self test |

are structured according to the particular filter implementation and filter state-vector formulation desired.

In an effort to reduce computational burden a two-step approach to satellite update is employed. As new satellites appear above the horizon, a full-precision ephemeris computation is carried out in 32-bit mantissa floating-point arithmetic. At the same time a matrix state-variable model representing the local satellite trajectory is initialized. This local model is then used to compute satellite position, velocity and clock offsets during the next 100 seconds; this enables a substantial savings since corrections are carried out in 16-bit signed integer arithmetic. Correction terms are added to the initial values forming ECEF whole-state values in 32-bit signed integer format.

Figure 2 shows the measurement incorporation strategy. A complete set of up to six satellites are incorporated at the 1Hz navigation update rate. For the i-th emitter the incorporation equation (for either delay or Doppler) is:

$$\underline{\Delta X_K} = \underline{K}^{(i)} \, [Z_K^{(i)} - h_i \, (\underline{\hat{X}_K} \, (-))]$$

and the updated state estimate is given by:

$$\hat{X}_K \, (+) = \hat{X}_K \, (-) + \underline{\Delta X_K}$$

With this scheme the satellite-user predicted pseudoranges are required at the measurement incorporation rate. However, matrix inversion is not required. An important byproduct is generation of positioning information for the demodulator. The 50 Hz demodulator tasks are free to request this information whenever lock monitor outputs fall below a predefined threshold. Such a capability is especially valuable during periods of maneuver-induced attenuation.

The next function is optimal gain computation, encompassing covariance extrapolation, gain update and covariance update. In this analysis we shall assume that filter gains are updated at the navigation update rate, 1 Hz. Using

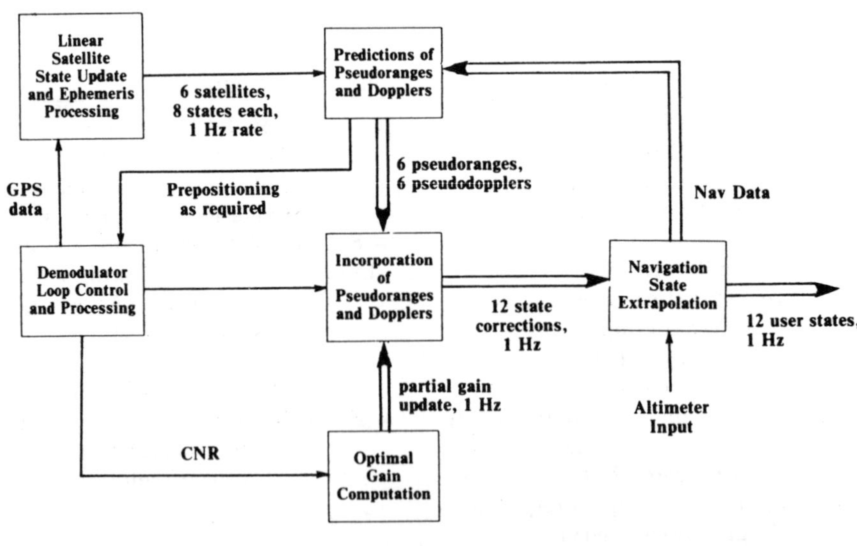


*Fig. 2—Fast multiplexed processor major functions.*

the emulator, loading sensitivity to gain update rate is easily investigated. The covariance extrapolation function is:

$$P_K(-) = \Phi P_{K-1}(+)\Phi^T + Q$$

where $Q$ is the state noise covariance matrix and $\Phi$ is the state transition matrix. Within the emulation array dimensions are variable and depend upon the number of navigation dimensions required for the particular application. The gain update equation, for the $i$-th satellite, is:

$$K_K^{(i)} = P_K(-) H_K^T [H_K P_K(-) H_K^T + R_K^{(i)}]^{-1}$$

Where $R$ is a scalar involving the $i$-th satellite CNR, and $H_K P_k H_K^T$ is scalar as well. Lastly, the covariance update following incorporation of the $i$-th emitter is computed:

$$P_k(+) = [I - K_K^{(i)} H_k] P_K(-)$$

Table 3 shows the loading estimates by task rate, for each microprocessor. Loading varies from a high of 500% to just under 50%. The biggest differences are seen in the 1Hz and .01Hz tasks. Here one finds the bulk of high-precision arithmetic. Among the 8-bit machines the 68b09 fares best with its limited 16-bit register capability and 8-bit multiply instruction.

In an effort to reduce loading a simplified tracker was also examined. The reduction possible with an Alpha-Beta tracker was modest.

Only the two most powerful processors were able to manage the speed needed in the fast-multiplexed processor. In practice it would be desirable to split the burden between two processors. For example a 6809 could be assigned to the 1

Table 3—Fast-multiplexed microprocessor loading

| Processor | Task Load—Seconds per Second | | | | Duty Percent |
|---|---|---|---|---|---|
| | 1 KHZ | 50 HZ | 1HZ | .01 HZ | |
| 6502 | .6 | 2.8 | 1.6 | .0047 | 500 |
| | | | .4* | | 380* |
| 8085A-2 | .78 | 2.9 | 1.7 | .0011 | 540 |
| | | | .53* | | 420* |
| 68b09 | .75 | .79 | .49 | .0010 | 203 |
| | | | .16* | | 170* |
| 680000 | .275 | .30 | .0875 | .00028 | 66 |
| | | | .024* | | 60* |
| 68020 | .22 | .20 | .060 | .00018 | 48 |
| | | | .012* | | 43* |

1. Navigation solution updated once/sec
2. 6 satellites tracked
3. 12 filter states
4. * indicates result for Alpha-Beta tracker

KHz tasks, which are primarily 8-bit I/O buffering and control with a 32/32 processor handling the remaining tasks.

## IV. ADVANCED INTEGRATED PROCESSOR

Many of the performance limitations of present radio-navigation systems are the result of compromises in the demodulation and filtering software. The timing emulator affords the designer the possibility of estimating the load for advanced filtering concepts. In this section we shall illustrate this for a combined demodulation and navigation architecture which is very close to optimum for the GPS system. The performance of such an integrated approach has been studied for applications involving signal attenuation and satellite failure[3].

The integrated processor uses its best overall estimate of user state to directly control the spread-spectrum correlation process. This enhances performance in several respects:

A) The demodulation process is less sensitive to signal amplitude changes.
B) The effective loop bandwidth of the demodulator is reduced, yielding higher accuracy.
C) Improved signal fault detection and fault isolation is possible.

Clearly, however, the integrated approach requires a substantial increase in computational speed. In particular many of the 1 Hz. tasks performed in the fast-multiplexed design must now be performed at 50 Hz. Unlike the fast-multiplexed design, in which the navigation observations are pseudoranges and Dopplers, the processor is now directly incorporating in-phase and quadrature statistics into the navigation solution. This requires faster projections of user and satellite states, as well as more complex non-linear incorporation equations.

Figure 3 depicts the new steps. These are carried out in parallel for all satellites in view. First the present user state, including clock state, is used to initialize the correlator system (the particular correlator implementation whether analog or digital is of no concern here). At the same time predicted signal correlation values are generated. Following the correlation interval the processor must use the six observed and predicted statistics to update user state. It is at this point that the processing method is quite different from the conventional Costas/delay-lock loop structure. In particular the integrated processor always attempts to make best use of all advanced/punctual/retarded correlation values. To accomplish this it first computes a maximum-likelihood estimate of signal amplitude and GPS telemetry data bits. Before the error residual is used to update the user state the residual magnitude is checked. On repeated evidence of large error the offending satellite channel is isolated from the overall integrated solution. Further isolation of the fault condition is then carried out. For intermediate values of error-residual the measurement is incorporated but with reduced gain, and estimates of GPS telemetry bits are ignored. This situation parallels the non-coherent mode found in present demodulators.

The overall task assignments to be computed are as shown in Table 4. The increased computational burden is reflected in Table 5. The 32/32 processor is now spending a very significant period in 50 Hz. tasks performing array computations to 32-bits accuracy. The emulator was employed to examine co-

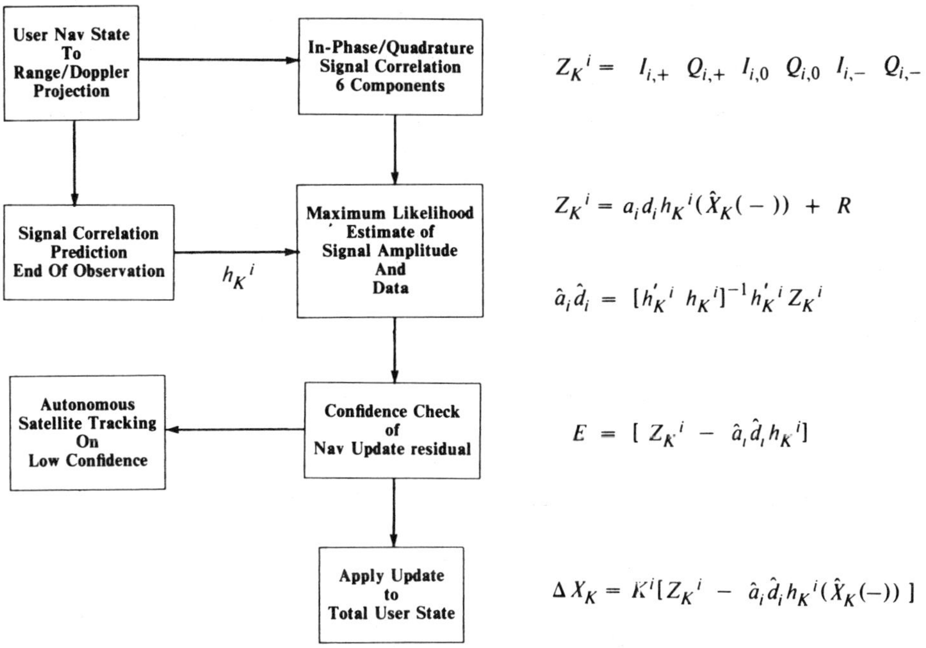


*Fig. 3—Advanced processor signal processing.*

Table 4—Advanced processor task summary

| | |
|---|---|
| 1 kHz RATE TASKS | 1 Hz RATE TASKS |
| —A/D reads | —Satellite slot assignments checking |
| —Code generator control | —Computation of RNAV along/ cross-track corrections |
| —Synthesizer control | |
| 50 Hz RATE TASKS | .01 Hz RATE TASKS |
| —Satellite state update | —Full ephemeris computation update |
| —Prediction of correlator statistics | —Orbit linearization |
| —Test statistic generation | —Hardware self test |
| —Amplitude/Data optimal estimates | |
| —Residual and fault detection check | |
| —Navigation state update | |
| —Test statistic processing | |
| —GPS data unpack | |
| —Computation of next complete set code/ synthesizer offsets | |
| —Keyboard service/display requests | |
| —Flight director service | |

processor speed requirements in this application. Postulated were devices that could perform 16 and 32-bit integer multiply and 32-bit floating point operations. Speeds were varied from four to one microseconds. The 4-microsecond specification is expected in soon-to-be announced co-processors. The fastest of the three reduces the load to about 75%. Parallel operation of several of the 32/32 processors would also be acceptable.

Table 5—Advanced processor microprocessor loading

| Processor | Task Load—Seconds per Second | | | | Duty Percent |
|---|---|---|---|---|---|
| | 1 KHZ | 50 HZ | 1HZ | .01 HZ | |
| 68b09 | .63 | 9.5 | 1.8 | .16 | 1200 |
| 68000 | .28 | 1.5 | .38 | .027 | 218 |
| 68020 | .22 | .81 | .27 | .018 | 132 |
| CP4* | .22 | .59 | .23 | .0086 | 104 |
| CP2* | .22 | .43 | .18 | .0043 | 83 |
| CP1* | .22 | .35 | .15 | .0022 | 72 |

CP1, CP2, CP4 augment 68020, performing 16 and 32 bit integer multiply as well as 32-bit floating-point arithmetic

Coprocessor speeds are 1, 2 and 4 microsecond respectively. This includes instruction and data fetch.

One can conclude from this that the next generation of microprocessors will allow optimum processors to be economically produced.

## V. SOFTWARE DEVELOPMENT SYSTEM

While considering needed refinements to the present emulator it became clear that the present structure contained the seeds of a highly portable and efficient operating system capable of carrying out both loading estimate calculations and real-time GPS processing. The key to this expanded role for the software is 1) reduction of the receiver and navigation tasks to highly structured multiplexed tasks operating at one of several synchronous rates and 2) use of an efficient and portable microprocessor language for coding all but primitive arithmetic utilities and input/output utilities.

The upgraded version of the emulator provides more detailed estimates of processor loading and also constitutes the bulk of the software in finally-implemented versions of GPS user configurations. A secondary goal was improvement of the user interaction during various modes of emulator operation, and to provide a friendly GPS software development environment within the widely available UNIX[4] operating system.

An overview of the upgraded implementation is shown in Figure 4. In addition to expanded versions of the three functions of Figure 1 are GPS processor configuration and run-time configuration routines. The upgraded versions of the GPS operating system model and subroutine model of Figure 1 contain operable versions of all functions, coded in C source code.

An important part of the development system is the operating system written in "C" which has the following features:

A) Multi-tasking
B) Memory resident
C) Multiple task priorities with queuing
D) Parent/child tasks
E) Interprocess communication
F) Memory allocation functions

These features were selected after a study of the typical multiplexing of tasks to be implemented in GPS user equipments.

In the run-time configuration routine the operator specifies the disposition

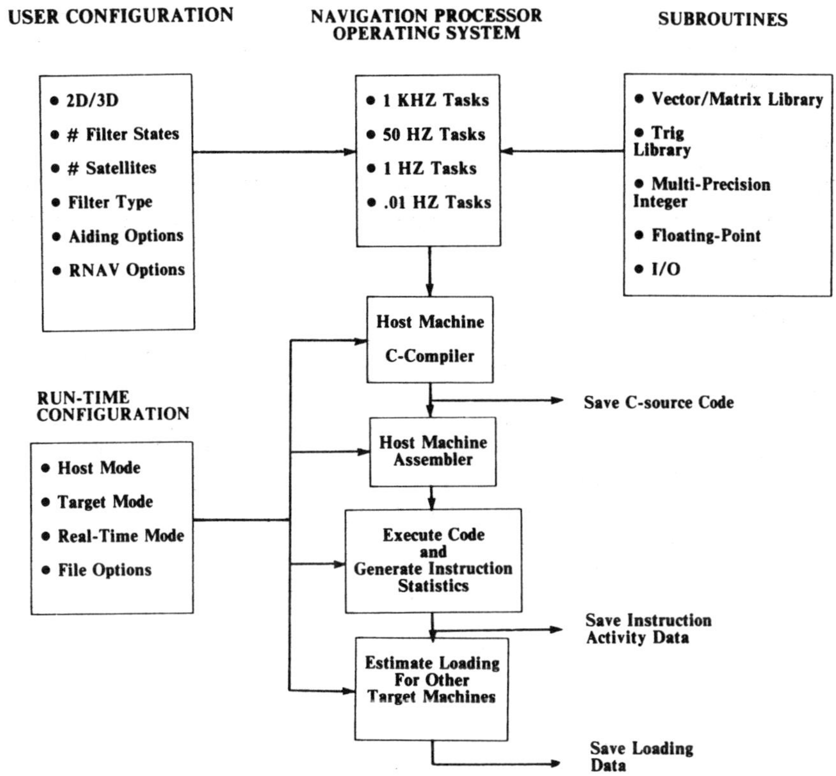


*Fig. 4—Software development system.*

of the source code. The "Host Is Target" mode generates loading statistics directly applicable to the machine on which the software development system/load-analyzer package is being run. The number of systems on which the complete package can be hosted is growing rapidly, with all the newer processors supported by one or more versions of the UNIX operating system. At Bradley the host machine is the DEC PDP-11.

In the non-host mode the operator working at the host machine wishes to develop timing estimates for some other processor. If the target machine hardware is available, along with a C cross-compiler and assembler, machine code is generated and down-loaded on that processor. Instruction statistics and timing are then generated and passed back to the host for further analysis.

## V. CONCLUSION

In this paper we have described a timing emulator and software development system aimed at radio-navigation and similar applications. The features of the timing emulator were illustrated for two different classes of GPS user equipment and five microprocessor types. An advanced integrated processor was postulated and timed. Augmentation of 32/32 processors with co-processors was examined. Lastly a portble GPS operating system incorporating multi-tasking capability was described, along with a development system for tailoring this operating system to meet the needs of present and future radio-navigation user equipments.

## ACKNOWLEDGEMENT

The author wishes to thank the U.S. Department of Transportation, Transportation Systems Center for their support. R. M. Kalafus was the technical monitor for the project.

## REFERENCES

1. Henderson, D. W. and Strada, J. A. "NAVSTAR Field Test Results," *Navigation,* Vol. 26,No. 1, Spring 1978.
2. Buige, A., La Frey, R. R. and Campbell, S. D., "The Design and Measured Performance of an Experimental GPS Navigation Receiver for General Aviation," PLANS '82 Conference Proceedings.
3. Sennott, J. W., "Fault/Maneuver Tolerance of Aided GPS Demodulation/Navigation Processors In The Non-Precision Approach Environment", Presented at National Aerospace Meeting, Institute of Navigation, March 1982.
4. UNIX is a trademark of Bell Laboratories.

# Sensitivity of GPS Acquisition to Initial Data Uncertainties

CAREY A. SMITH and KARL W. GRAVES
Aero Products Division, Litton Industries, Inc., Moorpark, CA
*Received January 1984*

## ABSTRACT

To acquire a GPS satellite signal, it is necessary to estimate the Doppler frequency shift due to the motion of the satellite relative to the receiver. This estimate is sensitive to uncertainties in the receiver position and in the GPS "Time-of-Week," which is used to compute the satellite's position and motion. Knowing this sensitivity and the uncertainties in the initial data, the receiver can determine the number of frequency bins that must be searched, which directly impacts the "Time-to-First-Fix." Ignoring this sensitivity and searching the wrong number of bins can result in missing the signal on the one hand or excessive delays on the other.

Since the Doppler shift is proportional to the range rate, its sensitivity to position errors is maximized by taking the gradient of range rate, and its sensitivity to time errors is given by the time derivative.

Somewhat surprisingly, the Doppler estimate is most sensitive to both kinds of initial errors when the satellite is directly overhead, implying that the common practice of acquiring the highest satellite first needs to be reexamined. Also, the elevation angle alone does not determine sensitivities, since it takes two angles to specify the orientation of the satellite's orbit to the receiver—even when earth rate is neglected.

In addition, numerical results for position and time sensitivities are tabulated for the entire spectrum of receiver/satellite orientations. These show that in the worst case, an error of one second in time causes nearly as large an error in estimated Doppler as does an error of one kilometer in position, roughly one hertz. One implication of this result is that a continuous, battery-powered "Real-Time-Clock" can significantly improve the Doppler estimate and reduce the acquisition time.

## 1.0 INTRODUCTION

This paper discusses the effects of errors in initial position and initial time on the estimated Doppler shift of the received satellite signals. Since the GPS user set will only acquire the satellite signal if its estimated Doppler is close enough to the true Doppler, the relationships derived herein allow the receiver to determine the number of frequency bins that must be searched, which directly impacts the "Time-to-First-Fix."

In Section 2.0, we derive an expression for range-rate and give its proportionality to Doppler shift. In Section 3.0, we compute the maximum errors in Doppler that can result from uncertainties in position for a given geometry by taking the gradient of range-rate. In Section 4.0, we calculate these errors for specific cases. In Section 5.0, we compute the errors in Doppler due to errors in initial time. Section 6.0 gives a derivation of the user's horizon and the mask angle in terms of the orientation of a satellite's orbit relative to the user; this helps interpret the calculations in Sections 4.0 and 5.0. Section 7.0 discusses other Doppler error sources. Section 8.0 is the conclusion.

## 2.0 EXPRESSION FOR RANGE-RATE AND DOPPLER

Figure 1 shows the geometry used for this analysis. This coordinate system is relative to the satellite's orbit and is independent of the earth-fixed latitude and longitude system. Since we will take the nominal $\lambda$ to be $\lambda_o = 90°$, $\theta$ can be thought of as the satellite's elevation above the horizon within its orbit. The angle $\phi$ is the angle between the user and the plane of the satellite. It can be useful to think of $\phi$ as latitude, but it must be remembered that there is no actual relationship between $\phi$ and latitude and that even for a fixed user position, $\phi$ will vary from satellite to satellite.

The satellite position can be seen to be:

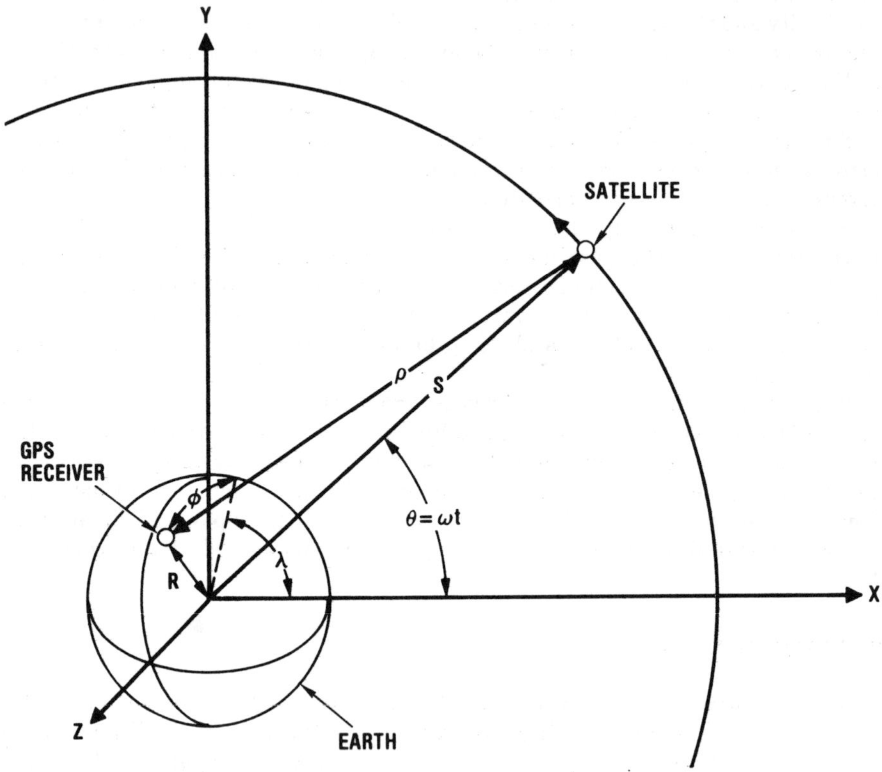


SATELLITE POSITION = $\boldsymbol{S} = (S{\cdot}\cos\theta, S{\cdot}\sin\theta, 0)$

USER POSITION = $\boldsymbol{R} = (R\cos\phi\cos\lambda, R\cos\phi\sin\lambda, R\sin\phi)$

RANGE = $\rho = \|\boldsymbol{S} - \boldsymbol{R}\|$

$S = 2.656{\cdot}10^7$m  $\lambda_O = 90°$

$R = 6.37 \cdot 10^6$m  $0° \leq \theta \leq 180°$(so that $|\theta - \lambda_O| \leq 90°$)

$\omega = 1.45 \cdot 10^{-4}$ rad/sec.  $|\phi_O| \leq 90°$

*Fig. 1—Definition of the coordinate system.*

$$S = (S \cdot \cos\theta, S \cdot \sin\theta, 0)$$
$$\text{where } S = 2.656.10^7$$

The GPS user position is:

$$R = (R \cdot \cos\phi \cdot \cos\lambda, R \cdot \cos\phi \sin\lambda, R \cdot \sin\phi)$$
$$\text{where } R = 6.37 \cdot 10^6.$$

The square of the range from the user to the satellite is:

$$\begin{aligned}\rho^2 &= \| S - R \|^2 \\ &= S^2 + R^2 - 2SR\cos\phi\cos(\theta - \lambda)\end{aligned}$$

Using $\theta = \omega t$, the range rate, $\dot{\rho}$ can be found from:

$$\begin{aligned}\frac{d}{dt}\left(\tfrac{1}{2}\rho^2\right) &= \rho\dot{\rho} \\ &= SR\omega\cos\phi\sin(\theta - \lambda) \\ \rightarrow \dot{\rho} &= \frac{SR\omega}{\rho}\cos\phi\sin(\theta - \lambda)\end{aligned}$$

Doppler shift, $\Delta D$, is related to $\dot{\rho}$ by the carrier frequency, $L_1 = 1.575 \cdot 10^9$ Hz, and $c = 2.998 \cdot 10^8$ m/s:

$$\Delta D = \dot{\rho}\left(\frac{L_1}{c}\right) \text{ and } \left(\frac{L_1}{c}\right) = 5.254 \text{ Hz/(m/s)}$$

## 3.0 THE GRADIENT OF RANGE-RATE WITH RESPECT TO ERRORS IN INITIAL POSITION

The gradient represents the vector of steepest ascent and its magnitude is the maximum magnitude of any directional derivative at that point. Interpreted as a Taylor series, let $u$ = a unit-vector displacement. The change in $\dot{\rho}$ due to a change of $u$ is:

$$\Delta\dot{\rho} = \dot{\rho}(x_o + u) - \dot{\rho}(x_o) \cong \nabla\dot{\rho}(x_o) \cdot u$$

This is maximized when $u$ is a unit-vector in the direction of $\nabla\dot{\rho}$. Then

$$\Delta\dot{\rho} = (\nabla\dot{\rho}) \cdot \frac{\nabla\dot{\rho}}{\|\nabla\dot{\rho}\|} = \| \nabla\dot{\rho} \|$$

Let $\delta\lambda$, $\delta\phi$ be infinitesimal displacements (in radians) of $\lambda$, $\phi$ from their true values $\lambda_o$, $\phi_o$. Then the corresponding distance displacements on the surface of the earth are:

$$\delta a = R \cdot \delta\phi$$
$$\delta b = R \cdot \cos\phi_o \cdot \delta\lambda$$

See Figure 2.

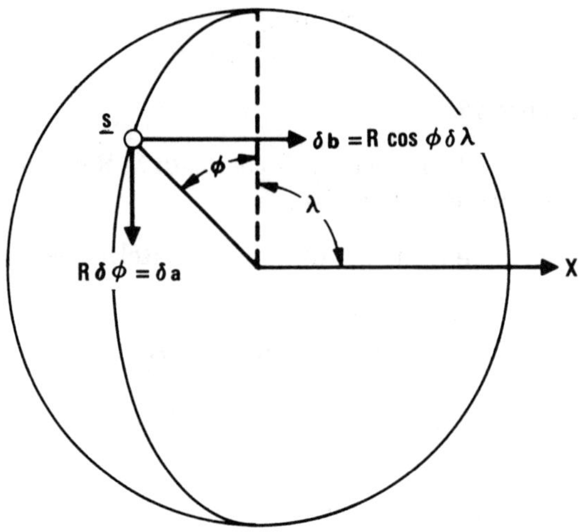


*Fig. 2—Small displacements in the φ,λ directions.*

Taking the gradient with respect to these distances, we get:

$$\nabla_{a,b}\dot{\rho} = \left(\frac{\partial\dot{\rho}}{\partial a}, \frac{\partial\dot{\rho}}{\partial b}\right)$$

$$\frac{\partial\dot{\rho}}{\partial a} = \left(\frac{\partial\dot{\rho}}{\partial\phi}\right)\left(\frac{\delta\phi}{\delta a}\right) = \left(\frac{\partial\dot{\rho}}{\partial\phi}\right)\left(\frac{1}{R}\right)$$

$$\frac{\partial\dot{\rho}}{\partial b} = \left(\frac{\partial\dot{\rho}}{\partial\lambda}\right)\left(\frac{\delta\lambda}{\delta b}\right) = \left(\frac{\partial\dot{\rho}}{\partial\lambda}\right)\left(\frac{1}{R\cos\phi_o}\right)$$

See Figure 3. (Note that the direction of the gradient is independent of our small displacements.)

We can now derive $\frac{\partial\dot{\rho}}{\partial\phi}$ and $\frac{\partial\dot{\rho}}{\partial\lambda}$ from our expression for $\rho\dot{\rho}$ as follows:

$$\frac{\partial}{\partial\phi}(\rho\dot{\rho}) = \left(\frac{\partial\rho}{\partial\phi}\right)\dot{\rho} + \rho\left(\frac{\partial\dot{\rho}}{\partial\phi}\right)$$

$$\rightarrow \frac{\partial\dot{\rho}}{\partial\phi} = \frac{1}{\rho}\left[\frac{\partial}{\partial\phi}(\rho\dot{\rho}) - \left(\frac{\partial\rho}{\partial\phi}\right)\dot{\rho}\right]$$

$$\frac{\partial}{\partial\phi}(\rho\dot{\rho}) = -SR\omega\sin\phi\sin(\theta-\lambda)$$

$$\frac{\partial\rho}{\partial\phi} = \frac{SR}{\rho}\sin\phi\cos(\theta-\lambda)$$

So

$$\frac{\partial\dot{\rho}}{\partial\phi} = -\frac{SR\omega}{\rho}\sin\phi\left[\sin(\theta-\lambda) + \frac{SR}{\rho^2}\cos\phi\sin(\theta-\lambda)\cos(\theta-\lambda)\right]$$

Likewise,

$$\frac{\partial \dot{\rho}}{\partial \lambda} = \frac{1}{\rho}\left[\frac{\partial}{\partial \lambda}(\rho\dot{\rho}) - \left(\frac{\partial \rho}{\partial \lambda}\right)\dot{\rho}\right]$$

$$\frac{\partial}{\partial \lambda}(\rho\dot{\rho}) = -SR\omega \cos\phi \cos(\theta-\lambda)$$

$$\frac{\partial \rho}{\partial \lambda} = -\frac{SR}{\rho}\cos\phi \sin(\theta-\lambda)$$

So

$$\frac{\partial \dot{\rho}}{\partial \lambda} = \frac{SR\omega}{\rho}\cos\phi\left[-\cos(\theta-\lambda) + \frac{SR}{\rho^2}\cos\phi \sin^2(\theta-\lambda)\right]$$

Evaluating $\rho$, $\frac{\partial \dot{\rho}}{\partial \phi}$ and $\frac{\partial \dot{\rho}}{\partial \lambda}$ at $\lambda = 90°$ and $\phi = \phi_o$, we get:

$$\rho_o = \rho(\lambda_o,\phi_o) = (S^2 + d^2 - 2SR\cos\phi_o \sin\theta)^{\frac{1}{2}}$$

$$\frac{\partial \dot{\rho}}{\partial \phi}(\lambda_o,\phi_o) = \frac{SR\omega}{\rho_o}\sin\phi_o(\cos\theta + \frac{SR}{\rho_o^2}\cos\phi_o\cos\theta\sin\theta)$$

$$\frac{\partial \dot{\rho}}{\partial \lambda}(\lambda_o,\phi_o) = -\frac{SR\omega}{\rho_o}\cos\phi_o(\sin\theta - \frac{SR}{\rho_o^2}\cos\phi_o\cos^2\theta)$$

Pulling this all together:

$$\begin{aligned}\|\nabla_{a,b}\dot{\rho}(\lambda_o,\phi_o)\| &= \left(\frac{1}{R^2}\left[\frac{\partial \dot{\rho}}{\partial \phi}(\lambda_o,\phi_o)\right]^2 + \frac{1}{R^2\cos^2\phi_o}\left[\frac{\partial \dot{\rho}}{\partial \lambda}(\lambda_o,\phi_o)\right]^2\right)^{\frac{1}{2}} \\ &= \frac{S\omega}{\rho_o}\left[\sin^2\phi_o\left(\cos\theta + \frac{SR}{\rho_o^2}\cos\phi_o\cos\theta\sin\theta\right)^2 + \left(\sin\theta - \frac{SR}{\rho_o^2}\cos\phi_o\cos^2\theta\right)^2\right]^{\frac{1}{2}}\end{aligned}$$

## 4.0 CALCULATION OF THE NORM OF THE GRADIENT FOR SPECIFIC CASES

When $\phi_o = 0°$, the norm of the gradient simplifies to:

$$\|\nabla\dot{\rho}\| = \frac{S\omega}{\rho_o}\,|\sin\theta - (SR/\rho_o^2)\cos^2\theta|$$

Considering only visible satellites, we have $0° \le \theta \le 180°$, so $\sin\theta \ge 0$. Since $(SR/\rho_o^2) < 1$ and $\cos^2\theta \ge 0$, the maximum clearly occurs at $\theta = 90°$. In this case,

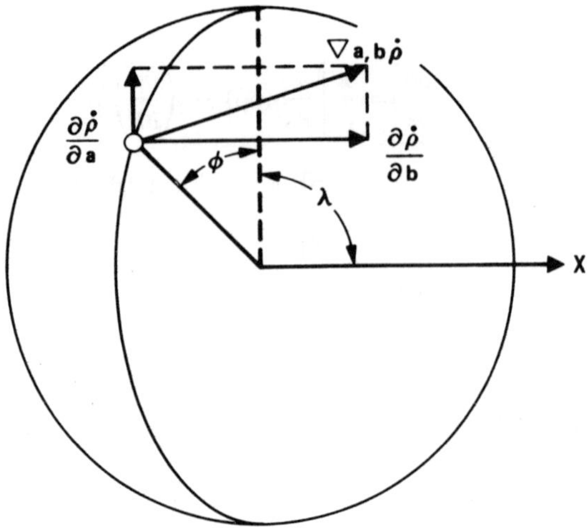


*Fig. 3*—$\nabla_{a,b}\dot{\rho}$.

the norm of the gradient is 1.002 hz/km. So, for a user in the plane of the satellite, a 1 km error in position can cause a 1 hz error in estimated Doppler, if the position error is in the direction of the gradient (which in this case would be along the "equator").

At $\phi_o = 90°$, the norm of the gradient is just:

$$\| \nabla\dot{\rho} \| = \frac{S\omega}{\rho_o}$$

This is independent of $\theta$, as it should be and is the same expression as we got for $\phi_o = 0°$ when $\theta = 90°$. However, the numerical value in this case is less (0.741 hz/km), since $\rho$ in this case is $\sqrt{S^2 + R^2}$, but in the previous case it was $(S-R)$.

As can be seen from Table 1, the first case ($\phi_o = 0°, \theta = 90°$) yields the largest Doppler sensitivity to position errors. Note also that for any $\phi$, the largest sensitivity occurs when the satellite is at its highest elevation relative to the user ($\theta = 90°$). However, two satellites at the same elevation but in different planes will not have the same sensitivity.

For example, $\phi = 0°, \theta = 0°$ and $\phi = 90°, \theta = 0°$ give the same elevation for two satellites in orthogonal orbits; but the sensitivity of the second is nearly five times the sensitivity of the first. So elevation angle alone isn't enough to determine which satellite will have the least sensitivity; the relative orbits need to be considered.

Section 6 computes the horizon and mask angle values of $\theta$ so that we can see which entries in Table 1 are below the horizon or below the mask angle.

Table 1—Values of Doppler error due to position errors (HZ/KM position error), as a function of $\phi_o$ and $\theta$.

| | $\theta \rightarrow$ | | | | | | | | | |
|---|---|---|---|---|---|---|---|---|---|---|
| $\phi$ | 0° | 10° | 20° | 30° | 40° | 50° | 60° | 70° | 80° | 90° |
| 0° | .168 | .050 | .085 | .236 | .400 | .571 | .735 | .874 | .968 | 1.002 |
| 10° | .210 | .145 | .168 | .279 | .426 | .586 | .741 | .874 | .964 | .996 |
| 20° | .299 | .272 | .296 | .375 | .491 | .624 | .757 | .872 | .950 | .978 |
| 30° | .398 | .392 | .419 | .479 | .568 | .672 | .777 | .867 | .929 | .951 |
| 40° | .493 | .499 | .526 | .574 | .640 | .716 | .792 | .857 | .900 | .917 |
| 50° | .578 | .589 | .613 | .650 | .696 | .748 | .798 | .841 | .870 | .880 |
| 60° | .647 | .659 | .678 | .704 | .733 | .765 | .795 | .820 | .837 | .842 |
| 70° | .698 | .708 | .721 | .736 | .752 | .768 | .783 | .795 | .803 | .806 |
| 80° | .730 | .735 | .741 | .747 | .754 | .759 | .764 | .768 | .771 | .772 |
| 90° | .741 | .741 | .741 | .741 | .741 | .741 | .741 | .741 | .741 | .741 |

Doppler error due to position = $\|\nabla\dot{\rho}\|$ $(L_1/C)$ in units of hz/km

## 5.0 SENSITIVITY OF RANGE-RATE TO ERRORS IN THE INITIAL TIME

Here we take the derivative of range-rate with respect to time. This gives us range acceleration, $\ddot{\rho}$:

$$\frac{d}{dt}(\rho\dot{\rho}) = \dot{\rho}^2 + \rho\ddot{\rho}$$
$$= SR\omega^2 \cos\phi_o \cos(\theta - \lambda)$$

Solving this equation for $\ddot{\rho}$, setting $\lambda = 90°$ and substituting for $\dot{\rho}$, we get:

$$\ddot{\rho} = \frac{S\omega}{\rho_o}\cos\phi_o\left[\sin\theta - \frac{SR}{\rho_o^2}\cos^2\theta\right](R\omega)$$

Again, we can see that $|\ddot{\rho}|$ is maximum at $\theta = 0°$, just as before. At $\phi_o = 0°$, max $|\ddot{\rho}| = 0.926$ hz/sec. of error in the initial time. Values of $|\ddot{\rho}|$ for other combinations of $\phi_o$ and $\theta$ are given in Table 2.

## 6.0 HORIZONS AND MASK ANGLES

In using Tables 1 and 2, it may be useful to know the value of $\theta$, given $\phi$, for a satellite on the horizon or at a 10° mask angle. These relations are derived here.

The horizon value of $\theta$ ($\theta_H$) can be found from Figure 4, as follows:

> The point *P* is at the intersection of the tangent plane to the earth at point *d* and the orbit of the satellite. The tangent plane intersects the orbit plane in a line parallel to the x-axis, and so the angle *A-P-O* is the same as $\theta_H$. In particular, the point A will be on this line. Since the *O-R-A* triangle is a right triangle with one angle $\phi$ and adjacent side R, the hypotenuse r is found from
>
> $$R = r \cdot \cos\phi$$

Then the right triangle *O-A-P* can be solved for $\theta$:

$$r = S\sin\theta_H$$
$$\Rightarrow \theta_H = \sin^{-1}\left(\frac{R}{S}\frac{1}{\cos\phi}\right)$$

To find the value of $\theta$ at the 10° mask angle ($\theta_\alpha$), we need to use a little bit of analytic geometry and Figure 5.

We construct a cone at *R* with axis along *R*. This axis will be our new Z-axis, and we use a new coordinate system with this Z-axis, the original X-axis, and a new Y-axis orthogonal to these two. The mask angle, $\alpha$, is the angle between the cone and the tangent plane at *R*. Then the equation of the cone is (in the new coordinate frame) is:

$$\tan\alpha(x^2 + y^2) = (Z - R)^2$$

Although the new coordinate system complicates the equation of the circular orbit, we can still write down the (X, Y, Z) coordinates of any point on the circle in terms of $\theta$:

$$(X, Y, Z) = S(\cos\theta, \sin\theta\sin\phi, \sin\theta\cos\phi)$$

Table 2.—Values of Doppler error due to time errors (HZ/Sec. of error) as a function of $\phi_o$ and $\theta$.

| $\phi$ \ $\theta \rightarrow$ | 0° | 10° | 20° | 30° | 40° | 50° | 60° | 70° | 80° | 90° |
|---|---|---|---|---|---|---|---|---|---|---|
| 0° | −.155 | −.046 | .078 | .218 | .370 | .527 | .678 | .807 | .894 | .926 |
| 10° | −.153 | −.045 | .077 | .215 | .364 | .518 | .666 | .791 | .876 | .906 |
| 20° | −.146 | −.043 | .075 | .206 | .347 | .491 | .628 | .744 | .821 | .849 |
| 30° | −.134 | −.038 | .070 | .191 | .319 | .448 | .569 | .670 | .737 | .760 |
| 40° | −.119 | −.033 | .064 | .170 | .280 | .391 | .492 | .575 | .630 | .649 |
| 50° | −.100 | −.026 | .055 | .143 | .234 | .322 | .402 | .466 | .508 | .523 |
| 60° | −.078 | −.019 | .044 | .112 | .180 | .246 | .304 | .349 | .379 | .389 |
| 70° | −.053 | −.012 | .031 | .077 | .122 | .164 | .201 | .230 | .248 | .255 |
| 80° | −.027 | −.006 | .016 | .039 | .061 | .082 | .099 | .112 | .121 | .124 |
| 90° | .000 | .000 | .000 | .000 | .000 | .000 | .000 | .000 | .000 | .000 |

Doppler errors due to time errors = $\ddot{\rho}$ ($L_1$/C)

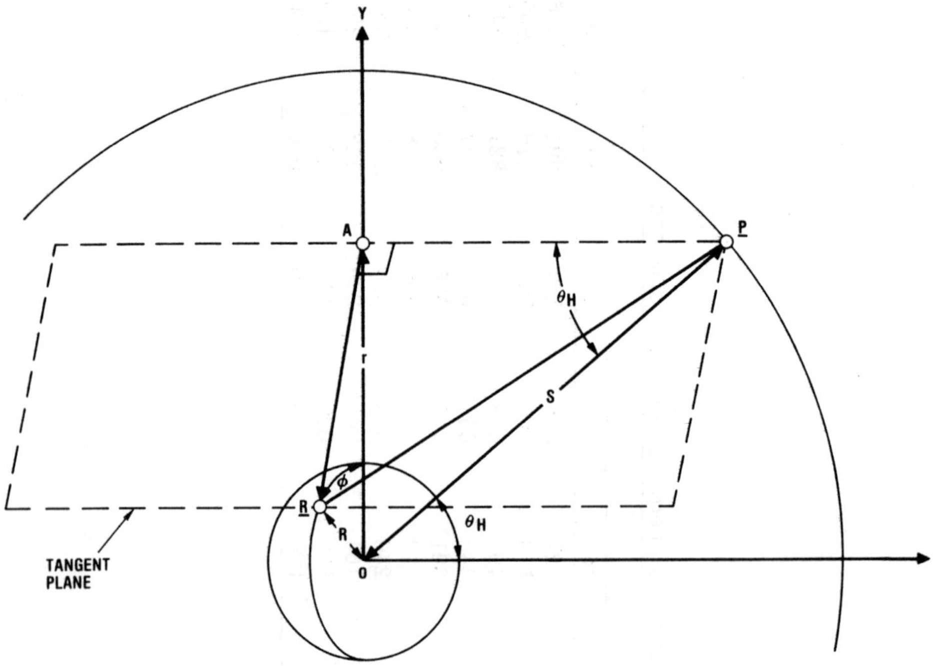


*Fig. 4—The user's horizon angle, $\theta_H$.*

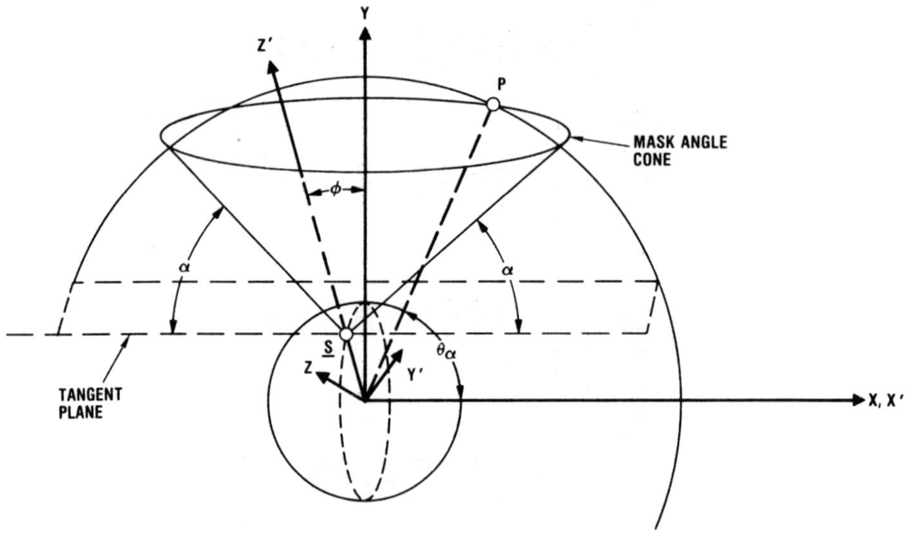


*Fig. 5—Calculation of the satellite orbit angle, $\theta_\alpha$, at the cone of the user's mask angle, $\alpha$.*

We find the point of intersection by substituting into the equation of the cone and simplifying. This yields:

$$\tan^2 \alpha - \tan^2 \alpha(\cos \phi \sin \theta_\alpha)^2 = (\cos \phi \sin \theta_\alpha)^2 - 2(R/S)(\cos \theta_\alpha \sin \theta_\alpha) + (R/S)^2$$

Rearranging terms, we get a quadratic equation whose solution is:

$$\cos \phi \sin \theta_\alpha = R/S \cos^2 \alpha \pm \sin \alpha(1 - (R/S^2)\cos^2 \alpha)^{\frac{1}{2}}$$

The root with the minus sign can be discarded, since this represents the intersection of the orbit with the extension of the cone below the tangent plane.

Numerical values for the horizon and mask angles are given in Table 3. Then, comparing this table with Table 1, we see that the entire first two columns and last two rows (plus a few other entries) represent cases below the horizon, and the first three rows and three columns represent cases below a 10° mask angle.

## 7.0 OTHER DOPPLER ERROR SOURCES

Two sources of Doppler errors not yet mentioned are velocity and earth-rate uncertainties. These are two sides of the same coin, since earth-rate is (locally) just a horizontal velocity, so let's first consider velocity uncertainties in general.

In the most general case when the velocity uncertainties are equal in all directions, the magnitude of the gradient of $\dot{\rho}$ with respect to user velocity is exactly 1, since the velocity uncertainty may be along the line-of-sight to the satellite. Note that the gradient in this case is independent of the orientation between the user and the satellite, so it has no influence on the order of acquiring the satellites. Less general cases have to be analyzed individually.

One special case of interest is the effect of uncertainties in the Earth-rate due to uncertainties in the user's position.

Let $\Omega$ = Earth-rate = 15°/hr = 7.27 $-10^{-5}$ rad/S. Then the local velocity vector due to Earth-rate is

$$V(\gamma,\mu) = \Omega R \cos \gamma \hat{T}(\mu) = 463 \text{ m/s} \cos \gamma \hat{T}(\mu)$$

where

$\gamma$ = user's latitude
$\mu$ = user's longitude
and $\hat{T}(\gamma,\mu)$ = the unit tangent to the latitude circle.

Table 3—Numerical values for horizon and mask angles

| $\phi$ | $\theta$ at the Horizon | $\theta$ at a 10° Mask Angle |
|---|---|---|
| 0° | 13.9° | 23.7° |
| 10° | 14.1° | 24.0° |
| 20° | 14.8° | 25.3° |
| 30° | 16.1° | 27.6° |
| 40° | 18.2° | 31.6° |
| 50° | 21.9° | 38.6° |
| 60° | 28.7° | 53.4° |
| 70° | 44.5° | — |
| 80° | — | — |
| 90° | — | — |

$\theta$ at the Horizon and at a 10° Mask Angle

An error in latitude either shrinks or expands $V(\gamma,\mu)$, without changing its direction, while an error in longitude tilts $V(\gamma,\mu)$ toward or away from the earth's spin axis.

Let

$$\delta q = R\delta\gamma \;=\; \text{a small distance displacement due to a change in latitude.}$$
$$\delta h = R\cos\gamma\delta\mu \;=\; \text{a small distance displacement due to a change in longitude.}$$

Then the sensitivity of $V$ due to displacements in latitude and longitude are:

$$\frac{\partial V}{\partial q} = \frac{\partial V}{\partial \gamma}\frac{\partial \gamma}{\partial q} = (-\Omega R \sin\gamma \hat{T}(\mu))\frac{1}{R}$$
$$= -\Omega \sin\gamma \hat{T}(\mu)$$
$$\frac{\partial V}{\partial h} = \Omega R\cos\gamma\left(\frac{\partial \hat{T}(\mu)}{\partial h}\right)$$

Now $\dfrac{\partial \hat{T}(\mu)}{\partial h}$ is the curvature of the latitude circle, $\kappa$, in the direction of the normal to the circle, $\hat{n}(\mu)$. So

$$\frac{\partial V}{\partial h} = \Omega R\cos\gamma\left(\frac{1}{R\cos\gamma}\hat{n}(\mu)\right)$$
$$= \Omega\hat{n}(\mu)$$

The Doppler error resulting from these earth-rate errors is the projection of the earth-rate errors onto the line-of-sight unit vector to the satellite, $\hat{\rho}$:

$$\Delta D(\gamma,\delta q,\delta h) = -\Omega\sin\gamma\delta q\hat{T}\cdot\hat{\rho} + \Omega\delta h\hat{n}\cdot\hat{\rho}$$
$$= \Omega(\delta q\sin\gamma\hat{T} + \delta h\hat{n})\cdot\hat{\rho}$$

At the equator, this becomes:

$$\Delta D(o,\delta q,\delta h) = \Omega\delta h\hat{n}\cdot\hat{\rho}$$

and the sensitivity is

$$\left|\frac{\Delta D}{\delta h}(o,\delta q,\delta h)\right| = |\Omega\hat{n}\cdot\hat{\rho}|$$

This is maximum when $\hat{\rho}$ is parallel to $\hat{n}$, that is, when the satellite is directly overhead, the same as in sections 3, 4 and 5 where Earth-rate errors were ignored. In this case the sensitivity is:

$$\max\left|\frac{\Delta D}{\delta h}\right| = \Omega$$

$= 7.27 \cdot 10^{-2}$ (m/s of relative velocity error)/(km of position error)

$=$ 0.382 hz of Doppler error/km of position error.

Comparing this with the sensitivities calculated in sections 4 and 5, we see that this is a secondary, but significant, effect.

As the latitude increases, the sensitivity to latitude errors also increases. In addition, note that $\delta q \sin \gamma \hat{T} + \delta h \hat{n}$ remains in the plane of the latitude circle. This plane is always parallel to the equator, so it appears to tilt toward the horizon as latitude is increased, until at the pole it becomes the tangent plane. Note also that the sensitivity of the Doppler estimate is maximized for satellites in the latitude plane and in the direction of $-\hat{n}$.

## 8.0 CONCLUSIONS

We have seen that a 1 km error in the user's position can cause up to 1.0 hz in the estimated GPS satellite Doppler frequency directly (ignoring earth-rate) and that the same error can cause an additional 0.38 hz error indirectly through mis-estimating earth-rate.

We have also seen that a one-second error in time can cause a 0.93 hz Doppler error.

The significance of these results is that large errors in estimated Doppler can cause the acquisition algorithms of GPS user sets to either be slow in acquiring the signal or to not acquire it at all. The design can be adapted in several ways to account for these errors.

First, the time error can be reduced to an insignificant level by using a battery-powered, continuous "Real-Time-Clock." While this is a common practice (to reduce the time errors and to save the pilot the effort of having to enter the time at each turn-on) the importance of using an accurate initial time has not been previously calculated. For example, an initial time error of 10 minutes could cause a Doppler error of 556 hz, which is more than half the PRN (pseudo-random-noise) code rate.

The Doppler errors due to position errors can not be eliminated so easily. Nevertheless, if one has some estimate of the position uncertainties, one can either design the receiver's acquisition algorithm to have an appropriate bandwidth for the resulting Doppler uncertainties, or design the receiver to search the appropriate number of frequency bins so that it neither misses the signal by searching too few nor wastes time searching too many.

Perhaps the most significant application is that it gives on a rational method of selecting the order of acquisition of the satellites to be tracked. In particular, the highest elevation satellite is generally the most sensitive to the position and time uncertainties discussed, and contrary to conventional wisdom, should not be acquired first. Additionally, choosing the lowest elevation satellite (above some mask angle) is generally not the best policy either: one must consider both the orientation of the satellite's orbit and its position within the orbit, even when earth-rate is ignored.

## REFERENCES

1. Kruh, P., "Satellite Constellation," Record of Symposium on GPS, 1983 Electronics Industries Association Spring Conference.
2. Sturza, Mark A., "Digital Signal Processing Techniques for GPS Receivers," IEEE National Telesystems Conference, November 1983.

# An Aspect of the Role of the Clock in a GPS Receiver

EDWARD M. COPPS

Intermetrics, Inc., Huntington Beach, CA
*Received January 1984*

## ABSTRACT

Analytic solutions for the least squares covariance and for steady state optimal estimate covariance have been derived. These solutions define and succinctly describe the relationship between clock performance, satellite geometry and system accuracy. It is seen that the clock attenuates the "bad geometry" effect associated with one-way ranging. The steady state solutions obtained depend on certain specializations of the system dynamics and on limit cases where the clock is either much better than or much worse than the position dead-reckoner.

## INTRODUCTION

A basic concept involved in the NAVSTAR Global Positioning System is frequently motivated by recognizing that measurements of time of arrival of transmissions from four (or more) geometrically diverse known points establishes the three-dimensional position of the receiver and time. Each measurement is described as a pseudorange measurement, meaning a measurement of time of arrival with an imperfect knowledge of time, such that a measurement contains a combination of distance and time error information. This note addresses the following issue. If a measurement to a satellite is expressed in the concept of pseudorange, which involves a time error, what is the role of the clock and, in particular, are good clocks better than bad clocks, and why? The note is limited to discussing performance accuracy while tracking.

## MATHEMATICAL PRELIMINARIES

### *Optimal Estimation*

The navigation process can be stated as a vector of linearized stochastic error differential equations: (Ref. 1, Ch. 12)

$$\dot{\boldsymbol{x}} = F\,\boldsymbol{x} + \mathbf{w} \tag{1}$$

with $\boldsymbol{x}$ the 4 vector of three-dimensional position error and time error. $\boldsymbol{w}$ is purely random process noise of power density $Q$ where the process noise variance is

$$E[\mathbf{w}(t)\,\mathbf{w}^{\mathrm{T}}(t+\tau)] = Q(t)\,\delta(\tau) \tag{2}$$

$\delta(\tau)$ is the Dirac delta function (Ref. 1, Ch. 11).

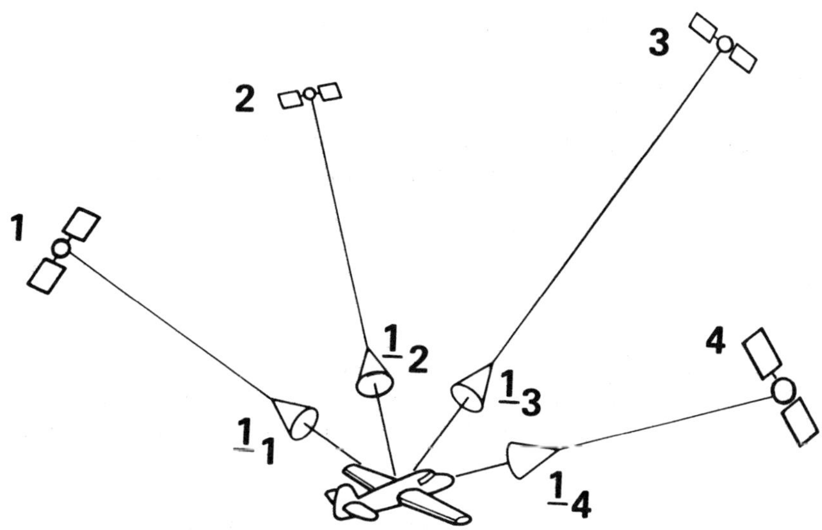


*Fig. 1*—$\mathbf{I}_n$—A Unit Vector Pointing at GPS Satellite *n*

Given a vector of continuous linear measurement: $\tilde{\mathbf{z}}$, (the pseudorange measurements)

$$\tilde{\mathbf{z}} = H\,\mathbf{x} + \boldsymbol{\nu} \tag{3}$$

where the measurement noise $\boldsymbol{\nu}$ is also purely random, with variance:

$$E[\boldsymbol{\nu}(t)\,\boldsymbol{\nu}^{\mathrm{T}}(t+\tau)] = R(t)\,\delta(\tau) \tag{4}$$

with $F$, $H$, $Q$ and $R$ known, the optimal estimate (Kalman filter) of all components of $\mathbf{x}$ is the solution of

$$\dot{\hat{\mathbf{x}}} = F\hat{\mathbf{x}} + PH^{\mathrm{T}}R^{-1}(\tilde{\mathbf{z}} - H\hat{\mathbf{x}}) \tag{5}$$

where $P$ is the solution of

$$\dot{P} = F\,P + P\,F^{\mathrm{T}} + Q - P\,H^{\mathrm{T}}\,R^{-1}\,H\,P \tag{6}$$

The covariance matrix $P$ defines the shape and orientation of the uncertainty ellipsoid of the estimate.

$$P = E[(\hat{\mathbf{x}} - \mathbf{x})\,(\hat{\mathbf{x}} - \mathbf{x})^{\mathrm{T}}] \tag{7}$$

### *Least Squares Estimation*

Another method of solution, least squares estimation, yields an estimate by the algorithm

$$\hat{\mathbf{x}} = (H^{\mathrm{T}}\,R^{-1}\,H)^{-1}\,H^{\mathrm{T}}\,\tilde{\mathbf{z}} \tag{8}$$

This minimizes the expected value of measurement residuals

$$E[(\tilde{z} - H\hat{x})(\tilde{z} - H\hat{x})^{\mathrm{T}}] \tag{9}$$

without accounting for information available from the process dynamics. The covariance of state vector errors in this case is

$$P = E[(\hat{x} - x)(\hat{x} - x)^{\mathrm{T}}] = (H^{\mathrm{T}} R^{-1} H)^{-1} \tag{10}$$

The matrix $R$, in this formulation describes the measurement error variance rather than its power spectral density.

When measurement errors are uncorrelated and have the same error statistics, which we will assume throughout this paper, the covariance simplifies to

$$P = (H^{\mathrm{T}}H)^{-1}\, r \tag{11}$$

where $R = r\,I$, $r$ scalar.

$(H^{\mathrm{T}}H)^{-1}$ describes the effect of measurement geometry on accuracy. The diagonal elements are the variances of the various components of the error state. Its trace is called the Geometric Dilution of Precision (GDOP). The trace of the upper left 3 × 3 partition is position error variance and is called PDOP, Position Dilution of Precision. The lower right element is time error variance, TDOP.

## CLOCK ERROR EFFECTS IN LEAST SQUARES PROCESSING

The linear relation between the three components of position error and time error to $n$ satellite measurements, $n \geq 4$, is

$$z = H\,x;\ H = \begin{bmatrix} \mathbf{1}_1^{\mathrm{T}} & 1 \\ \mathbf{1}_2^{\mathrm{T}} & 1 \\ \cdot & \cdot \\ \cdot & \cdot \\ \mathbf{1}_n^{\mathrm{T}} & 1 \end{bmatrix} \tag{12}$$

It follows that

$$H^{\mathrm{T}}H = \begin{array}{c} \begin{array}{cc} \longleftarrow 3 \longrightarrow & \longleftarrow 1 \longrightarrow \end{array} \\ \left[\begin{array}{cc} \sum_{j=1}^{n} \mathbf{1}_j\, \mathbf{1}_j^{\mathrm{T}} & \sum_{j=1}^{n} \mathbf{1}_j^{\mathrm{T}} \\ \sum_{j=1}^{n} \mathbf{1}_j^{\mathrm{T}} & n \end{array}\right] \end{array} \begin{array}{c} \updownarrow 3 \\ \updownarrow 1 \end{array} \tag{13}$$

Call the upper left 3 × 3 matrix $M^{-1}$, if $M$ exists, and the vectors in the upper right and lower left partitions $\ell$. Then, the inverse of $H^{\mathrm{T}}H$ is

$$[H^{\mathrm{T}}\, H]^{-1} = \begin{bmatrix} M + M\,\ell\ell^{\mathrm{T}}M/(\mathrm{n} - \ell^{\mathrm{T}}M\ell) & -M\ell/(n - \ell^{\mathrm{T}}M\ell) \\ -\ell^{\mathrm{T}}M/(n - \ell^{\mathrm{T}}M\ell) & 1/(n - \ell^{\mathrm{T}}M\ell) \end{bmatrix} \tag{14}$$

The upper left 3 × 3 partition of $[H^{\mathrm{T}}H]^{-1}$ is the position error covariance. It is composed of $M$—which is seen from equation 13 to be the position error covariance that would occur if the time error state was somehow suppressed, as it could be with a perfect clock (or as it is in two-way or RHO-RHO distance measurement systems). The additive term is the result of the necessary inclu-

$$\sum_{j=1}^{n} \underline{1}_j^t \, \underline{1}_j^{\,t} \equiv M^{-1}$$

$$\sum_{j=1}^{n} \underline{1}_j \equiv \underline{\ell}$$

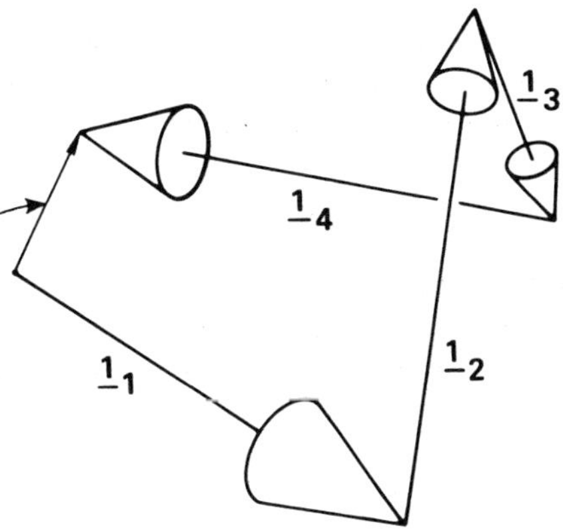


*Fig. 2—Definition of M and ℓ.*

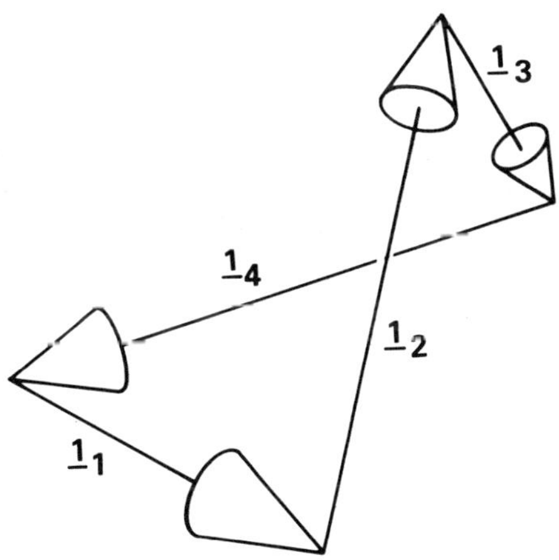


*Fig. 3—An illustration of the condition ℓ = 0.*

sion of time error in the formulation. Position error covariance using the least squares algorithm is the upper left 3 × 3 partition

$$P_{UL} = (M + M\,\ell\ell^T M/(n - \ell^T M\,\ell))\, r \tag{15}$$

Therefore, the second term is the position accuracy penalty paid in least squares estimation for not having a perfect clock. When $\ell = O$, the necessity of carrying the time error state has no effect on accuracy. Figure 3 is a sketch of four unit vectors for which $\ell = 0$. This figure is not rigid. It can be manipulated onto the edges of a tetraheydron, or flattened into two, or even one dimension. As long as $\ell = 0$, equation 15 tells us that the addition of the time error state does not change the solution if it exists.

## CLOCK ERROR EFFECTS IN OPTIMAL PROCESSING

We now examine the effect of clock errors when an optimal estimator is used. The error covariance matrix is governed by the quadratic matrix differential equation, eq. 6, repeated here

$$\dot{P} = F\,P + P\,F^{T} + Q - P\,H^{T}R^{-1}H\,P \tag{16}$$

Under certain circumstances, the steady state value of $P$ can be analytically evaluated. We assume that $F = 0$ and that $H$, $Q$ and R are constants. $Q$ is diagonal and $R$ is diagonal with all diagonal elements equal.

$$R = r\,I,\ r \text{ scalar} \tag{17}$$

As stated, we have the following set of equations, in terms of error state variables

$$\begin{aligned} \dot{x} &= w_p \\ \dot{y} &= w_p \\ \dot{z} &= w_p \\ c\dot{t} &= w_c \end{aligned} \tag{18}$$

That is, the three dimensions of position error are independent random walk processes with the same process noise power density, and clock error is also a random walk process with different noise power density. Time error is scaled by the speed of light $c$.

The power density of the noise $w_c$ is the measure of the inaccuracy of the clock.

The $Q$ matrix is diagonal

$$Q = \begin{bmatrix} q & & & 0 \\ & q & & \\ & & q & \\ 0 & & & q_c \end{bmatrix} \tag{19}$$

where $q$ is the noise power density describing the decay of knowledge of position, and $q_c$ is the noise power density describing the clock drift.

In order to support the mathematics, we choose to rescale the time error variable $t$ to make the diagonal elements of $Q$ equal. This is accomplished by

$$t' = \sqrt{q\,/\,q_c}\;t \tag{20}$$

With this redefinition, the noise power density matrix becomes

$$Q' = q\,I \tag{21}$$

The system of equations describes a process whose states are independent and identical random walks, but which are strongly coupled due to the measurements. In order to accomplish this, the time error state has been scaled by a factor related to the square root of the ratio of position process noise to clock process noise.

Our purpose is to produce useful analytic solutions which show how clock errors affect position error. To accomplish this, we construct the $H$ matrix for these state variables.

$$z = \begin{bmatrix} 1_1^T & k \\ 1_2^T & k \\ \cdot & \cdot \\ \cdot & \cdot \\ 1_n^T & k \end{bmatrix} x' \tag{22}$$

where $k = \sqrt{q_c / q}$

The factor $k$ occurs because of the rescaling of time error. The unit vectors $1_j, j = 1, n$ point at the $n$ satellites.

It follows that (see equation 13)

$$H^T H = \begin{bmatrix} M^{-1} & k\ell \\ k\ell^T & k^2 n \end{bmatrix} \tag{23}$$

By rescaling the time error variable, the matrix Ricatti equation (eq. 16), simplifies to

$$\frac{\dot{P}}{q} = I - P \begin{bmatrix} M^{-1} & k\ell \\ k\ell^T & k^2 \mathrm{n} \end{bmatrix} P / (qr) \tag{24}$$

If a steady state for $P$ exists, it will be

$$P = [\ \sqrt[S]{H^T H}\ ]^{-1} \sqrt{qr} \tag{25}$$

where the notation defines the positive symmetric square root of the $H^T H$ matrix. To the author's best knowledge, it is not possible to find an analytical solution for the symmetric square root of a symmetric matrix. Numerical solutions can be obtained by diagonalizing the matrix. The symmetric square root has the same eigenvectors and has eigenvalues that are the square root of the matrix eigenvalues.

However, in this case, it is possible to find approximate solutions at the following limits: $k \to 0$, and $k \to \infty$.

Since $k = \sqrt{q_c/q}$ this corresponds to a clock which is very good compared to position error build-up, and to a clock which is very poor compared to the same criterion.

For $k \to 0$ (a good clock), the symmetric square root is (terms of $o(k)$ retained)

$$\sqrt[S]{H^T H} \quad \cong \quad \begin{bmatrix} \sqrt[S]{M^{-1}} & \sqrt[S]{M}\, k\, \ell \\ \ell^T \sqrt[S]{M}\, k & k\sqrt{(n - \ell^T M \ell)} \end{bmatrix} \tag{26}$$

Inverting equation 26 and multiplying by $\sqrt{q\ r}$ yields the covariance matrix $P$.

$$P \quad \cong \quad \begin{bmatrix} \sqrt[S]{M} & \\ +\ M\,\ell\ell^T\, Mk/\sqrt{n - \ell^T M\ell} & -\ M\ell/\sqrt{n - \ell^T M\ell} \\ -\ \ell^T M/\sqrt{n - \ell^T M\ell} & 1/(k\sqrt{n - \ell^T M\ell}) \end{bmatrix} \sqrt{q\ r} \tag{27}$$

Transforming back to the original variables by unscaling the time error variable yields the covariance matrix for position error and time error $-\ ct$.

$$P_{UL} \cong \sqrt[S]{M}\,\sqrt{q\ r} + (M\,\ell\ell^T M\ /\ \sqrt{n - \ell^T\, M\ell})\,\sqrt{q_c r}$$

position error covariance

$$P_{UR} \cong -\ (M\ell\ /\ \sqrt{n - \ell^T M\ell})\,\sqrt{q_c r} \tag{28}$$

position-time error covariance

$$P_{LR} \cong (\ 1/\ \sqrt{n - \ell^T M\ell})\,\sqrt{q_c r}$$

time error covariance

This distinctly describes the effect of clock process noise on navigation and time accuracy for good clocks. Namely, the steady-state position error covariance is the sum of a term proportional to the square root of position process noise and another term proportional to the square root of clock process noise. The term proportional to the square root of position process noise is scaled by the *square root* of the *"RHO-RHO"* PDOP. The term proportional to the square root of clock process noise is similar to the additive term which differentiates RHO type from RHO-RHO type navigation, using least squares processing. The vector $M\ell$ defines the direction in which a singularity can occur.

When $k$ approaches infinity, corresponding to a clock whose error build-up is much faster than the build-up of position error, the approximate solution for steady state covariance is

$$P_{UL} = \sqrt[S]{M + M\ell\ell^T M\ /\ (n - \ell^T M\ell)}\ \sqrt{q\ r} \tag{29}$$

position error covariance

$$P_{UR} = -\ (\ \sqrt[S]{M + M\ell\ell^T M\ /\ (n - \ell^T M\ell)}^{\,-1}\ M\ell/(n - \ell^T M\ell)\ \sqrt{q\ r}$$

position-time covariance

$$P_{LR} = ((1 - \ell^T M\ (M + M\ell\ell^T M\ /(n - \ell^T\ M\ell))^{-1}\ M\ell)/\ (n - \ell^T M\ell))^{\frac{1}{2}}\ \sqrt{q_c\ r}$$

time error covariance

This indicates that position error is not affected by clock error if the clock is "poor." Another point of view is that with a poor clock, the "bad geometry" term

$$M\,\ell\ell^{T}M\,/\,(n-\ell^{T}M\,\ell) \tag{30}$$

is not attenuated.

## THE SINGULARITY CONDITION

If the unit vectors to all of the satellites being used lie on a cone of arbitrary half angle, then the unit vectors do not span the four-dimensional space. The direction in position space where position error is not separable from time error is the axis of the cone. Stated another way, if it is possible to lay a plane on top of the tips of all the unit vectors, then a singularity exists perpendicular to the plane. This condition is equivalent to the condition $n - \ell^{T}M\ell = 0$, in the equations developed.

When such a singularity exists, if the clock process noise is not zero, position error along the axis of the cone will grow without bound. It is therefore important to relate clock inaccuracy to the rate of growth of position uncertainty. The following simplified reasoning describes this effect. We confine ourselves to the two-dimensional problem of position error along the axis of the singularity cone. Along this axis, position error is undistinguishable from time error. For each measurement, the strength of the measurement gradient in this direction is $\cos\alpha$ where $\alpha$ is the cone half angle. Choosing the linear combination

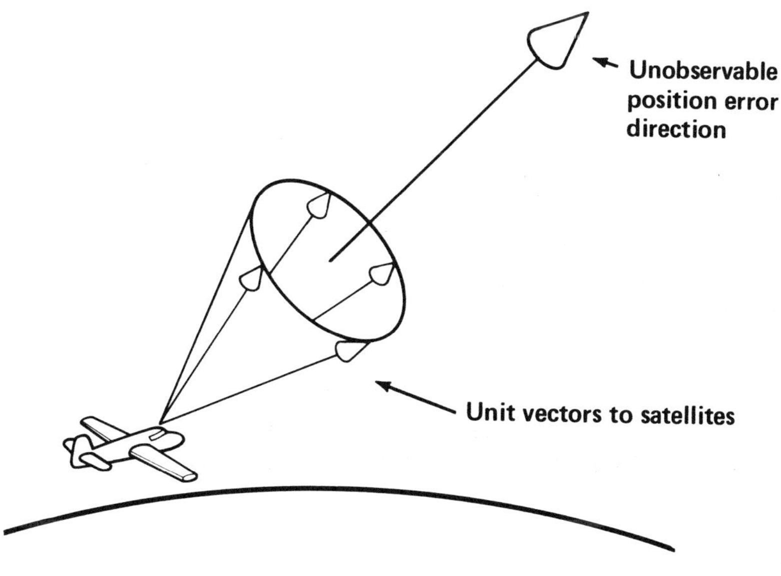


*Fig. 4—Bad geometry.*

of position error $x$ and clock error $ct$ that are respectively observable and unobservable, we have for $n$ satellites,

$$z = n \cos \alpha \, [1 \quad 0] \begin{bmatrix} x + ct \\ x - ct \end{bmatrix} \tag{31}$$

defining $x_1 = x + ct$
$x_2 = x - ct$

we have

$$\dot{P}' = Q' - P' H^T H P'/r$$

$$\text{where } Q' = \begin{bmatrix} (q + q_c) & (q - q_c) \\ (q - q_c) & (q + q_c) \end{bmatrix} \tag{32}$$

$$H^T H = \begin{bmatrix} n^2 \cos^2 \alpha & 0 \\ 0 & 0 \end{bmatrix} \tag{33}$$

The following set of scalar equations results

$$\dot{P}'_{11} = q + q_c - P'^2_{11} n^2 \cos^2\alpha \,/r$$

$$\dot{P}'_{12} = q - q_c - P'_{11} P'_{12} n^2 \cos^2\alpha \,/r \tag{34}$$

$$\dot{P}'_{22} = q + q_c - P'^2_{12} n^2 \cos^2\alpha \,/r$$

$P'_{11}$ and $P'_{12}$ reach steady state values, after which $P'_{22}$ grows linearly with time

$$P'_{22} = \frac{4 \, q \, q_c}{q + q_c} \, t \tag{35}$$

Transforming back to position error and time error yields the following relationship for position variance $\sigma_x^2$

$$\sigma_x^2 = (P'_{11} + 2P'_{12} + P'_{22}) \,/\, 4 \tag{36}$$

Thus, after the other terms have reached steady state, position variance in the singular direction grows as

$$\sigma_x^2 = \frac{q \, q_c}{q + q_c} \, \mathrm{t} \tag{37}$$

When $q_c$ is small with respect to $q$; that is, when we expect to depend on the clock to yield position information, we have, for position variance

$$\sigma_x^2 \cong q_c \, t \tag{38}$$

## SUMMARY

The analytic solutions for least squares estimates, and the steady state solutions of the covariance equation help to define and describe the complex relationships between clock accuracy, geometry, and system performance. The $M$ matrix, which one can analogize to two-way ranging (RHO-RHO) geometry, and the $\ell$ vector are the quantities that characterize GPS geometry.

Besides the general improvement in visualization of these effects that this paper may provide, the formulas may be particularly useful in choosing satellites and in mapping the expected accuracy of a GPS receiver as a function of satellite constellation geometry. The tendency to ignore the contribution of the receiver clock may yield too pessimistic predictions of GPS performance.

## ACKNOWLEDGMENTS

Reference 2 was used to develop the inversion process for a partitioned matrix. They refer to references 3 and 4 as their source. The author also wishes to acknowledge the support provided by Dr. Lawrence O. Lupash of Intermetrics, Inc. in discussing procedures for obtaining symmetric square roots and for reviewing the manuscript; and my thanks to Ms. Shirley Kalle for typing the manuscript.

## REFERENCES

1. Bryson, A. E., and Ho, Yu-Chi *Applied Optimal Control,* John Wiley & Sons, 1975.
2. Athans, M., and Schweppe, F. C., "Gradient Matrices and Matric Calculations," TN 1965-53, Lincoln Laboratory, Lexington, MA (1965).
3. Bodewig, E., *Matrix Calculus,* North Holland Publishing Co. and Interscience Publishers, NY, (1956).
4. Bellman, R., *Introduction to Matrix Analysis,* McGraw-Hill Book Co., NY (1960).

# GPS Doppler Processing for Precise Positioning in Dynamic Applications

JAVAD ASHJAEE

Private Consultant, Sunnyvale, CA*

*Received January 1986*

## ABSTRACT

In this paper we show that the judicious use of all the information available from C/A code can provide results better than the reported accuracy of P code, with much lower cost. This performance, however, is limited to low-dynamic, low-jamming environments which are typical of civilian applications. The elegance of P-code is primarily manifested in high dynamic vehicles (jets and missiles) and in jamming environments.

Results are provided in stationary and low-dynamic cases based on code-phase, carrier-phase, doppler, and integrated doppler measurements from TRIMBLE 4000 receivers. In particular, it is shown that the relative or differential accuracy of better than 10 centimeters can be achieved in low-dynamic applications. Position fixes are obtained every second, after a short time of initial calibration and tracking.

The improved accuracy is based on aiding the code-phase measurements with integrated doppler. Each one second position fix, which is provided in both polar and Cartesian coordinates, is based on a complete and independent set of measured data, with no hypothetical modeling, interpolation, or extrapolation.

## INTRODUCTION

Even a brief survey of the growing body of GPS literature leaves little doubt that GPS will indeed replace all existing positioning and timing systems in the near future. The attractiveness of GPS is not only due to its technical superiority, but also to economic considerations. It is interesting to observe that, even in this early stage of the GPS program, production of low cost GPS receivers is a reality. Today C/A code GPS receivers can be manufactured, in large volumes, at a lower cost than LORAN receivers.[1]

Of the two classes of GPS signals, C/A and P code, only C/A code will be available for civilian applications[2]. This restriction might lead one to conclude that the precision positioning and timing capabilities of GPS will not be available to civilian users. This is a false conclusion.

Judicious use of all the information available from C/A code can provide results equivalent to P code, with much lower cost. In this paper we prove this claim by providing results based on code phase, doppler, integrated doppler, and carrier phase measurements from TRIMBLE C/A code receivers. It is our belief that even if P code were available for civilian applications it would add very little, if any, to the precision that is available from C/A code. The elegance

*Formerly with Trimble Navigation, Sunnyvale, CA.

of P code is mostly manifested in military applications such as in high dynamic vehicles (jets and missiles) and in jamming environments.

In this paper we distinguish between four levels of accuracy that can be obtained from the C/A code system:

Level A = Accuracy of about 10 meters,
Level B = Accuracy of about 1 meter,
Level C = Accuracy of about 10 centimeters,
Level D = Accuracy of about 1 centimeter.

Each one of these accuracy levels can be achieved with a different type of C/A code information processing, and with a different operational requirement. Furthermore, a user set may be optimized for a specific accuracy level, to avoid unnecessary complications and cost overhead. Of course, it is desirable that a user set be easily upgradable to a higher level. The receivers discussed in this paper do have this feature.

The cost and the procedure for obtaining each one of the accuracy levels, depends primarily upon the deterministic (bias) and random (noise) errors of the receiver. Random errors are mostly generated by the tracking loops and by the internal oscillator of the receiver. They can readily be removed in a multichannel receiver which allocates a tracking module to each satellite. This allows *continuous* and *simultaneous* measurements which results in reduction of random errors in the following three ways:

1. Simultaneous measurements from each tracking module causes total cancellation of errors that are generated by the oscillator. Note that due to this error cancellation, a multichannel receiver using an inexpensive oscillator, performs better than a single-channel receiver even with an atomic clock.
2. Continuous measurement provides at least four times more data which results in 6 db of reduction in random errors.
3. Continuous tracking of satellites eliminates the need for reacquiring. This enables the tracking loops to maintain narrower bandwidth which has direct effect on reducing the random errors. Furthermore, eliminating reacquisition allows continuous tracking under lower signal-to-noise conditions.

In addition to the advantages mentioned above, in a multichannel receiver, continuous observation of the health status and synchronization flag of NAV data[3] inhibits the inadvertent use of unhealthy satellites.

Although there is universal agreement regarding the advantages of multiple channel design, it has been argued that the "interchannel differences" reduce its attractiveness. This concern is valid and if not carefully considered in the design of a receiver, will introduce deterministic errors. This issue is discussed in the next section and it is shown that the advantages of multiple channel design can be retained without introducing any significant bias.

Also described is the doppler processing technique for smoothing code phase measurements. Results in obtaining accuracies of levels A and B are presented. It must be noted, however, that the data presented in this paper is based on the current performance of GPS.

## INTERCHANNEL BIASES

The interchannel differences affect code phase, doppler, integrated doppler, and carrier phase measurements. Theoretically these differences can be eliminated by locking all channels to the same satellite and determining the interchannel differences as part of a power-on automatic receiver calibration. The advantage of this method is that it does not impose any overhead (except for about 100 seconds of initial calibration). It is effective only for the receivers whose characteristics do not change noticeably with time. In receivers that lack this property, calibration must be performed periodically either by locking all channels to the same satellite, or by providing an extra calibrating channel.

The design of time-invariant channels is possible, and next we present data to show that the interchannel biases can be removed with a short automatic calibration time.

The interchannel biases for code phase, doppler, and integrated doppler measurements in a Trimble 4000A receiver are presented in Figures 1, 2, and 3, respectively. These figures show about two hours of data while the receiver was exposed to high temperature gradient. Interchannel bias for carrier phase, which only concerns the accuracy of Level D (4000S Surveyor receiver), is generally less than 0.1 millimeter.

Inspection of Figures 1 through 3 reveals that after about 100 seconds of initial calibration, the residual calibration errors are less than 5 centimeters for code phase, less than 0.0001 Hz for doppler, and less than 3 millimeters for

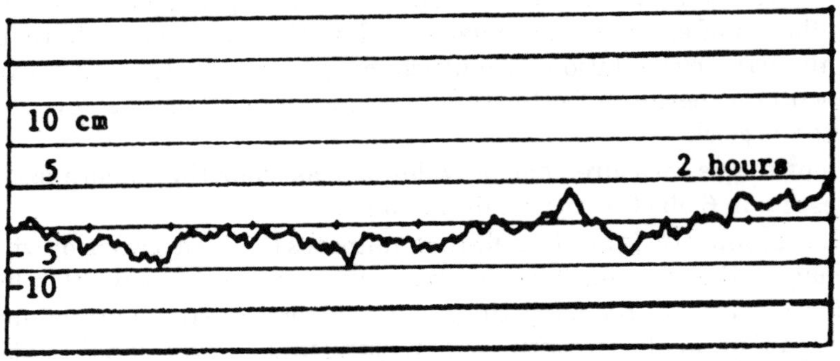


*Fig. 1—Interchannel Biases for Code Phase Measurements for 100 Sec. Averaging*

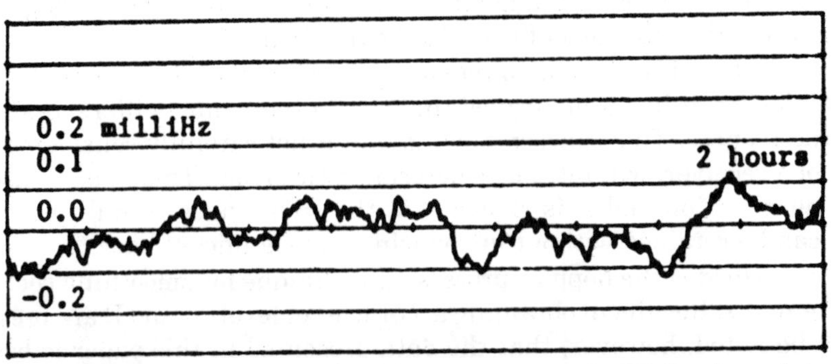


*Fig. 2—Interchannel Biases for Doppler Measurements for 100 Sec. Averaging*

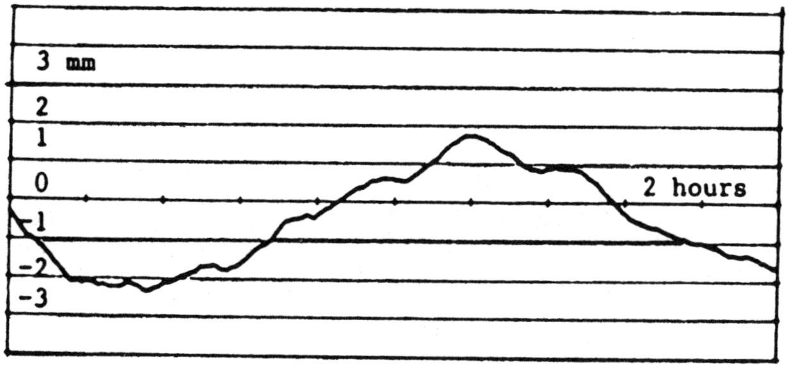


*Fig. 3—Interchannel Biases for Integrated Doppler Measurements for 100 Sec. Averaging*

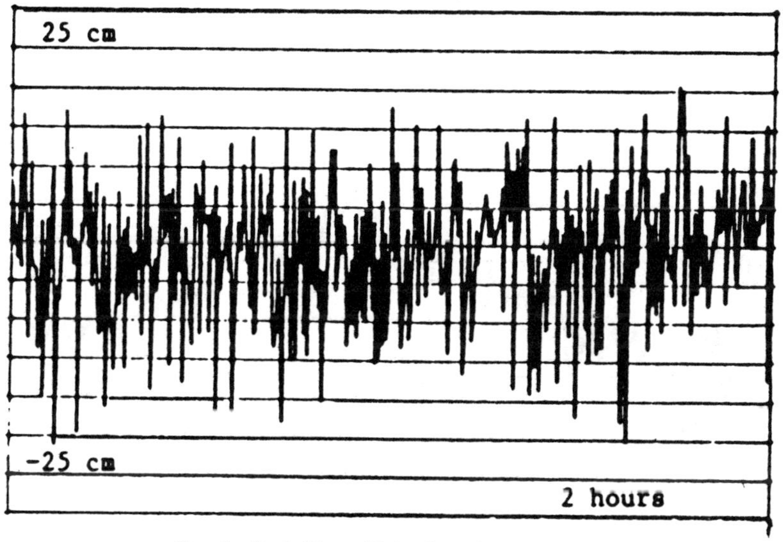


*Fig. 4—Code Phase Noise for 5 Sec. Averaging*

integrated doppler measurements. Consistency of these values during about two hours of tracking demonstrates the time invariant property of the receivers. These empirical results do indeed match the theoretical expectations derived from worst case analysis of hardware delays.

The above investigation reveals that the advantages of multiple channel design can be utilized without suffering from errors due to interchannel differences, and without introducing any significant operating overhead time. Note that the residual calibration errors can further be reduced by increasing the power-on calibration time to more than 100 seconds if higher accuracy is required.

Figures 4, 5, and 6 represent the levels of uncorrelated noise in code phase, doppler, and integrated doppler measurements, respectively. Each point in these plots is an average of 5 seconds of data.

Notice that the noise levels in doppler and integrated doppler measurements, are about 0.005 HZ and 1 millimeter, respectively. This high resolution is

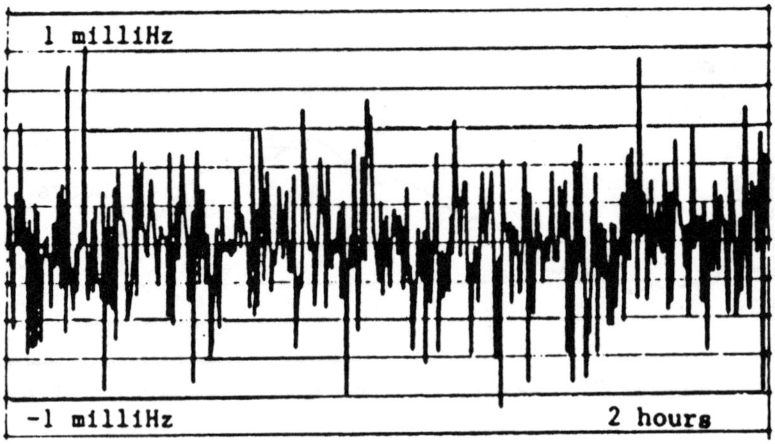


*Fig. 5—Doppler Noise for 5 Sec. Averaging*

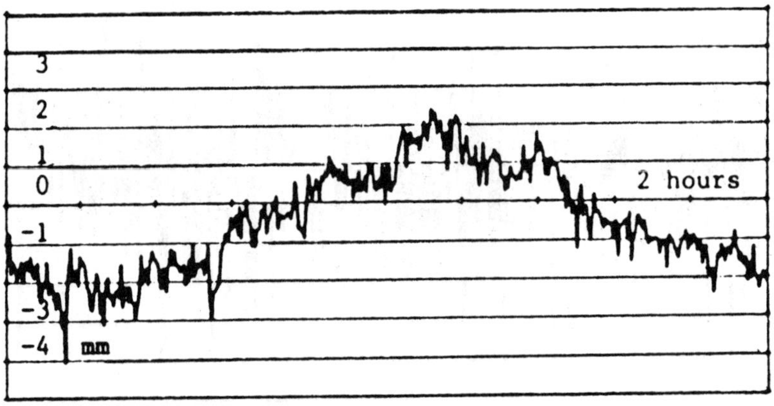


*Fig. 6—Integrated Doppler Noise for 5 Sec. Averaging*

mostly due to the multichannel design which allows simultaneous measurements from satellites and results in cancellation of errors that are caused by the short term oscillator instability. This, in turn, allows inexpensive oscillators to be used in precision measurements.

The scatter in code phase measurements is about 0.5 meter, as shown in Figure 4. For calibration this noise was reduced by simple averaging. Long-time averaging cannot be done in position fix mode since the channels are locked to different satellites and the receiver may be in a dynamic environment. Besides, 100 seconds of averaging is unacceptable in many operational applications.

The higher level of noise in code phase measurements of C/A code is considered a disadvantage, as compared with the P code. In the next section we show that by taking advantage of the fine resolution of integrated doppler we can smooth the code phase measurements of the C/A code to levels much better than conventional resolution of the P code.

## AIDING CODE PHASE WITH INTEGRATED DOPPLER

Let us first define the following variables and notations.

$C_i$ = Code phase measurements from a satellite at time $T_i$, corrected for errors mentioned below.
$D_i$ = Integrated doppler measurement from a satellite during the time interval from $T_o$ to $T_i$, corrected for the same errors.
$B_{ci}$ = Deterministic residual bias in $C_i$.
$B_{di}$ = Deterministic residual bias in $D_i$.
$V_{ci}$ = Noise in $C_i$.
$V_{di}$ = Noise in $D_i$.

A "bar" above a variable indicates the true value of the variable.

A "hat" above a variable indicates the estimated (smoothed) value of the variable.

Considering the above definitions we have

$$\overline{D}_i = \overline{C}_i - \overline{C}_o \tag{1}$$

$$C_i = \overline{C}_i + B_{ci} + V_{ci} \tag{2}$$

$$D_i = \overline{D}_i + B_{di} + V_{di} \tag{3}$$

Now define and compute

$$\begin{aligned} C_{oi}' &= C_i - D_i \\ &= \overline{C}_i - \overline{D}_i + B_{ci} - B_{di} + V_{ci} - V_{di} \\ &= \overline{C}_o + B_{ci} - B_{di} + V_{ci} - V_{di} \end{aligned} \tag{4}$$

The major error sources that affect $B_{ci}$ and $B_{di}$ are satellite clock error, ionospheric variation, tropospheric variation, relativistic effect, rotation of the earth, and receiver clock error.

Except for the ionospheric variation, the above sources of error have the same impact on code phase as on integrated doppler[4] such that their effects in (1) cancel out. Hence they shall not be considered in the following formulation.

Let $I_i$ be the residual error in the GPS ionospheric correction model[5] at time $T_i$. Therefore,

$$\begin{aligned} C_{oi}' &= \overline{C}_o + I_i - (I_o - I_i) + V_{ci} - V_{di} \\ &= \overline{C}_o + 2\, I_i - I_o + V_{ci} - V_{di} \end{aligned}$$

Next define

$$\begin{aligned} \hat{C}_{oi} &= \frac{1}{i+1} \sum_{j=0}^{i} C_{oj}' \\ &= \frac{1}{i+1} \sum_{j=1}^{i} (V_{cj} - V_{dj}) + C_o + \frac{2}{i+1} \sum_{j=0}^{i} (I_j - I_o) \end{aligned} \tag{5}$$

Let us examine the terms of the above equation. The term

$$\frac{1}{i+1} \sum_{j=0}^{i} (V_{cj} - V_{dj}) \tag{6}$$

represents the average of random errors in code phase and integrated doppler measurements. Note that $V_{dj}$ (less than 1 millimeter) is much smaller than $V_{cj}$ (on the order of 0.5 meter). The noise average will vanish as the number of measurements increases. On the other hand, the term

$$\frac{2}{i+1}\sum_{j=0}^{i}(I_j - I_o) \tag{7}$$

changes from the initial value of $I_O$. This term is deterministic, and changes very slowly with time. Note that for small values of i, the contribution of (6) dominates, and as the number of measurements increases this term vanishes and the drift of (7) becomes more visible.

The plots of (5) for about 2 hours of tracking of two channels are displayed in Figure 7. In Figure 7 the plot of one channel is slightly shifted to show the consistency of the two channels. The actual difference of the two channels for (5) is shown in Figure 8, which shows the two channels agree to better than 2 centimeters after a few minutes of tracking.

Now, assuming (6) is negligible after several measurements, we have a noise free estimate for $\overline{D}$ and we can compute a noise free estimate of the current code phase measurement as:

$$\hat{C}_i = \hat{C}_{oi} + D_i = \hat{C}_{oi} + \overline{D}_i - I_i + I_o + V_{di}$$

Using (5) and ignoring the small value of $V_{di}$ (about 1 millimeter), we obtain

$$\hat{C}_i = \overline{C}_o + \overline{D}_i + \frac{2}{i+1}\sum_{j=0}^{i} I_j - I_i \tag{8}$$

Figure 9 shows the plot of (8), which are code phase measurements aided

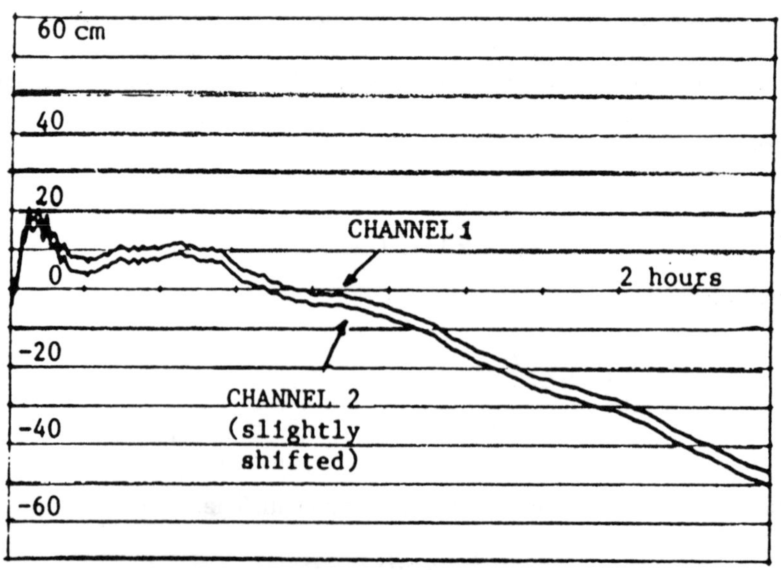


*Fig. 7—Smoothed $\hat{C}_{oi}$ as a Function of i (Time)*

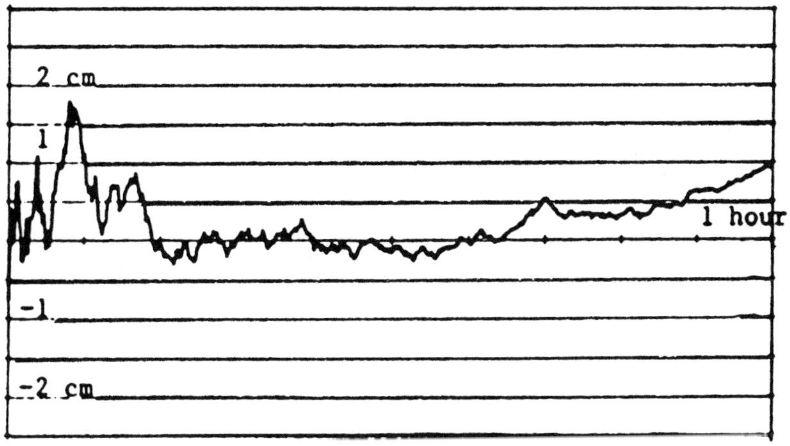


*Fig. 8—Differences Between Two Channels in Smoothing $\hat{C}_{oi}$*

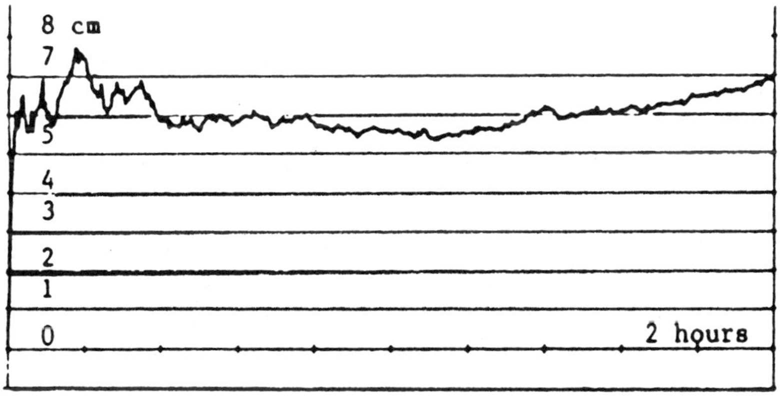


*Fig. 9—Code Phase Measurements Smoothed with Integrated Doppler*

(smoothed) with integrated doppler measurements. Note that the smoothing is applicable in both static and dynamic environments.

The plot of Figure 9 consists of a slowly changing component, which varies of the order of 1 centimeter after a few minutes of tracking (and can be removed with differential mode), and a random component. Figure 10 shows the random component of the aided code-phase measurements. As can be seen in this figure, the magnitude of the random component is about 3 millimeters.

The noise free and slowly changing nature of (8) is useful in relative and in differential positioning, especially in dynamic environments where the effect of noise in code-phase measurements cannot be reduced by averaging the position fixes. In static conditions, although position fixes can be averaged, smoothing the code-phase measurements yields better results, because the noise is reduced before the multiplicative effect of GDOP can amplify its impact.

It is important to note that the code phase smoothing described in this section requires very accurate integration of doppler. In particular, the occurences of carrier cycle slips causes significant degradation of accuracies. In the receivers

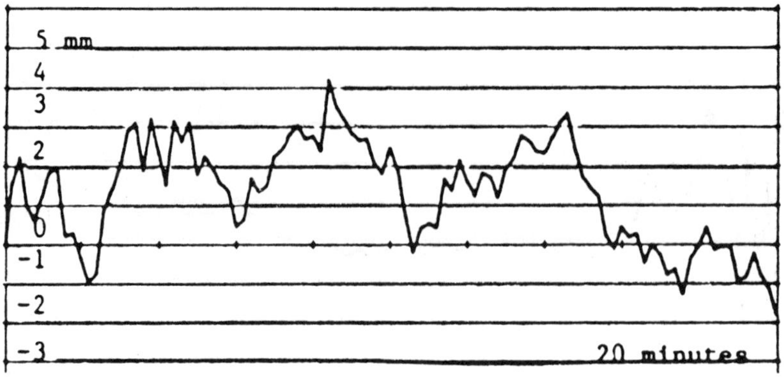


*Fig. 10—Noise Component of Smoothed Code Phase Measurements*

mentioned in this paper, cycle slips are automatically detected and edited by the receivers during doppler integration.

In the next two sections the results on level A and level B accuracies are presented. Note that the 4000S Surveyor receiver, which provides Level C and Level D accuracies, does not use code phase for translocation, hence resolution of code phase is irrelevant.

## LEVEL A: SINGLE-RECEIVER POINT POSITIONING USING AIDED CODE PHASE

In this section the accuracy of level A in static and dynamic conditions is discussed. This level of accuracy can be achieved efficiently by using a single C/A code receiver capable of providing code phase, doppler, and integrated doppler measurements. The effectiveness of integrated doppler-aiding, especially in the dynamic case, is demonstrated by the results obtained from the implementation of this technique in a Trimble 4000A Locator receiver.

### *Static Case:*

Position computation in this case is based on four code-phase measurements, which provide instantaneous results. The error sources affecting the accuracy, and methods of removing them are as follows:

1. Satellite orbit and clock estimation errors. Magnitude of these errors, during several hours after the upload, is about 1 meter. They affect C/A code as well as P code. The errors can be removed by post-processing of the measured data using the knowledge of the magnitude of errors, or via differential mode as discussed in the next section.
2. Ionospheric and tropospheric effects. Most of these two effects are removed via mathematical models. The model used for tropospheric delay (for low altitudes) is

   $$2.4225 \,/\, (\sin E + 0.025) \text{ meters}$$

   where E is the elevation angle. The model used for the effect of the ionosphere is presented in Reference 5. The residual uncorrected errors, during the past several months of observations (in Mountain View, CA), were about 2 meters for elevation angles of greater than 15 degrees.

3. Resolution of C/A code phase. The noise due to the coarse granularity of the C/A code is of the order of a few meters. This error source can be reduced by (a) averaging the code phase measurements from each satellite and/or averaging the computed position fixes, or (b) aiding the code phase measurements with integrated doppler as presented earlier. The latter technique provides PDOP times better improvement without introducing averaging error due to nonlinear motion of the satellites (most of this nonlinearity error is modeled and removed).

Figure 11 shows a plot of static position fixes based on code phase measurements smoothed with integrated doppler. In this figure, the drift in position is due to the error sources mentioned above. Note that the noise component in the plot of Figure 11 is very small. The center of the graph has 5 meters of offset with respect to WGS-72 coordinates. Note that most of the offset and drift

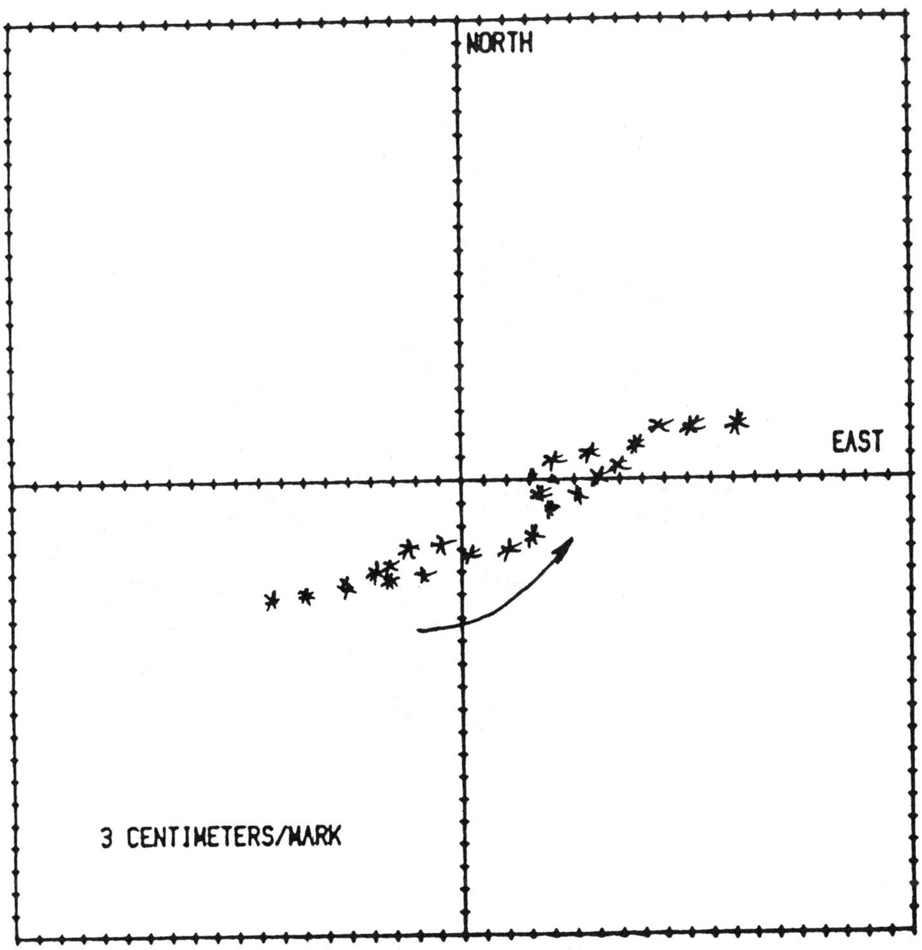


*Fig. 11—Instantaneous Static Position Fixes Using Aided Code Phases*

can be removed in differential mode as will be discussed in the next section on Level B accuracy.

*Dynamic Case:*

The error sources and the methods of removing them are similar to those in the static case, except that noise in code phase measurements cannot be removed by averaging the position fixes. Here the integrated doppler-aiding technique must be used. In the dynamic case, however, a new source of error is introduced. This error is a function of nonlinear motion of the vehicle during the measurement time. If the motion of the vehicle during every 0.25 second is linear (no sudden change in velocity or direction) the contribution of this error is negligible.

To show the effectiveness of integrated doppler-aiding in dynamic conditions, the phrase "C/A CODE GPS" was scripted on the ground, with each letter 3 by 5 meters. The GPS antenna then was moved along the letters. Figure 12 shows

*Fig. 12—Instantaneous Dynamic Position Fixes Using Aided Code Phases*

the result. The offset of the origin in this plot is about 5 meters with respect to WGS-72 coordinates.

The ability of the 4000A to measure doppler accurately, results in velocity estimates with an accuracy of better than 5 millimeters/second (0.01 Knots), as presented in Reference 6.

## LEVEL B: DIFFERENTIAL POSITIONING WITH AIDED-CODE-PHASE

In this section the accuracy of Level B in static and dynamic environments is discussed. This level of accuracy can be achieved by using a pair of receivers similar to Level A, but used in differential mode. The position fixes can be obtained in real time if a communication link between the two receivers is available. If not, the position fixes are recorded at each site and compared later. The errors determined by the receiver at a pre-surveyed site are removed from the data provided by the remote receiver. The enhanced resolution of the code phase measurements is the key factor in providing this level of accuracy. A pair of 4000A Locator receivers were used in the tests reported in this section. The accuracy of differential positioning is based on the fact that most of the systematic (deterministic) errors are common between the two receivers. The residual errors due to satellite orbit and clock estimation and ionosphere and troposphere effects, are predicted to be typically of the order of 1 centimeter per kilometer of separation between the two receivers.[7] Much better results have been reported in practice.[8,9,10]

Although the errors mentioned above are easily removed in differential computation, the noises due to the coarse granularity of the code phases are totally uncorrelated between the two receivers, and cannot be removed by the translocation technique. In fact, the accuracy of translocation is limited by this noise. Removal of receiver noise in static and dynamic translocation is presented next.

### *Static Case*

As in the case of single receiver position fixes, the code phase noise can be reduced by (a) averaging the measurements in both receivers or (b) employing the doppler aiding technique. The doppler aiding technique, is not affected by GDOP nor by the nonlinear motion of the satellites. This technique yields better results than does averaging the position fixes.

The static translocational accuracy of two Trimble 4000A Locator receivers, using aided code-phase measurements, is shown in Figure 13. Shown in this figure are the plots of latitude, longitude, and altitude components of about 100 position fixes, obtained synchronously from the two receivers labeled “R1” and “R2”. A high degree of correlation can be observed in this figure. Figure 14 shows a two dimensional view of the same data. The plots of position fixes provided by each receiver are marked with “R1” and “R2”, and the translocation results are marked with “T”. Figure 15 shows the error in the altitude components of translocation.

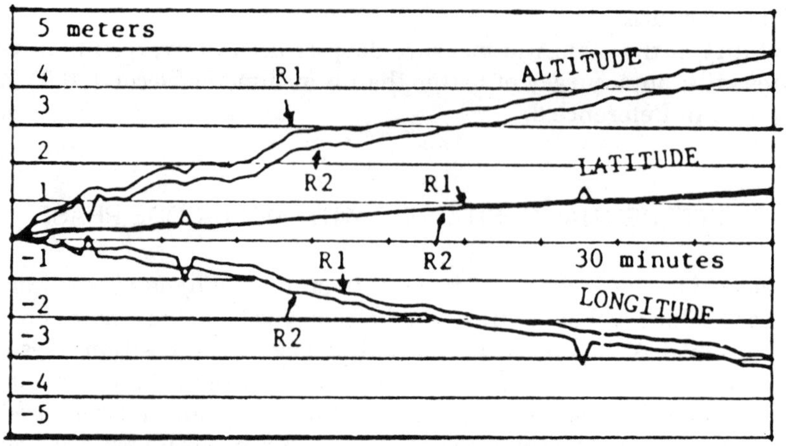


*Fig. 13—Position Components of 100 Static Fixes Obtained Synchronously From Two Receivers*

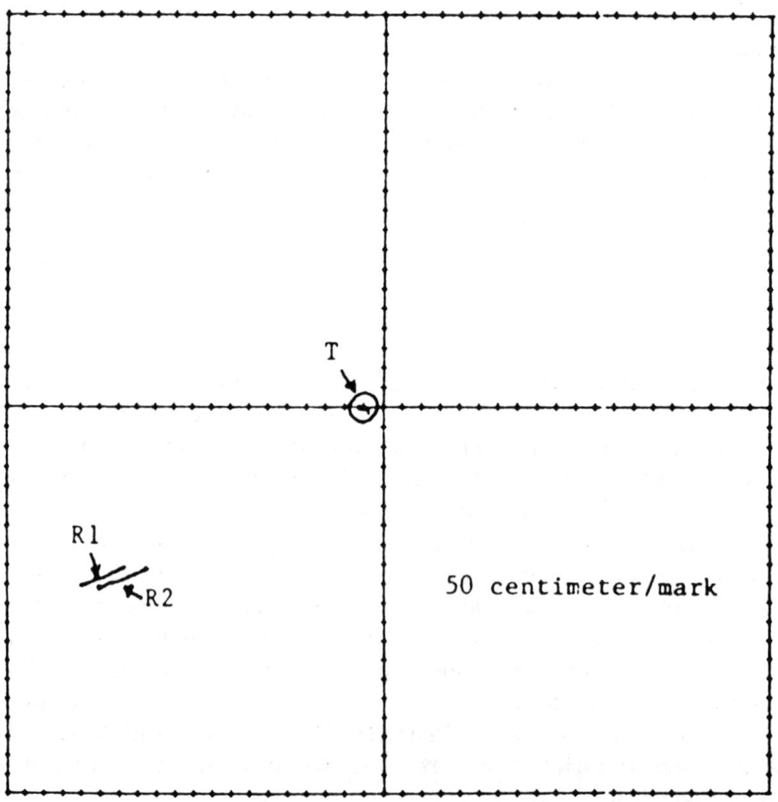


*Fig. 14—Two-Dimensional View of Position Fixes of Figure 13*

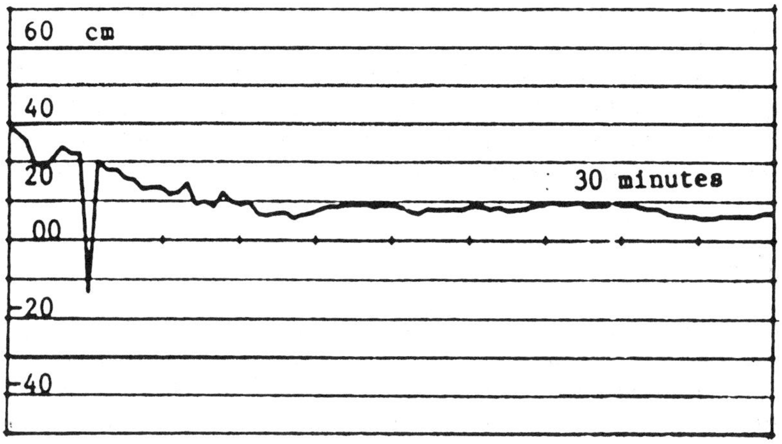


*Fig. 15—Error in Altitude Component of Translocation*

*Dynamic Case:*

The result of translocation in this case is very similar to the static case, provided that the motion of the remote receiver can be approximated to be linear during every .25 second time interval.

Since, in this case, the motion of the remote receiver inhibits code phase and/or position averaging, integrated doppler aiding provides an elegant solution for obtaining sub-meter real-time differential accuracy. Plots of position fixes in Figures 12 through 15 show that the differential accuracy of about 50 centimeters can be achieved by using C/A code phase measurements aided with integrated doppler.

## ACCURACIES OF LEVELS C AND D

These levels of accuracy are briefly described in this section. They are partly discussed in Reference 11.

Level C accuracy can be obtained differentially by using a pair of C/A code receivers capable of measuring carrier phase, in addition to those quantities of levels A and B. Online results can be provided if a data communication link between the two receivers is provided. If such a link is not available, the measurements must be recorded and post-processed. In either case, real time results at each site provide assurance of the proper operation of the receiver while data is being recorded. This level of accuracy is available only under static conditions and can be achieved with about 15 minutes of tracking. Results demonstrating this level of accuracy were obtained using a pair of Trimble Navigation 4000S Surveyor receivers.[11]

The accuracy of Level D is obtained by using a pair of C/A code receivers similar to Level C. The increased accuracy in this case is achieved by using the integrated Doppler, carrier phase and code phase ($IC^2$) information processing technique.[11] The initial estimate of translocation is obtained by the method of Level C and then further improved by processing the carrier phase measurements using single and/or double difference interferometric algorithms, simplified due to the good (less than 1/2 carrier wavelength error) initial estimate obtained by the receivers.

Accurate timing is very crucial in obtaining all levels of accuracies discussed in this paper. It is interesting to observe that, due to the time transfer capability of GPS, there is no need for highly stable oscillators. Accurate timing can be provided by using the accuracy of the GPS master clock in a time transfer mode.

## CONCLUSION

In this paper we demonstrated that the integrated Doppler aiding technique significantly improves the resolution of the code phase measurements of the C/A code system. The enhanced resolution was shown to be significantly better than the resolution of the conventional code phase measurements of the P code. This degree of enhancement is achievable in low dynamic cases which are typical of civilian applications.

The enhanced resolution was shown to be the key factor in providing submeter accuracy in differential positioning, especially in low dynamic applications.

The major factors in obtaining the presented accuracy are *continuous* tracking of satellites via separate physical channels and *simultaneous* tracking that eliminates the effect of the short term instability of the local oscillator.

We also showed feasibility of low cost design of multiple channel receivers with complete parallel tracking capability, which do not have the drawback of interchannel biases. The overhead of the design was shown to be about 100 seconds of automatic calibration during the power-on procedure. The advantages of design are due to a high degree of digital signal processing.

## REFERENCES


1. Braisted, P.; Eschenbach, R., and Sutherland, S.; *$1000 GPS Receiver Technology,* The Institute of Navigation, Forty-First Annual Meeting, Annapolis, MD, June 1985.
2. Baker, P., Zedeck, F., *Civil Use of the Global Positioning System Precision Positioning Signal (PPS)*, Proceedings of First International Symposium on Precise Positioning with GPS, NOAA, Rockville, MD, April 1985, pp. 33–35.
3. Anon., *NAVSTAR GPS Space Segment/Navigation User Interfaces, ICD-GPS-200,* Rockwell International Space Operations and Satellite Systems Division, September 1984.
4. Klobuchar, J., *Ionospheric Effects on Earth-Space Propagation,* Air Force Geophysics Laboratory Document AFGL-TR-84-0004, Hanscom AFB, Dec. 1983.
5. Ashjaee, J., *Improving the Accuracy of C/A Code Systems,* Proceedings of The National Technical Meeting of The Institute of Navigation, San Diego, CA, January 1985.
6. Ashjaee, J., and Helkey, R., *Precise Positioning Using a 4-Channel GPS Receiver,* Proceedings of the IEEE PLANS '84, San Diego, CA, November 1984, pp. 236–244.
7. Kalafus, R. *et al. NAVSTAR GPS Simulation and Analysis Program,* DOT-TSC-RSPA-83-11, Dec. 1983.
8. Collins, J., *Precise Baseline Determination With a New Codeless GPS Receiver,* same as Reference 2, pp. 201–205.

9. MacDoran, P. F.; Miller, R. B.; Buennagel L. A., and Whitcomb, J. H.; *Codeless Systems with Positioning with NAVSTAR-GPS,* same as reference 2, pp. 181–190.
10. Ladd, J. W., Counselman, C., *The Macrometer II Dual-Band Interferometric Surveyor,* same as reference 2, pp. 175–180.
11. Ashjaee, J., *New Results On The Accuracy of The C/A Code GPS Receiver,* same as 2, pp. 207–214.

# NAVSTAR Global Positioning System Collins User Equipment: An Evolutionary Assessment

GREGORY J. HUDAK

Collins Government Avionics Division, Cedar Rapids, IA
*Received January 1986*

## ABSTRACT

This paper describes the evolution of Collins GPS receiver architecture as it has matured from Phase I. The growth in functional capabilities required of the receiver over the course of the GPS program has necessitated the development of various processes and techniques to ensure that User Equipment (UE) could be produced at low cost while meeting the imposed performance requirements. The results of Life Cycle Cost (LCC) trade studies became the driving force in determining receiver architecture as the receiver design progressed from Phase I. Phase I, Concept Validation was structured to prove three major tasks. These were: (1) GPS validity as a navigational system, (2) verification of the engineering concept chosen for implementation, and (3) demonstration of military utility. These tasks necessitated the development of various types of User Equipment (eg, GDM, High and Medium Dynamic sets, Manpacks) tailored to provide certain test data without, necessarily, optimizing for costs or producibility. Phase 2, Full Scale Development, demonstrated the integration of GPS Receivers into a select group of eight Host Vehicles (HVs) representing a wide range of HV types across low, medium and high dynamic applications. The emphasis in this phase was to utilize available technology to produce receivers (in a factory environment) that provided the required performance and survived the Host Vehicle environment. LCC results played the key role in determining the Phase IIB architecture—the production baseline for Phase III. Phase III is concentrating on optimizing the integration concepts to take full advantage of the increased functional capabilities of the GPS receiver. Improvements in Hardware (HW), as well as an inventive interface processing approach in Software (SW) have allowed the receiver to offer the basic navigational data in a wide range of functional "flavors" while minimizing the HW and SW changes from Phase IIB. Finally, a glimpse of improvements planned for future receivers will be shown.

## INTRODUCTION

The deployment of a successful military operational capability generally has its conception in a technology program. In the case of NAVSTAR GPS the program was TIMATION. TIMATION provided the demonstration and verification of many of the baseline concepts and processes that became the core of NAVSTAR GPS. Ranging techniques, Ionospheric (IONO) correction process, downlink frequency selection, and Pseudorandom Code modulation are just a few. The launch of NTS-1 (third in the TIMATION series) provided the marrying of TIMATION and the Air Force 621B program into the NAVSTAR GPS program. Significance of NTS-1 was the incorporation of two Rubidium clocks and the addition of the Pseudorandom Code L Band downlink. This launch heralded the close of the TIMATION program and the inception of the NAVSTAR GPS Program. The subsequent launch of NTS-2 as the first satellite designed and built under the NAVSTAR Program provided a baseline constellation under which the GPS User Equipment development could occur. This intensive development continues today with a much expanded constellation.

## PHASE I

Phase I provided an extensive test environment that evaluated numerous different receiver architectures. During this period, User Equipment was fielded by Collins Government Avionics Division, Texas Instruments, and Magnavox Advance Products Division. Table 1 identifies the diverse nature of the receivers tested during this period.

During Phase I, the overall requirements for GPS Receivers led to the partitioning of receiver types into four distinct types: high performance, medium performance, manpack, and low cost. These types were arrived at by a detailed analysis of requirements that had to be met. The manpack application is a good example of the process where high performance needed to be traded off to

Table 1. Phase I Receivers Tested.

| SET | DESCRIPTION/ FUNCTION | CONTRACTOR |
|---|---|---|
| GDM | 5 Channel/Hi Anti-Jam and Dynamics | Collins Government Avionics Division |
| XU | 4 Channel Unaided | Magnavox Advanced Products Division |
| XA | 4 Channel INS Aided/ Weapon Delivery | Magnavox Advanced Products Division |
| YU | 1 Channel | Magnavox Advanced Products Division |
| YA | 1 Channel Aided | Magnavox Advanced Products Division |
| HDUE | 5 Channel | Texas Instruments, Inc. |
| MVUE | 1 Channel Manpack, Vehicular | Texas Instruments, Inc. |
| MP | 1 Channel Manpack, Vehicular | Magnavox Advanced Products Division |

achieve size, weight, and power goals. Each of the contractors fielding equipment during Phase I utilized the concepts of functional partitioning in the design of their equipment and, in varying degrees, other techniques to minimize cost and increase flexibility to future requirements. Each of the receiver types were designed to provide certain test data required to verify the concept validation goals of Phase I.

The primary function defined for the Collins Generalized Development Model (GDM) was to determine the maximum performance capable in high dynamic and high jamming environment with only limited system constraints on size, weight, and power. This required a design that provided flexible test capabilities for evaluation of receiver performance. The GDM shown in Figure 1 and the functional block diagram in Figure 2 illustrate Collins' approach.

At a high level, three major functional areas were developed: Receiver/Processor, Navigation (data processing), and Interfaces. Realization of test flexibility was achieved with a design that provided signal processor channel configuration under control of SW, SW tracking loops, and use of a HOL in both the receiver and data processor. HW commonality, although not a driving force in this application, is reflected in the common receiver channels (performing a common task) which are under control of the receiver processor function HW/

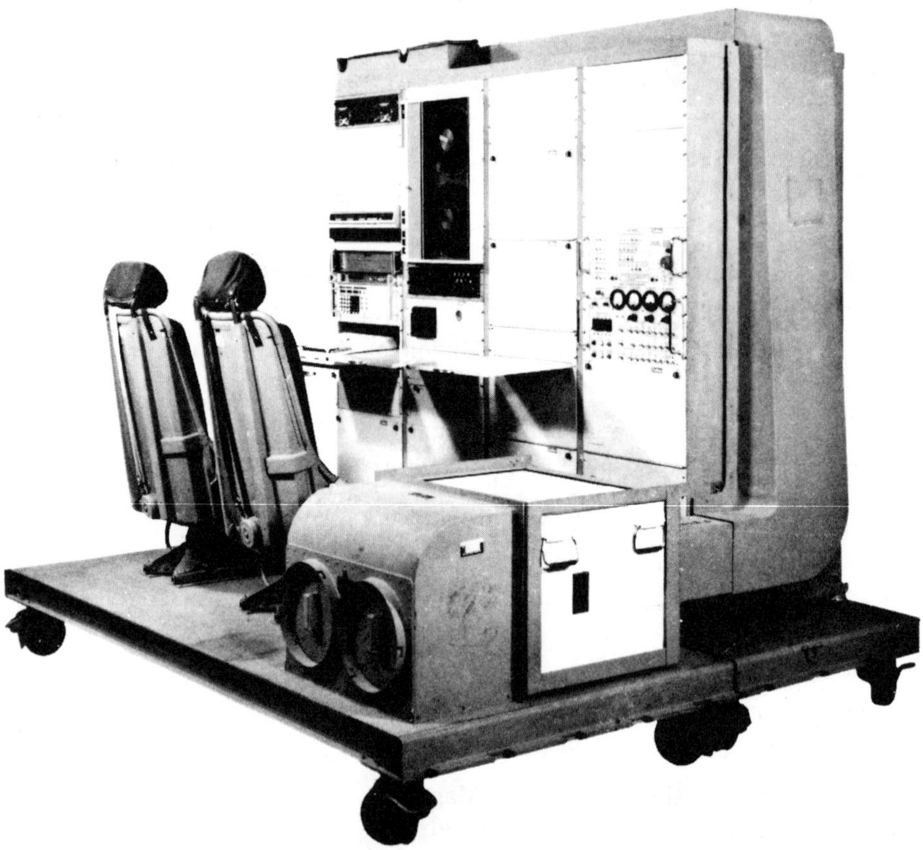

*Fig. 1—Collins/AFAL NAVSTAR GPS Generalized Development Model*

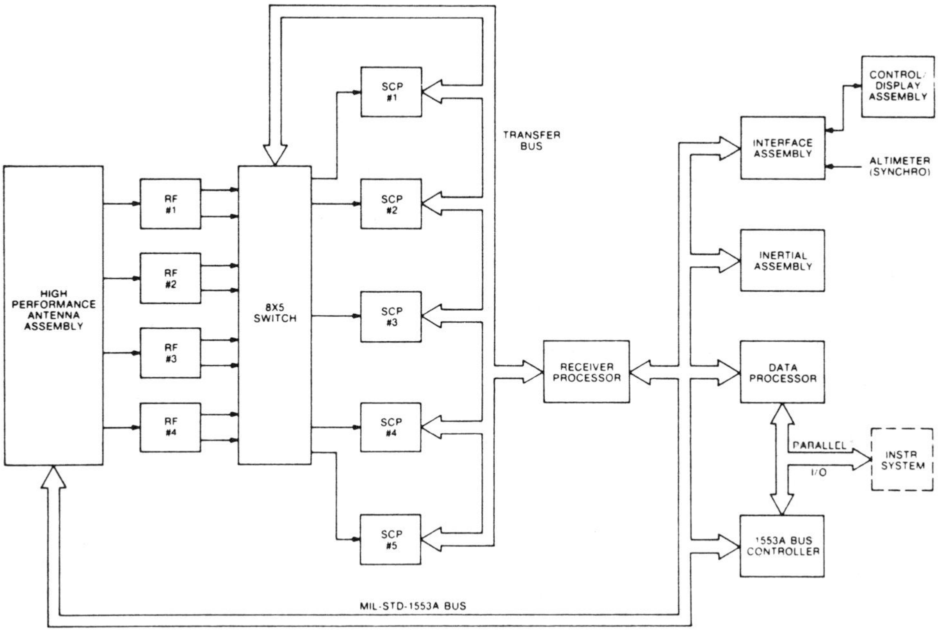


*Fig. 2—GDM Functional Block Diagram*

SW. The 1553 interface, as well as the interfaces to the Antenna Subsystem, Control Display Unit (CDU) and INS also constitute distinct functional partitioning with the Data Processor providing centralized control.

## PHASE II

Without a doubt the single most important driver in the selection of the Phase IIB architecture was Life Cycle Cost (LCC). The key to producibility was deemed to be affordability, not only for the 50 odd preproduction prototypes that constituted Phase IIB but also in achieving an architecture that could provide an effective transition to full scale production in Phase III. In addition, the severe tri-service requirements of performance, size, weight, and power across the spectrum of User Equipment types necessitated a design approach that fully utilized any advantages realized from the results of the LCC trade studies. Many of the decisions early in the Phase II design process were therefore dictated by LCC. For example, the deletion of high risk LSI (RF) and Bubble Memory (processor) but the inclusion of custom LSI for code generators (Figure 3) was based on LCC results showing that inexpensive UE is derived by low Unit Production Cost (UPC), minimal unique circuitry and high reliability, low risk technology. Given the importance of the LCC studies performed we will look at the role they played in determining the receiver architecture Collins chose for Phase IIB.

### *LCC*

The LCC studies utilized for Phase IIB receiver architecture determination were largely performed as part of Phase IIA. Eighty-one trade studies were

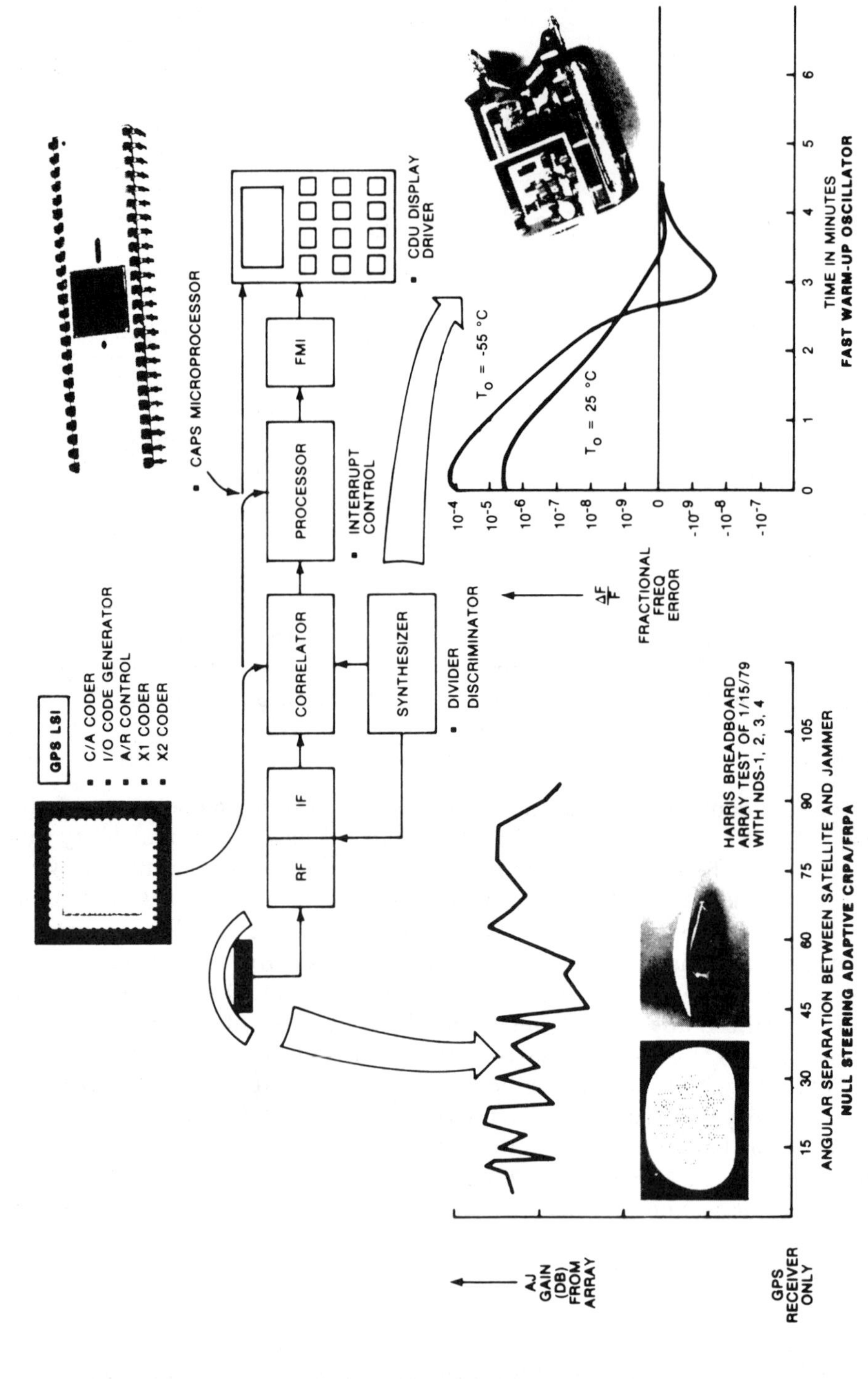


*Fig. 3—LSI Use in Phase IIB Receivers*

initially started, of which fifty-eight were finally determined to be applicable. Of these fifty-eight, thirty nine were specifically applicable and their results fundamental to the final receiver hardware and software design chosen. The studies were conducted in a variety of ways, from use of a government furnished cost model to one based on cost estimate only. Most, however, fell somewhere in between these methods. An initial output of preliminary LCC work indicated that minimum LCC was arrived at by paying close attention to functional commonality, requirements partitioning, and selection of the correct implementation technique. It was also shown that about 70% percent of LCC was driven by UPC.

With the broad spectrum of requirements imposed by SS-US-200 it was necessary to determine an architecture that met the goals of performance, cost savings, reliability, and with design flexibility sufficient to meet the vagaries of current and future Host Vehicle integrations. Five major trade studies were conducted by Collins to achieve an architecture that met these goals. These were: (1) Receiver/Processor Unit (R/PU)-Flexible Modular Interface (FMI) partitioning, (2) Receiver channels required, (3) Processor selection, (4) Manpack application, and (5) Packaging selection. The outcome of each of these played an integral role in the final architecture that Collins designed, built, and tested in Phase IIB. In all, twenty-five different HV configurations were used in the LCC data base.

## *R/P-FMI Partitioning*

With weight and size goals dictating one Line Replaceable Unit (LRU) (minimized connectors, wiring, etc.), permitting simplified HV installation and reduced LRUs to be maintained, two alternative approaches were considered: (1) the receiver could be uniquely designed for each HV application or (2) one receiver could be designed for all applications. Method two clearly implied a complex, expensive approach that greatly restricted adaptability and, at the same time, required all users to pay the price of unwanted capabilities. Method one eliminated commonality at the LRU level and since LCC is most strongly influenced by LRU production volume this method did not offer the minimum LCC. Although this increases common Shop Replaceable Unit (SRU) production cost savings, it is done at the expense of increased labor intensive assembly and test cost. The outcome was a design that utilized the positive features of both methods. The partitioning was done at the R/P-FMI subassembly level

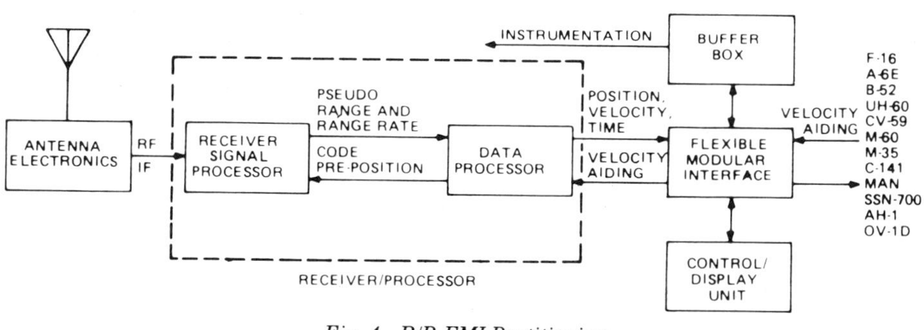


*Fig. 4—R/P-FMI Partitioning*

(maximizing adaptability and subassembly commonality) which, when bolted together forms one LRU (maximizes LRU production commonality). Figure 4 shows this basic, but fundamental, partitioning. This architecture provided a high level of commonality (in such areas as nav and receiver processing) located in one assembly, thereby allowing the benefits of a large production base to be realized. The effects of HV unique requirements ae then localized to the FMI where the advantages of FMI SRU commonality can be realized were possible (eg, CDU and Instrumentation Port HW and SW).

### *Receiver Channels*

Another significant study was to determine the number of tracking channels required. The 5-channel receiver could certainly meet all performance requirements while the single-channel Manpack provided a low weight and power approach with diminished performance over dynamics. However, there existed a category of medium performance host vehicles that could possibly use something less than 5 channels—if this approach was LCC effective. It was determined that a 2 channel receiver (as opposed to a three or four) could meet the medium dynamic requirements. Three different approaches were studied to assess the LCC effect of adding this third receiver type. They were: (1) Use the 5 channel for all, (2) Use a 2 channel for surface ship category only, and (3) use a 2 channel for ships and helicopters. The LCC showed that by adding the helicopter to the surface ship grouping, LCC went down due to the addition of a large volume of helicopter receivers. As a collateral benefit, the 2 channel architecture could meet the future needs of transport type of aircraft. Table 2 provides the data on delta LCC for each configuration analyzed.

### *Processor Selection*

This study was initiated by first specifying that the SW architecture (instruction set, data types, and address space) be defined and then hardware implementation was selected to meet those needs. Five different architectures were evaluated on the basis of various assessment criteria. The study results showed that the Collins Adaptive Processing System (CAPS) family provided a 20 to 60 percent memory savings over the other contenders. The generic processor

Table 2. Receiver Channels LCC Results.

| CONFIGURATION | ΔLCC<br>$ XM |
|---|---|
| ■ 5-CHANNEL ONLY | -- |
| ■ 5- & 2-CHANNEL MIX<br>(2-CH IN SURFACE SHIP ONLY) | +18 |
| ■ 5- & 2-CHANNEL MIX<br>(5-CHANNEL IN F-16, A-6E, B-52, SSN) | –6 |

implementation chosen required two processors that, used alone or in pairs, met all present and expected applications, minimized memory, lowered LCC, and gave high performance across the range of UE types.

### Manpack

The stringent power, weight, and size requirements imposed on the manpack (MP) required that producibility dictate the approach chosen. An initial approach to use hybrid microcircuits for the MP application and then adapt them to the 2 and 5 channel applications was dropped in favor of a method that featured producibility by laying out the required circuit boards and then deciding where weight savings could be obtained. Results of trade studies performed on the various MP circuit cards showned that the best results were obtained when the processor, wide-band amp, and clock were implemented using miniature circuitry. At this time further weight reduction was still necessary either by designing a unique correlator card for the MP or reducing chassis weight. The unique correlator was not the preferred approach because it negated commonality with the 2 and 5 channel correlator cards increasing its UPC. By going with a nylon chassis it was possible to reduce chassis weight, thereby meeting performance, environment, and weight requirements while retaining the commonality of the correlator, frequency synthsizer, and CAPS 10 with the air and shipborne applications.

### Packaging

Packaging constraints were those of temperature and dynamics for aircraft, shipboard shock, MP immersion, and size and power. Maximizing commonality among assemblies and piece parts such as PC connectors and SRU mechanical parts addressed some LCC objectives. Others were addressed by using plug-in SRUs instead of bolted-on modules, thereby providing the advantages of easy

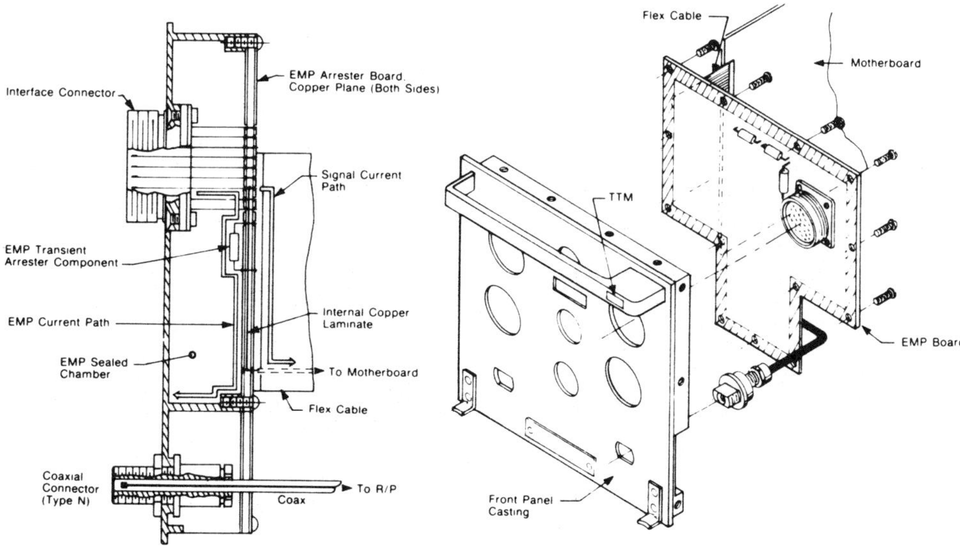


*Fig. 5—Connector/EMP Board Implementation*

replacement, minimum wiring to a motherboard, increased cooling efficiency, mechanical integrity, and maintainability. In addition, an acceptable production method for interfacing Series III connectors to the internal receiver circuitry was developed. This approach (Figure 5) eliminates the problems, both in manufacturing and reliability that are inherent with ribbon cable while at the same time reducing overall receiver length and providing a very effective location for necessary EMP circuitry. Typical receiver packaging is shown in Figure 6.

### *Realized Receiver Design*

The resulting receiver architecture was a Design to Life Cycle Cost (DTLCC) based design that benefited from an ordered design approach. That is, implementation candidates were identified with evaluation and selection was based on engineering knowledge and LCC results. The process manifested itself in a Generic Receiver configuration (Figure 7) which extends HW commonality across LRUs, SRUs, and piece-parts while realizing all requirements in a factory producible design (where 80% of LCC control exists).

In addition, the receiver consisted of low risk current parts technologies, and minimal custom circuitry, enhancing future piece-part improvements. These combinations resulted in a 75 percent HW commonality figure with SW commonality extending across 84 percent of all HV applications (Figure 8).

From a HW perspective this commonality meant that eighteen FMI SRUs satisfied the requirement of the eight defined Phase IIB HVs with only three additional SRUs required to extend to seventeen HVs. The all-digital design (except or specific HV analog interfaces) combined with a common frequency synthesizer (common 1, 2, and 5 channel frequency plan) increases producibility features and improves maintainability performance. SRU layout was driven by functional commonality, interconnection and structural advantages along

*Fig. 6—Typical Receiving Package*

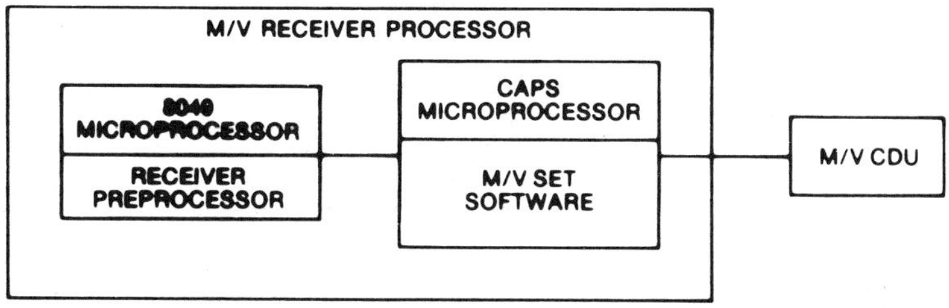


**M-V Processor Configuration.**

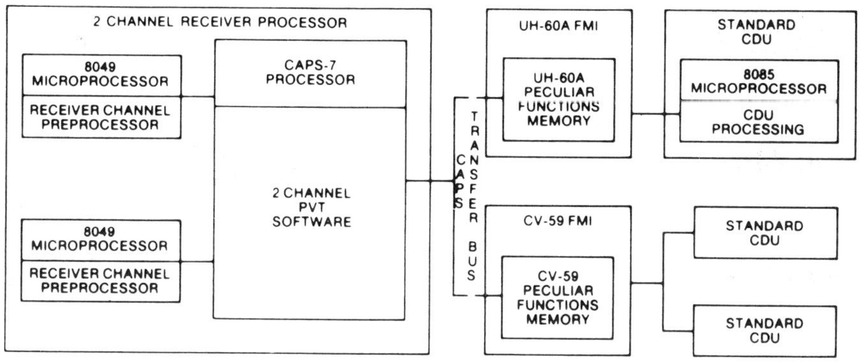


**2-Channel Processor Configuration.**

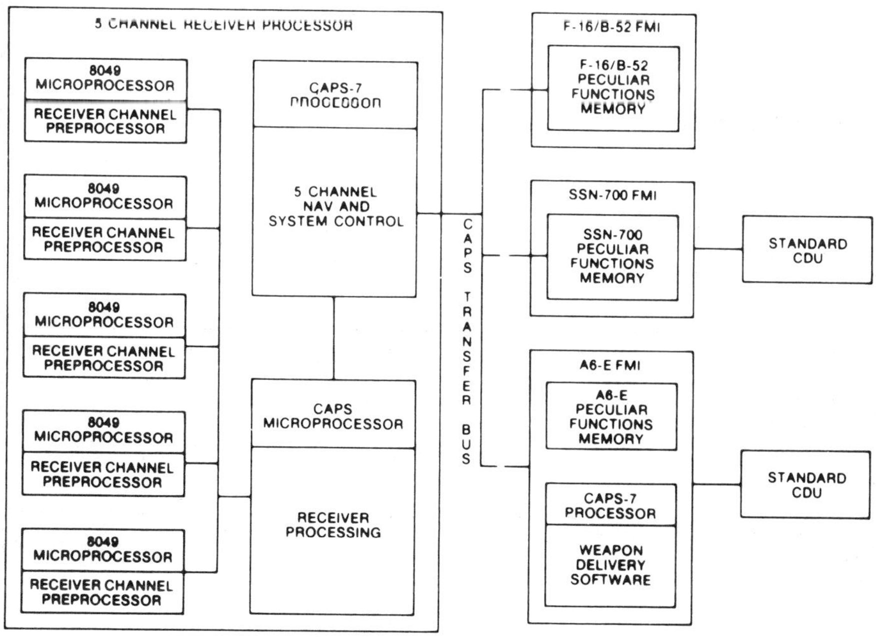


**5-Channel Processor Configuration.**

*Fig. 7—Generic Receiver Configurations*

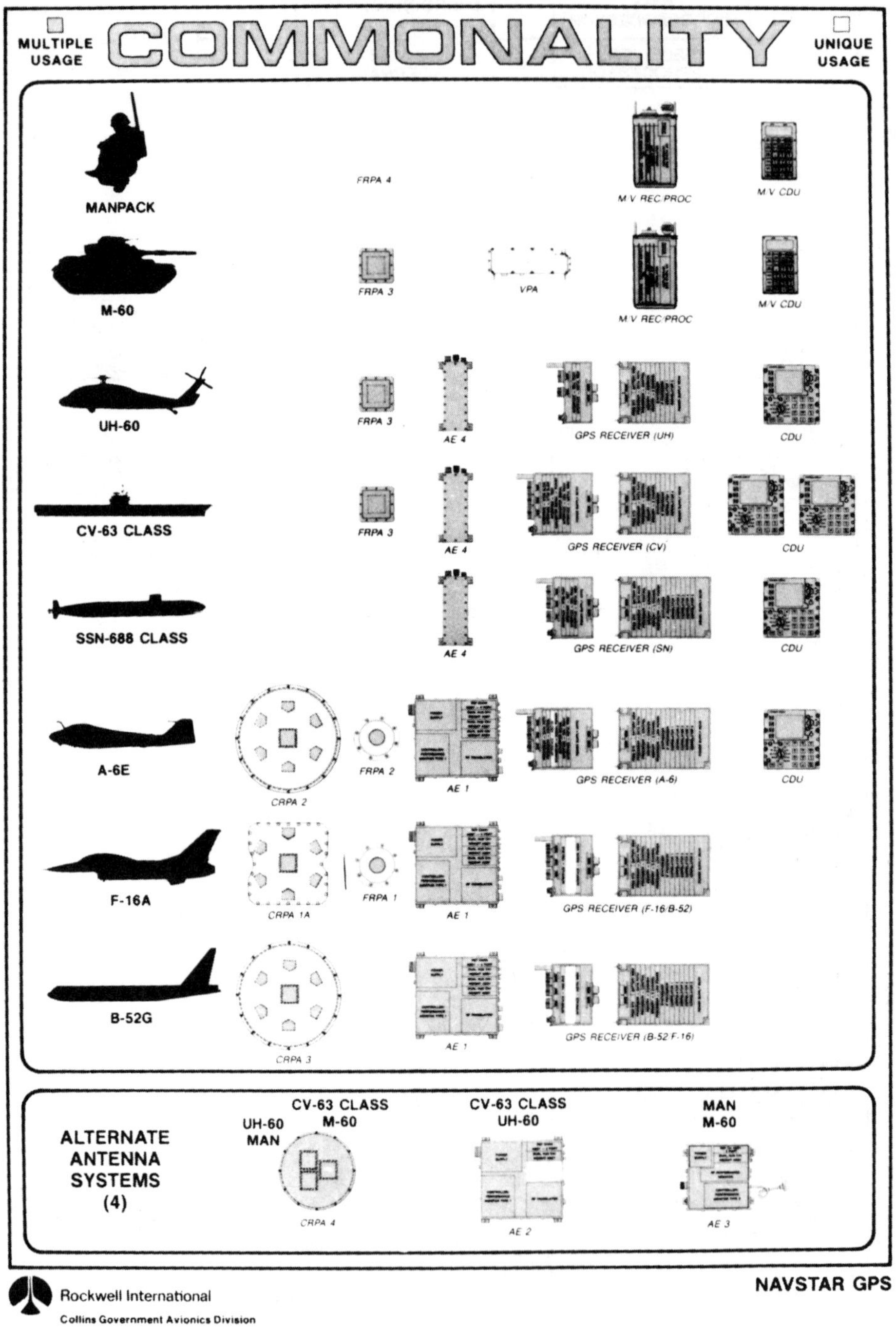


*Fig. 8—Phase IIB Receiver Commonality*

with an optimal mix of microminiature circuitry, and discrete parts to meet environmental and producibility (fab, auto insert, test) requirements.

The SW implementation was realized in a design approach that minimized HW required (low UPC), utilized an efficient, capable, proven HOL (reduced maintenance costs), and minimized logistics (common SW partitioning). In addition, the design featured the advantages of modularity (maximizes SW commonality across HVs) and partitioning (requirements allocated to discrete functional areas). Figure 9 demonstrates this design.

A collateral benefit with this design was the repeated verification that naturally occurred as each of the receiver types was tested, thereby providing a large data base on the performance of common functions.

A multiple computer bus (common and local) facilitated the transfer of data between the R/P and the FMI and allowed the use of FMI SRU space for the storing of program memory. Furthermore, the ability to interface any R/P with any FMI as a result of the FMI common bus implementation shown in Figure 10 provided great flexibility for future applications as well as an optimized use of available SRU space. This approach provided a limited number of general function SRUs with application controlled by SW. By locating "Interface" RAM and ROM on the FMI SRUs and executing the code in the R/P, additional HV applications can then be addressed by primarily an R/P SW change only.

## PHASE III

The Phase IIB architecture, verified during extensive in-plant and field testing, provided a solid production baseline from which to implement the government requested changes for Phase III, i.e., Engineering Change Proposal (ECP) flexibility was built into the Phase IIB architecture. From an architecture perspective, the only major change required by Phase III was the capability to accommodate the requirements of one-hundred-twenty candidate host vehicles for integration. A necessary ingredient was to standardize interfaces as much as possible (e.g. MIL-STD-1553) while at the same time allow sufficient flexibility to adapt to HV unique requirements. In addition, a requirement was levied that no changes be made to the Receiver/Processor functions. In each of the cases the existing Phase IIB architecture was ideally suited for adaptation.

### *HW Changes*

In general, HW changes (Figure 11) were minimal, but very important, in providing an even more producible and reliable receiver. Piece-part improvements allowed by technology advances permit direct substitution of ICs, increasing memory from 64 to 256K within the same device family as well as decreasing access time from 450 to 250 nanoseconds. These features combined to facilitate a memory map reorganization which aided the SW changes required for the additional HW interfaces. Also, the increase in parts efficiency enabled the deletion of program memory on the receiver common bus which increased system throughput. In addition, semicustom LSI provided a one chip drop-in replacement code generator for five Phase IIB chips.

Of significance in the HW mechanical area was the decision to go to a single chassis fabrication while still retaining the overall FMI-R/P SRU functional partitioning. In part, this approach was dictated by the need for extra space in

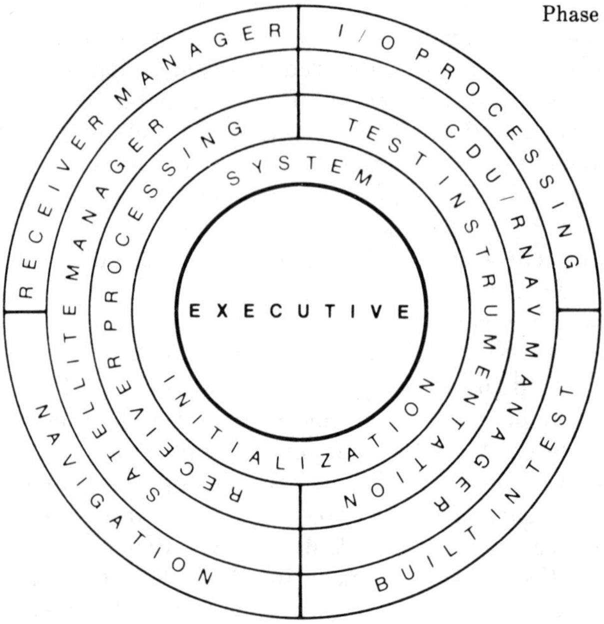


| | |
|---|---|
| EXECUTIVE | Provides realtime, multitasking operations with flexible user interfaces and error management. |
| SYSTEM INIT | Sets up those SW functions used in particular applications. |
| NAVIGATION | Contains background and foreground modules processing GPS measurements and HV sensor inputs. |
| RCVR PROCESSOR | Processes satellite signals to provide PR and DR measurements to NAV. |
| RCVR MANAGER | Controls continuous and sequential acquisition and tracking activity on a per channel basis. |
| SATELLITE MNGR | Establishes data base for down-link data and satellite position estimates and recommends optimum GDOP combinations. |
| CDU MANAGER | Provides CDU Interface, Waypoint navigation, and coordinate conversion functions. |
| TEST INSTR | Provides Interface to Instrumentation Port and data formatting. |
| BUILT IN TEST | Provides a sequence of HW and SW testing and real-time performance monitoring. |
| I/O PROCESSING | Provides the interface to Host Vehicle sensors and subsystems. |

*Fig. 9—Software Functional Modules*

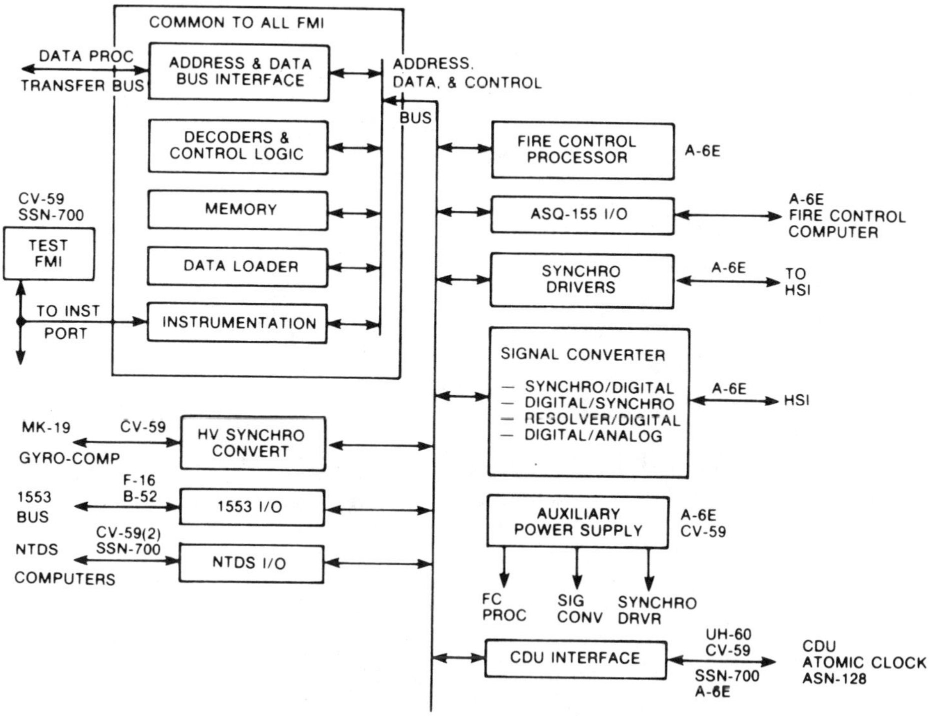


*Fig. 10—FMI Common Bus Implementation*

the airborne receiver (RCVR 3A) required for the extra interface SRUs (Precise Time and Time Interval Flight Instruments). This extra space (enough for two SRUs) was achieved by eliminating the Phase IIB R/P-FMI Interface connector which also provided roughly a two-pound weight savings as well as a cost reduction of about $1K per LRU. Also, Shipborne LRU weight savings were achieved by going to a card cage arrangement, increasing similarity with the RCVR 3A motherboard layout.

## SW Changes

The only significant SW change for Phase III concerned modifying the interface implementation approach to achieve compatibility with the one hundred odd HVs expected. The effective, though somewhat limited approach of Phase IIB was replaced with one that structured the interface SW design into two functional areas: Interface Manager to handle general HV I/O, and Interface Descriptor to define the parametrics of the interface data. This structure was made possible by the HW device changes mentioned above. Figure 12 shows the HV I/O implementation transition from Phase IIB to Phase III. Adaptability to the Phase III I/O required only a six percent change to the Phase IIB receiver LRU SW. As can be seen in Figure 13, from a receiver-level SW perspective this only required updating the I/O Processing Module to reflect requirements of new interfaces. By using a connector strapping scheme on the receiver to identify available interfaces, a means was provided to allow the installation to

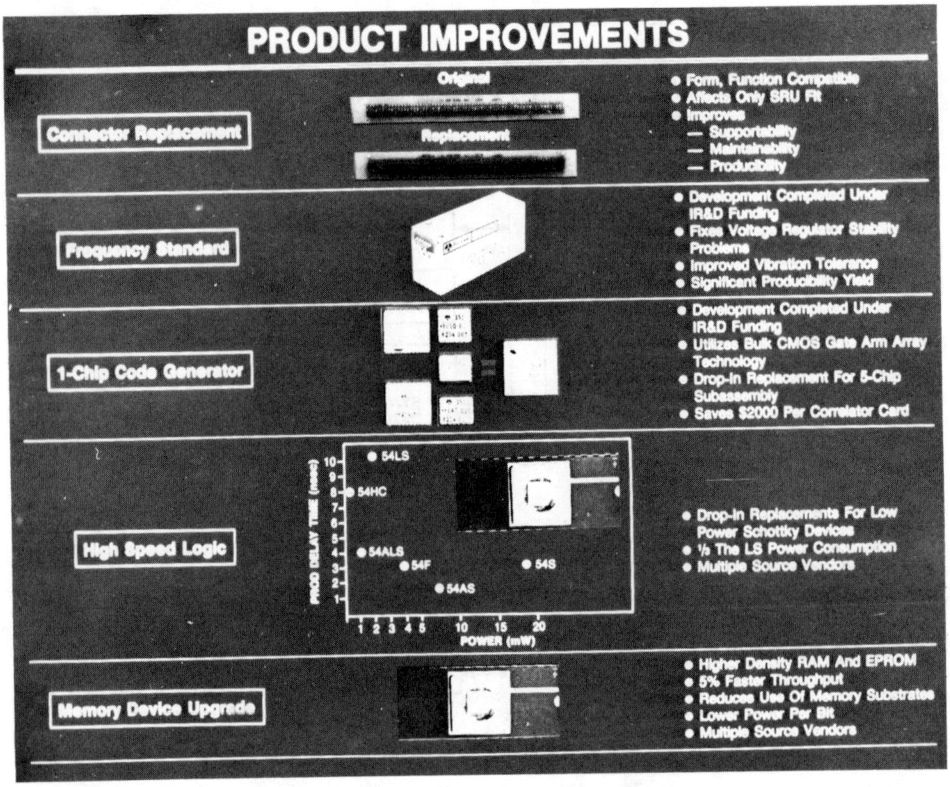


*Fig. 11—Phase III Hardware Producibility Improvements*

define electrical characteristics, data types and protocols, etc, thereby activating any or all receiver interface capabilities available to the user.

### Manpack

The Manpack remains nearly unchanged from Phase IIB. The replacement of the SRU Blueline connectors will increase environmental reliability and the modification of SW to provide a more Human Factors related (user friendly) operation will increase military effectiveness.

In summary, the minimal changes in Phase III were made possible by the aggressive engineering approach that was undertaken during Phase IIB to design a Receiver that was producible, flexible to future requirements, and met performance goals. As in Phase IIB, the architecture once again meets the flexibility and commonality required across the spectrum of users (Figure 14).

## FUTURE IMPROVEMENTS

Certainly one of the most pleasing challenges to future architectures will be the necessity to reduce volume and weight while retaining, if not increasing, the current level of functional capabilities. Receivers in the 3/4 ATR short and 3/8 ATR long-form factors are already being requested by users who are continually short on equipment space. A reduction in the number of correlators

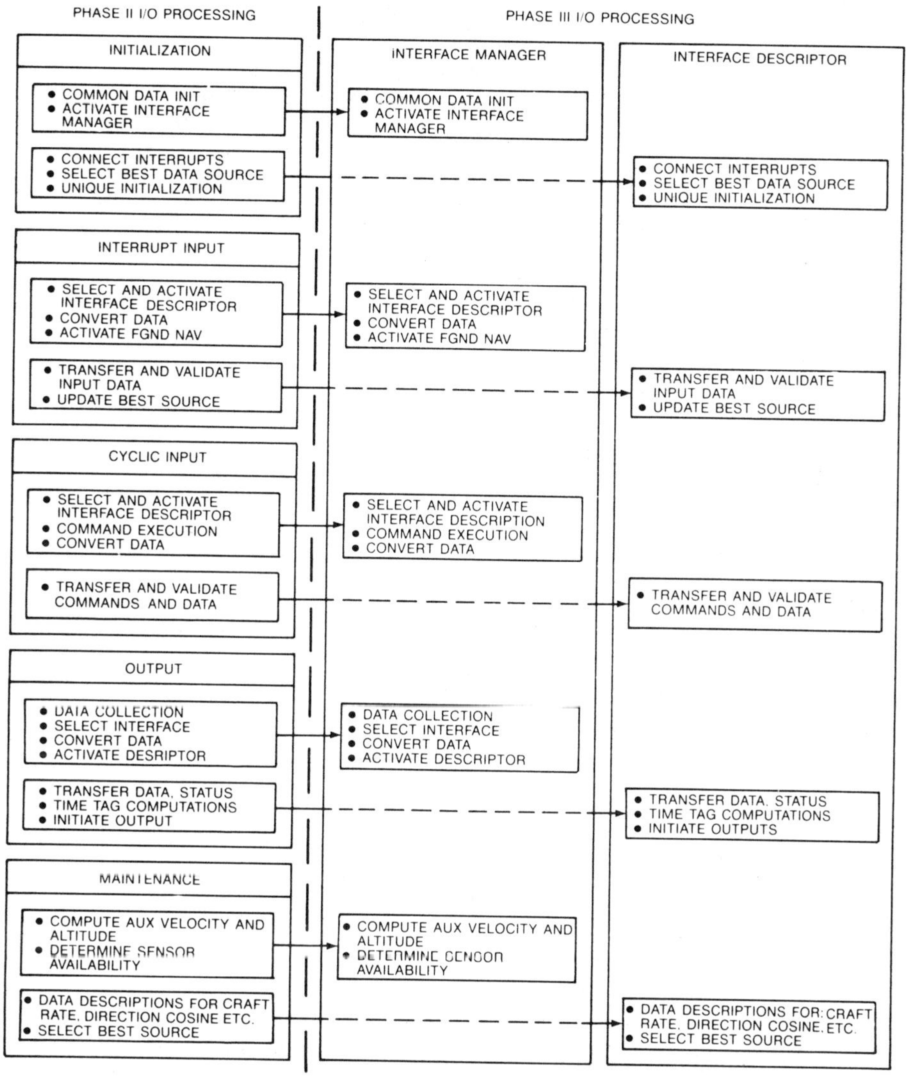


*Fig. 12—HV I/O Processing Implementation*

required from five to two while still retaining the maximum performance advantages of the 5 channel receiver will provide the first step towards this reduction. A prime candidate for a reduced size GPS Receiver will be in the Integrated Communications, Navigation, Identification Avionics (ICNIA) program where potentially sixteen distinct avionics packages will be integrated into one unit with a planned weight and volume savings of fifty percent.

Future decreases in the manpack size may take two paths. The first, a weight reduction in the fifty percent range, is presently available by a repackaging of current device technology. However, this approach requires the use of expensive, high density parts thereby incurring production risk as well as extensive certification testing. The second path will be in the realization of the effort Collins is performing under the auspices of DARPA to provide a small (less

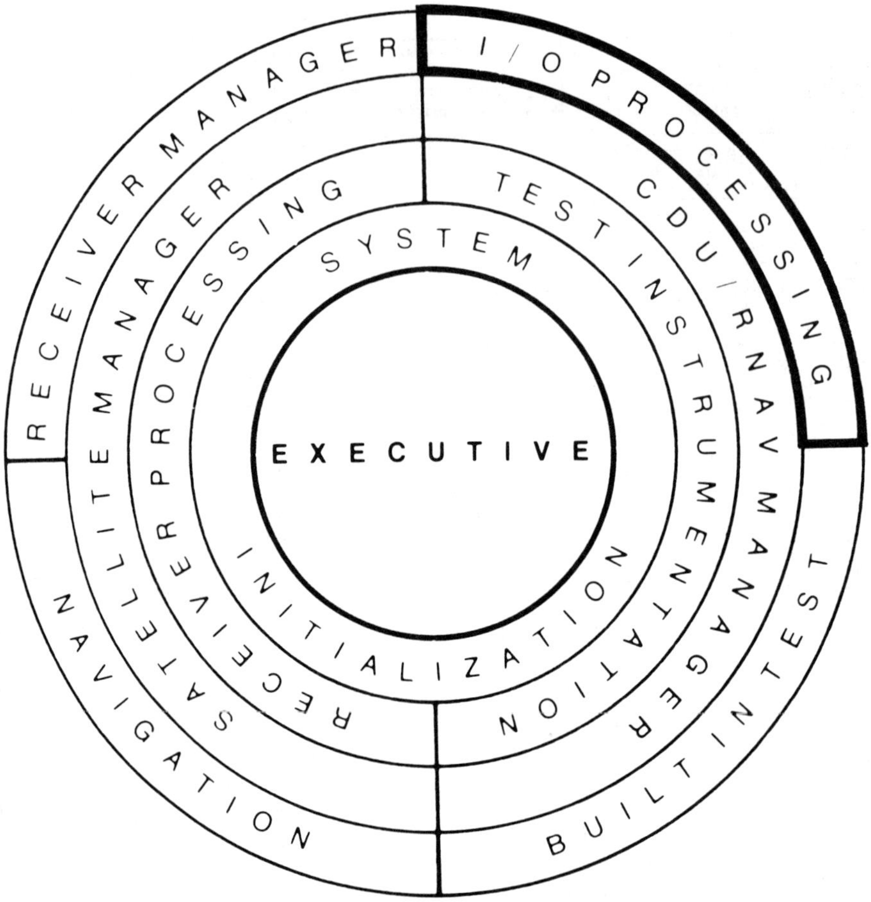


*Fig. 13—Phase III Software Modularity*

than one hundred cubic centimeters), hand-held receiver with an integrated CDU in the four-pound range with capabilities nearly matching those of the present manpack and equivalent or better navigation performance (Figure 15). This advanced receiver architecture will require the use of VLSICs and digitizing the receiver functions to as close to the antenna as possible. However, it will utilize the existing (Phase IIB/Phase III) SW. Without a doubt the driver for these improvements will be in the electrical device area where smaller size, power, higher memory, and throughput capabilities are continually being obtained.

## CONCLUSION

The utility and effectiveness of GPS has certainly been well documented and field proven since Phase I. Similarily, the ability of the User Equipment contractors to build and field a receiver compatible with all the user requirements was demonstrated in Phase IIB and certified in the award of the Phase III production contract. In the next few years the capabilities of GPS will be spread far and wide as the evolution of the receiver architecture continues. Advances

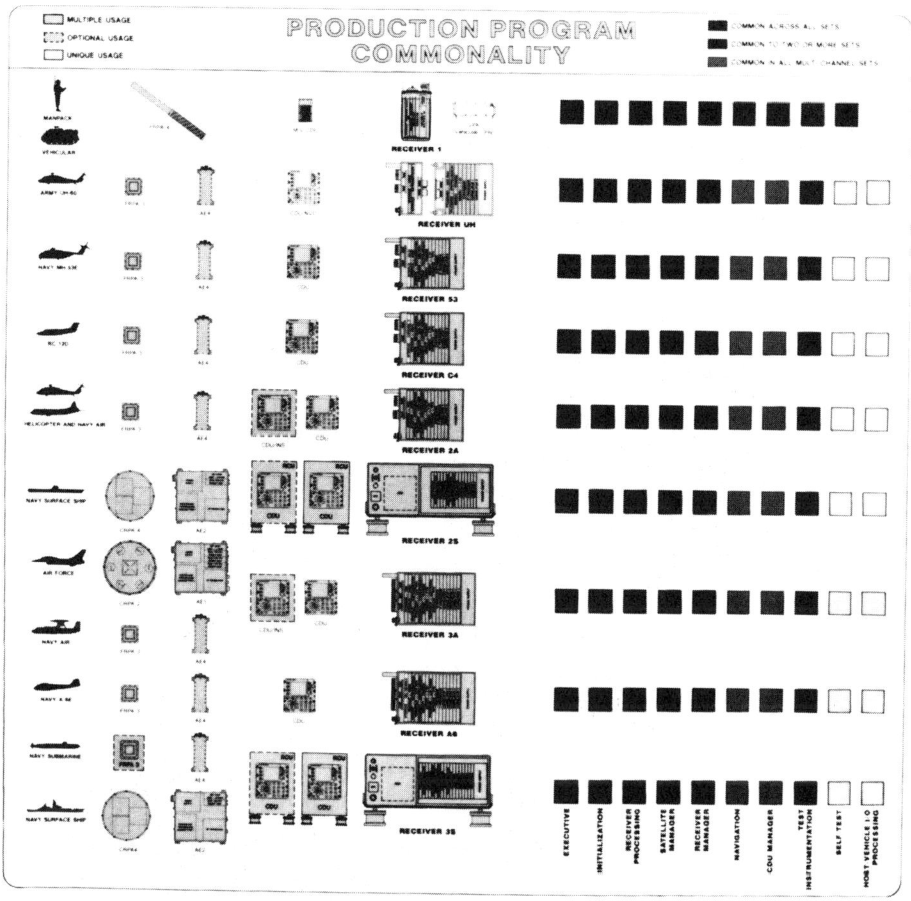


*Fig. 14—Phase III Receiver Commonality*

in SW programming techniques and technological improvements in HW components and packaging will make the receiver available to any user, regardless of his most stringent requirements.

## BIBLIOGRAPHY

1. Easton, R. C., *The Navigation Technology Program,* NAVIGATION Vol. 25, No. 2, Summer 1978, pp 15–20
2. Denaro, et al, *GPS Phase I User Equipment* NAVIGATION, Vol. 25, No. 2, Summer 1978, pp 125–131
3. Krishnamurti, G.; Harshbarger, S. A. and Smith, T. N.; *The Design and Performance of GPS Phase II User Equipment Navigation Software,* NAVIGATION, Vol. 32 No. 3, Fall 1985, pp 263–281

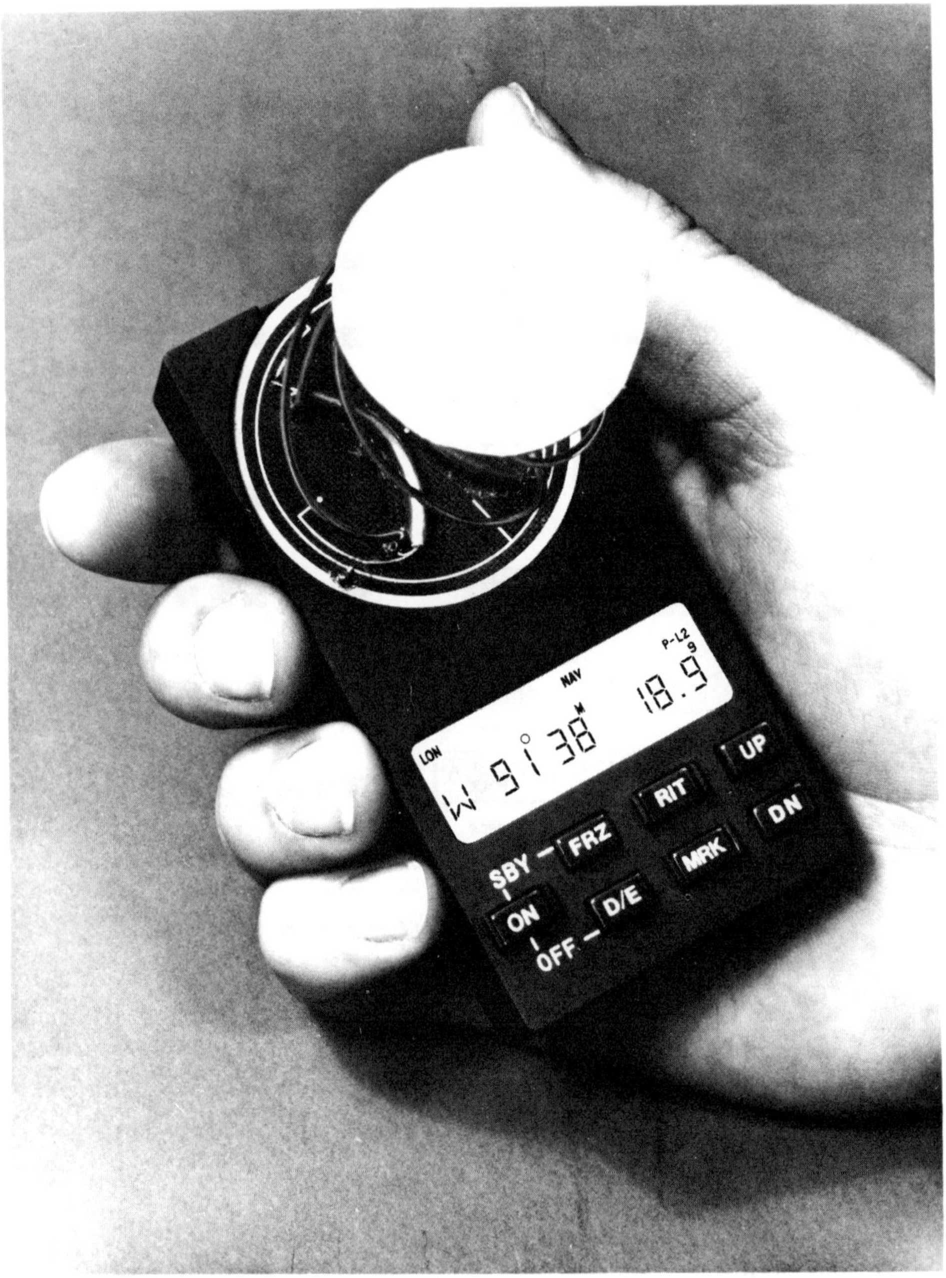


*Fig. 15—DARPA Technology Receiver*

# Overview of Differential GPS Methods

EARL G. BLACKWELL

SRI International, Menlo Park, California
*Received January 1985*

## ABSTRACT

A local GPS reference receiver (RR) can be used to eliminate common errors in the GPS navigation solution of other nearby receivers. A threefold improvement in remote GPS position accuracy can be achieved by this technique under optimum conditions. Several methods of accomplishing this process are presented.

## INTRODUCTION

The NAVSTAR GPS under development by the Department of Defense (DoD) will provide highly accurate position, velocity, and time anywhere in the world to users equipped with suitable GPS receivers. The GPS operates by the use of a constellation of 18 satellites in 12-hour (19,000-km) orbits that continuously broadcast their identification, position, and time using specially coded signals. The typical accuracies quoted for user equipment operating in P code are: positioning to 10 m RMS in each axis; velocity to 0.1 m/s RMS in each axis; and time to 0.1 μs.[1] Although these accuracies are considered exceptional for navigation systems, test and training ranges and other applications can require even more accurate positioning information.

One technique for improving the performance of GPS positioning that is receiving increasing interest is the use of differential GPS processing. This concept involves the use of data from a GPS RR in the vicinity of the GPS receiver-equipped participants that allows certain errors common to both receivers to be removed from the participant's position measurements. Several schemes to accomplish differential GPS have been identified; this paper discusses and contrasts some of these schemes.

### *GPS Error Sources*

The GPS receiver computes its position from the receiver-to-satellite ranges it has determined through signal reception time measurements and from satellite orbital information and time transmitted by the satellite. Each satellite transmits its identity number, the time (resolved to the nearest nanosecond), and orbital ephemeris corrections, in addition to a host of other data (such as health, clock errors, drift rates, etc.). The receiver basically obtains the range to the satellite vehicle (SV) by noting the difference in the time the signal is received (using the receiver's clock) and the transmitted time. Because the receiver's clock error may be large relative to the ranging accuracy required (feet), the receiver's SV ranges are termed pseudoranges (PRs), and four such ranges from separate SVs are used to provide an overdetermined 3-D solution, thus eliminating the receiver's clock error. The location of each SV is obtained using the SV's ephemeris message to allow the receiver to calculate its position, which is defined in terms of earth-centered, earth-fixed coordinates.

Several sources of error are associated with this process, however, and our interest is with those which can be eliminated or significantly reduced because they are common to both participant and reference receivers. Fundamentally, these errors are composed of the satellite's clock error, errors in the satellite's broadcasted ephemeris data, and signal propagation delays that are not accounted for by the receiver's measurements or modeling. These errors will be seen over the relatively short intervals of a test or training operation (about 10 to 20 min) as relatively constant bias errors in the receiver's position data. When two receivers in the same vicinity (within about 100 nmi) are using the same set of four satellites (up to seven or eight can be above the local horizon at one time), the above errors will be common to both and can be removed or essentially eliminated by differential techniques.

Another source of receiver position error is selective availability (SA), which will be imposed by DoD and will limit C/A code users to around 250-m accuracy in the absolute navigation mode.[2] Only authorized users will be able to access P code. Differential GPS, however, will enable civilian sector C/A-code receivers to substantially eliminate SA effects in a local region near a reference receiver as well as remove other common errors. Other error sources in the positioning process that are not correlated between two receivers are: receiver interchannel biases, receiver noise and quantization errors, and local multipath effects.

Table 1 tabulates a typical error budget for nondynamic GPS (absolute mode) user equivalent ranging error (UERE) and subsequent position determination for a nominal carrier to noise ratio ($C/N_o$) of 38 dB-Hz. The satellite clock error and the control segment ephemeris errors (up-loaded for SV retransmission by the satellite control segment) are common to both P-code and C/A-code operation. The ionospheric delay is determined by dual frequency measurements ($L_1$ & $L_2$) for P code, but is estimated by an algorithm when in C/A-code operation

Table 1—GPS Static Error Budget

| Error Source | Predicted Error (ft) | |
|---|---|---|
| Satellite clock error | 10 | |
| Ephemeris error | 8.6 | |
| | P code | C/A code |
| Ionospheric delay error | 1.3 | 21.0 |
| Tropospheric delay error | 1.3 | 1.3 |
| Receiver noise/quantization errors | 0.8 | 8.0 |
| Receiver interchannel bias | 0.5 | 2.0 |
| Multipath | 4.0 | 10.0 |
| UERE (RMS) | 13.9 | 27.9 |
| Resulting Position Accuracy | | |
| RMS horizontal (x,y) position error (assume HDOP = 1.5) | 21 | 42 |
| RMS vertical (z) position error (assume VDOP = 2.5) | 35 | 70 |

Source: Reference 4

since only $L_1$ is used and is consequently less precise. The tropospheric delay error is estimated by an algorithm for C/A code and P code and is the same for both. Receiver noise and quantization errors are noncorrelated and are seen to be ten times less for P code than for C/A code because of the 10:1 bandwidth differences between the two codes. Receiver interchannel biases can be considered as those PR differences between multichannel receiver measurements if all four channels were tracking the same satellite. Multipath errors are unique to receiver location and antenna installation and are, therefore, noncorrelated between two receivers.

### *Differential GPS Error Budgets*

Figure 1 illustrates the principle of common error removal by differential GPS. A reference receiver at a known (surveyed) location measures the range to a NAVSTAR satellite. This measured range includes the actual range to the satellite plus several errors. These errors are taken as additive (for illustrative convenience) and include the satellite clock error, the tropospheric delay, and the ionospheric delay. Since the reference receiver's coordinates are known, the satellite's ephemeris message allows the range to be calculated; however, an ephemeris error can exist in the calculated range, so the difference between the measured range and the calculated range includes all previously discussed errors. If this range correction is applied to the measured range of a remote GPS user, a "corrected" range is obtained that will allow the remote receiver to remove the error in its uncorrected position. For clarity, this basic diagram assumes the receiver's clock bias term has been removed from the satellite range measurements and does not show noncommon receiver-unique errors such as receiver noise and quantization errors or local multipath errors.

Considering that two receivers are using the same set of SVs, the satellite clock and ephemeris errors are obviously common to both and to a large degree the ionospheric delays are common due to the high altitude of the ionosphere (25 to 250 nmi).

With increasing separation distance between remote and reference receivers, a slight decorrelation occurs with the ionospheric delays and satellite ephemeris errors due to difference in satellite viewing angles. Ephemeris error decoupling between two receivers has been estimated by Beser and Parkinson[2] for receiver separations of 100 nmi and a worst-case situation where all the ephemeris error is in the along-orbit direction. This error is of the order of 0.5 m for each 100 m of along-orbit ephemeris error. Ionospheric delay decoupling with receiver separation is more difficult to estimate due to diurnal and seasonal variations in the ionosphere. Static differential field test measurements have been reported by Fickas and Wadsworth[3] using Phase I GPS equipment. These data were taken for receiver separations of 0.5 nmi up to 280 nmi. The bulk of these data was obtained at night (when the ionosphere is stable), although some day-night transition measurements were included. Dispersions (i.e., spread) in the horizontal and vertical RMS data (approximately $\pm 1$ m in horizontal P-code data, and $\pm 3$ m in the vertical data) were masking any trends of error growth over the 280-nmi separation, although an error growth of $\sim 1$ m could have been easily embedded in the data. Thus, it may not be unreasonable to consider employing differential GPS for up to 300-nmi separations.

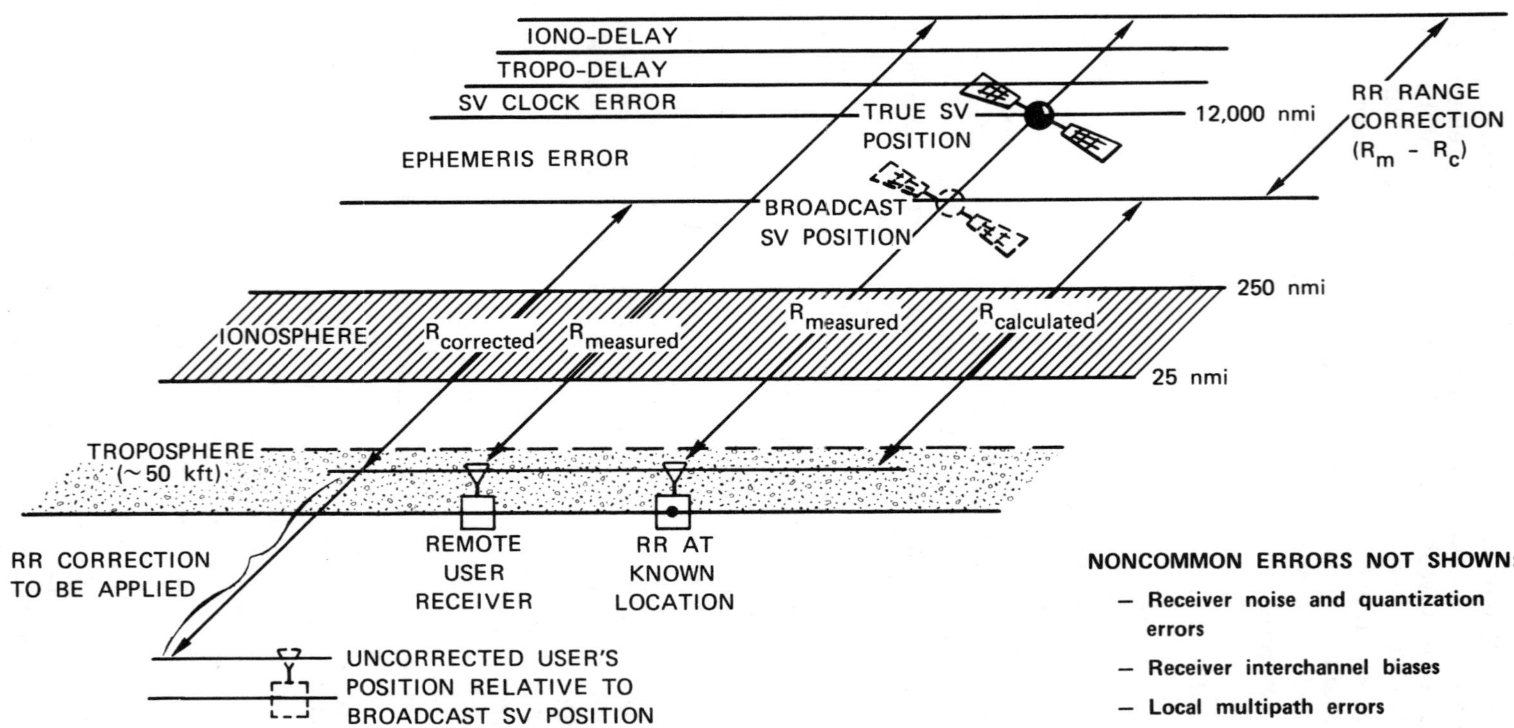


*Fig. 1—Basic diagram for common error removal by differential GPS*

Tropospheric delays, however, are only expected to be common when both receivers are at the same altitude and in relatively close (~50 mi) proximity (90 percent of the tropo delay occurs below 30 kft). Nominal error budgets for differential GPS have been estimated (see Table 2) and appear in the Air Force GPS Range Applications Joint Program Office (RAJPO) specification[4] for a high-dynamic medium-volume GPS Instrumentation Set (HDIS) for co-altitude receivers at a separation of 50 nmi. The iono and tropo values are considered to be the residual noncommon mode errors for combined iono and tropo delays remaining after defferential processing. As with Table 1, receiver noise and quantization as well as multipath error budgets are the same.

The UERE values in Tables 1 and 2 represent the RMS error in PR measurement to each satellite, assuming a nominal SV elevation angle about the local horizon. To obtain estimates of the position solution accuracy, one multiplies the pseudoranging accuracy by the expected horizontal and vertical geometric dilution of precision (HDOP and VDOP). For the purposes of estimation, values for HDOP and VDOP are taken as 1.5 and 2.5, respectively. The resulting estimates for horizontal and vertical static position accuracies are summarized at the bottom of Tables 1 and 2. It should be noted that the error budget estimates presented in Tables 1 and 2 are for a receiver $C/N_o$ condition of 38 dB-Hz. At poorer signal conditions (such as loss of lock) the receiver noise contributions will increase by approximately a factor of four, which increases the values in Table 1 for ionospheric delay measurements for P code and the receiver noise/quantization error for both C/A and P code. However, at poorer signal conditions, only receiver noise/quantization errors will be increased for the differential mode error budgets in Table 2, since the iono delay need not be measured because it is common. In the differential GPS methods described in this paper, it is assumed that neither a remote participant receiver nor the reference receiver will remove ionospheric delays from their data since they are assumed to be common. Tropospheric delays, however, are

Table 2—Differential GPS Error Budget*

| Error Source | Predicted Error (ft) | |
|---|---|---|
| Residual satellite clock error | 0 | |
| Residual ephemeris error | 0 | |
| | P code | C/A code |
| Residual ionospheric/ tropospheric delay error | 0.5 | 0.5 |
| Receiver interchannel bias | 0.5 | 2.0 |
| Receiver noise | 0.8 | 8.0 |
| Multipath | 4.0 | 10.0 |
| UERE (RMS) | 4.2 | 13.0 |
| Resulting Position Accuracy | | |
| RMS horizontal position error (assume HDOP = 1.5) | 6.3 | 19.5 |
| RMS vertical position error (assume VDOP = 2.5) | 10.5 | 32.5 |

Source: Reference 4

*Assume 50-nmi separation between remote and reference receivers.

assumed to be estimated by each receiver, since differences in altitude can cause these delays not to be common.

### *Methods for Implementing Differential GPS*

Two basic techniques can be used to effect differential GPS between a GPS-equipped user vehicle and a reference receiver; corrections can be determined and made in the measurement domain (PR data) or in the solution domain (position data). Both methods should result in essentially the same final position accuracy, and they differ principally in the specific data that must be transferred between the participant vehicle and the reference receiver, and whether uplink or downlink (or both) communication paths are used between the reference receiver and the participant vehicles. All schemes discussed accommodate the possibility that participants may independently select sets of four satellites and that multiple participants can be involved. Four methods are considered which are combinations of either corrections of PRs or of the navigation solutions, and whether the corrections are made on board the participant or at the reference receiver.

**Method 1: Uplink of pseudorange corrections for participant on-board processing**—This method requires that the position of the RR is known and that the RR measures the PRs to all visible satellites (typically seven) and computes the differences between the calculated range to the SV and the measured range. The differences are taken after the receiver clock bias has been removed and the calculated range to each SV uses the satellite's position as defined in the ephemeris message. Because the satellites are in motion and SA may be invoked, the rate of change of PR corrections should also be computed for each SV and uplinked for use by the participants. Another reason for including the rate of change of PR corrections is that uplinked correction data need only be transmitted about every six seconds. In this method, the RR does not need to know which set of four satellites a given participant is using since PR corrections are broadcasted for all visible SVs. Each participant thus selects the appropriate set of corrections and applies them to its position processing. Figure 2 illustrates this method.

The data that must be uplinked in this method include:

- PR corrections (each visible SV)
- Rate of change of PR corrections (each SV)
- Age of ephemeris data (AODE) used by RR (each SV).

The AODE term is included to assure the use of the same ephemeris and clock corrections data by both reference and remote receivers, since one receiver may read and apply newly uploaded data before the other. The advantages of this method are that any number of participants can receive and use the differential correction data, and the corrected position data are available on board the vehicle for precision navigation purposes. Disadvantages are that an uplink capability is required and provisions are needed for on-board incorporation of the correction data by the participant's GPS receiver.

**Method 2: Uplink of position corrections for participant onboard processing**—This method is similar to Method 1 in that the reference receiver position is known, but the correction data are computed based on differences in the reference receiver's surveyed position coordinates and GPS measured

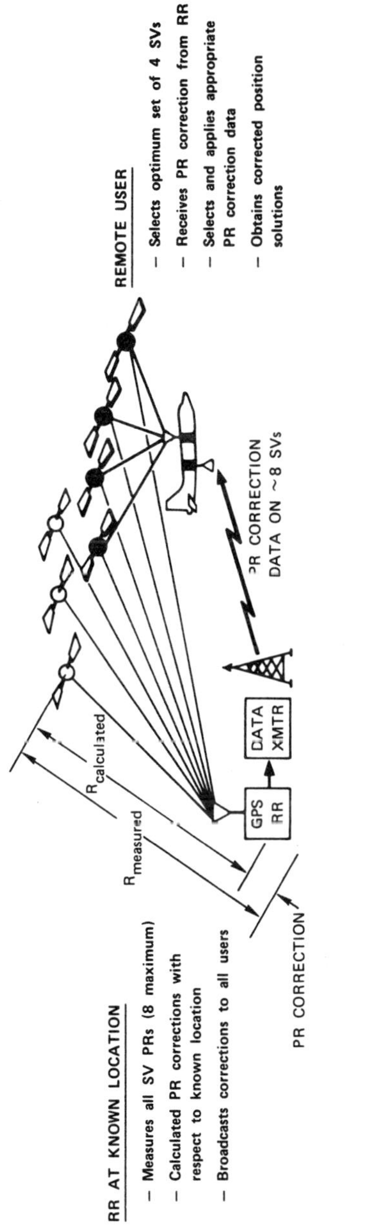


*Fig. 2—Differential GPS—Method 1*

coordinates. Conceptually, these coordinate errors ($\Delta X$, $\Delta Y$, $\Delta Z$) can be uplinked to a participant for correcing its navigation solution. Of course, the position corrections must be derived for the same set of satellites the participant is using. Method 2 is illustrated in Figure 3. For Method 2, the uplinked data will include:

- $\Delta X$, $\Delta Y$, $\Delta Z$ corrections for each user's SV set
- Rate of change of the corrections ($\Delta\dot{X}$, $\Delta\dot{Y}$, $\Delta\dot{Z}$)
- AODE (each SV)
- Participant's address.

The advantage of this method (as with Method 1) is that the corrected position data are available at the participant vehicle. Disadvantages of Method 2, however, outweigh its advantage. First, it obviously requires an uplink capability as well as correction processing provisions at the participant. Second, the participant may need to downlink its selected set of 4 SVs. More importantly, when multiple participants are involved (and it is impractical to require that all participants use the identical set of four SVs because of vehicle structural masking of the antenna), up to 70 combinations of four satellites are possible when eight are above the horizon. Thus, the designer must require that either (1) all participants use the same set of four SVs, (2) each participant downlinks the set of four SVs it uses so the RR produces the appropriate correction terms for each user, or (3) the RR ground system broadcasts corrections for all possible SV combinations.

**Method 3: Downlink all participants' raw-measured data for ground differential processing**—This method involves differential corrections in the measurement domain, differing from Methods 1 and 2 in that the differential processing is performed at the reference receiver's location. This method was conceived to take advantage of a ground computer's larger capacity to perform more sophisticated processing than is typically possible in a remote participant's navigation processor. Also, this method is applicable when no uplink is available and when corrected position data are not required on board the participant vehicle. The data to be downlinked for this method include inertial sensor data, which are usually required on board high-dynamic vehicles to preserve accuracy and to accommodate position data integrity during satellite obscuration during maneuvers. Figure 4 illustrates Method 3. The data that must be downlinked from each participant are:

- Four PRs and PR rates
- Four satellite ID numbers
- AODE (each SV)
- Receiver ID
- User time
- Measurement quality estimates
- IRU attitude tests (3)
- IRU accelerations (3)
- Attitude estimate (direction cosines)
- Accurate time-tags for all measured data.

The reference receiver produces the same PR corrections for all visible satellites, as in Method 1. These data are made available to the ground-based position processing system along with the downlinked data from each participant.

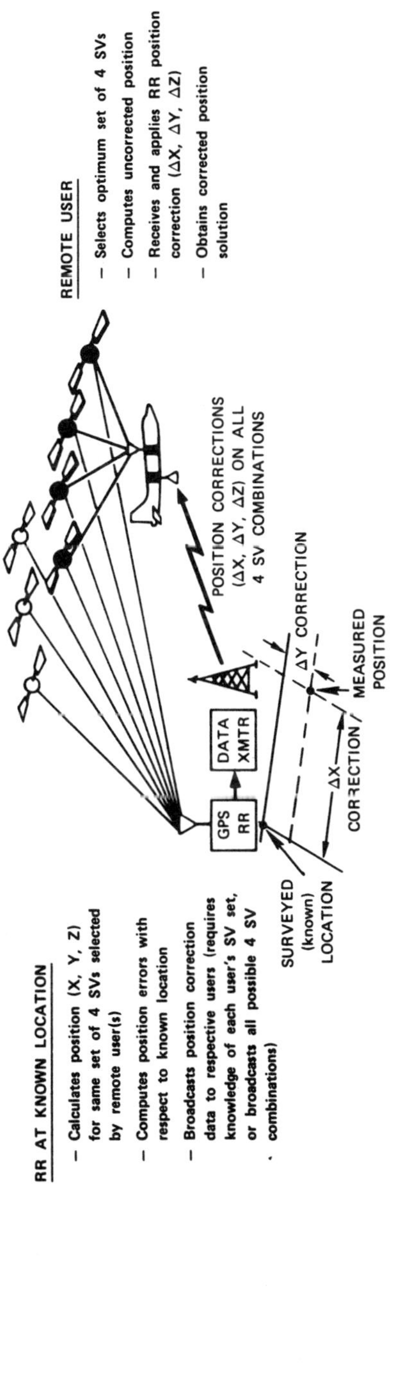


*Fig. 3—Differential GPS—Method 2*

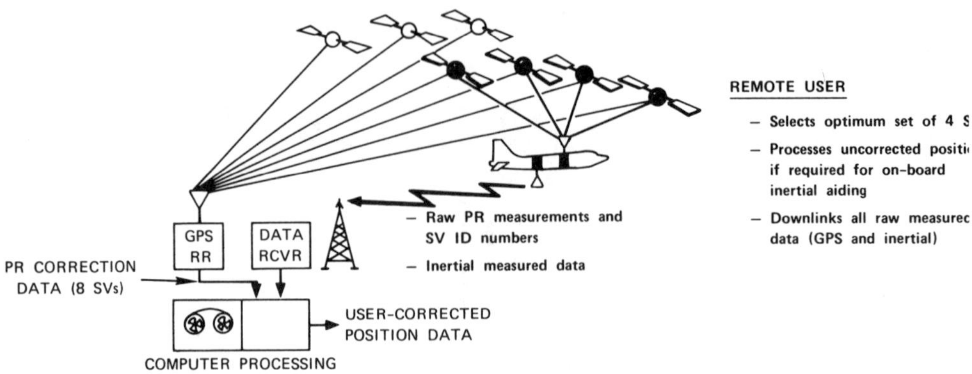


*Fig. 4—Differential GPS—Method 3*

Advantages of Method 3 are as noted earlier; no uplink is required and more extensive ground data processing resources can be used for the differential navigation solution. Disadvantages are seen in the large amount of raw data to be downlinked for ground-based differential processing. Additionally, if the user requires the corrected data on board, then an uplink would also be required.

**Method 4: Downlink uncorrected participant position data for ground differential processing**—This method performs differential corrections in the solution domain and can also be used with or without the position of the RR being known. Method 4 is used (1) for applications that do not require the corrected position data to be on board the remote user platform, (2) where an uplink and on-board differential processing are not available, and (3) when large amounts of raw measurement data are not easily downlinked. Method 4 requires each participant to downlink its uncorrected position data along with the satellite set being used. The RR (when it's position is known) can calculate the ($\Delta$X, $\Delta$Y, $\Delta$Z) corrections (using the same set of satellites as the participant). These corrections are then applied to the respective participant's (X, Y, Z) position data to obtain the differentially corrected position data. Of course, a separate set of reference receiver ($\Delta$X, $\Delta$Y, $\Delta$Z) corrections must be made for each unique participant set of four satellites. Rather than having a large number of separate navigation processors working with the reference receiver's eight sets of satellite measurements, it has been suggested that the set of eight PR corrections can be used with a simple matrix inverter to derive each set of four ($\Delta$X, $\Delta$Y, $\Delta$Z) corrections to correspond with each participant's choice of satellites. Method 4 is illustrated in Figure 5.

When the RR position is not accurately known (e.g., with an at-sea application), one may use Method 4 by downlinking uncorrected participant positions (including SV numbers) and simply difference these data from the RR's measured position as determined by using the participant's set of satellites in separate navigation processors for each satellite set. This process has been termed *relative* GPS because the processed participant's location is not tied to absolute position on the earth, but is relative to the reference receiver, which may be taken as the origin of a local tangent plane. The data that must be downlinked from each participant are:

- Uncorrected participant position (X, Y, Z)

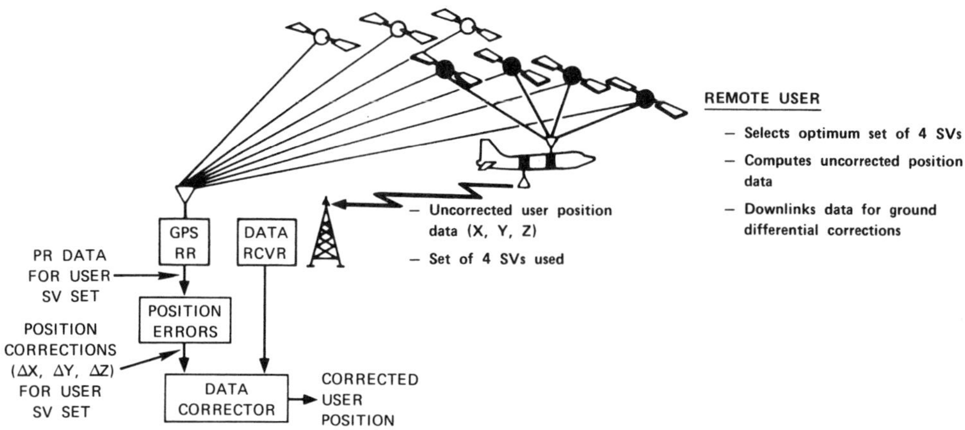


*Fig. 5—Differential GPS—Method 4*

- Satellite (vehicle) ID numbers
- Data measurement time
- AODE (each SV).

## Summary and Conclusions

Of the four differential methods discussed, Method 1 (broadcasting PR corrections) and Method 2 (uplinking position corrections) will provide corrected position data on board the remote user's vehicle. This is basic to any application where precise vehicle navigation is required. Method 1 is the preferred method when large numbers of user vehicles must be accomodated, since the PR correction data for ~8 satellites is sufficient regardless of the SV set chosen by a given user. Method 2 is least preferred when multiple users that may use different satellite sets are involved.

Methods that involve strictly ground processing of downlink user vehicle GPS data (Methods 3 and 4) are most applicable to situations where precise position data are required at a central RR facility (e.g., test range target control applications or post-flight processing to obtain higher trajectory accuracies). Method 3 (downlink of all new PRs and inertial data if aiding is used for high-dynamic players) will be the most demanding to implement due to the need for high data-link integrity (to preserve the ground navigation processing filter continuity) and the need for multiple ground processors when multiple objects are involved. Method 4, although reducing the downlink data load requirements, will still require separate differential solution processors when multiple objects are using different satellite sets.

The advantage of differential GPS for DoD authorized receivers can yield up to a threefold accuracy improvement over absolute navigation modes. For the civilian community using C/A code receivers, the advantage is even more dramatic in that up to tenfold accuracy improvements are possible in local regions when selective availability is involved.

## REFERENCES


1. *GPS NAVSTAR User's Overview,* YEE-82-009, NAVSTAR GPS Joint Program Office (September 1982).
2. Beser, J. and Parkinson, B. W., *The Application of NAVSTAR Differential GPS in The Civilian Community,* Journal of the Institute of Navigation, Vol. 29, No. 2 (Summer 1982).
3. Fickas, E. and Wadsworth, I. *GPS Field Test Measurement Program,* International Telemetering Conference, Las Vegas, Nevada (October 1984).
4. *High Dynamic Instrumentation Set (HDIS) Specification,* YISP-83-14, United States Air Force (AD/YI), Eglin AFB, Florida (July 1984).

# Special Committee 104 Recommendations for Differential GPS Service

RUDOLPH M. KALAFUS

DOT Transportation Systems Center, Cambridge, MA

AL J. VAN DIERENDONCK

Stanford Telecommunications, Inc., Santa Clara, CA

NEVIN A. PEALER

Systems Control Technology, Inc., Washington, D.C.

*Received June 1986*

## ABSTRACT

The Radio Technical Commission for Maritime Service (RTCM) established Special Committee 104 on Differential NAVSTAR/GPS Service in November 1983 to address the methods to be used to provide differential GPS service and to develop standards for the data format. The Special Committee has finished the first phase of its deliberations, which involves data message and format standards, and recommendations for communications bands and a proposed design for pseudolite operation. This paper describes the recommendations of the RTCM SC-104. The standard message provides range and range-rate corrections, along with the satellite and station health, and issue of data (which provides satellite message timing). Auxiliary messages are defined, to be interposed occasionally, which provide ground station location, constellation health, tropospheric model parameters, and further range corrections to adjust for satellite message changes. A special message type is defined to support surveying applications. This message provides high-precision Doppler counts using carrier phase measurements, which make possible centimeter resolution for stationary platforms. Navigation and communication bands are identified for the provision of public and private radionavigation and radiolocation. The use of the marine radiobeacon band, in particular, appears attractive for marine navigation in harbor/harbor approach and restricted waterways.

## BACKGROUND

RTCM Special Committee 104 on Differential NAVSTAR/GPS Service was established in November 1983 to develop recommendations on standards for the broadcast of differential corrections to users of the Global Positioning System (GPS), and to define the data link between the reference station and the users. Recognizing that differential GPS has potential for other than marine use, the Committee's terms of reference call for maintaining liaisons with the Radio Technical Commission for Aeronautics and other organizations, and for ensuring that the standards and recommendations do not restrict differential operation to marine usage. The three working groups and their chairmen are:

Data Message and Format Working Group—Dr. Al J. Van Dierendonck
Communications Working Group—Nevin A. Pealer
Pseudolite Working Group—Thomas A. Stansell, Jr.

The Data Message and Format recommendations include several standard messages which are considered to be in final form. Other messages are defined provisionally, but it is expected that it will be several years before either ground station or user equipment will require them. The first message includes the pseudorange and range-rate corrections for each satellite in view of the ground reference station, along with satellite ID, issue-of-data (which identifies the time of the satellite data update), and satellite and station health. The second message provides further pseudorange corrections to account for the possibility that the user may be using old satellite data in its pseudorange computations. The third message provides essential auxiliary data such as the location of the reference station, location of differential broadcast transmitters, and tropospheric correction factors. The fourth message is for surveying applications, and has highly accurate carrier phase measurements along with pseudorange corrections and issue-of-data indications. Other useful auxiliary messages have been defined as well, and some have been deliberately left undefined so that service providers can fashion their own messages to their clientele.

The Communications Working Group identified a number of navigation and communications bands which could be used to provide radionavigation and radiolocation services. In particular, the use of the Marine Radiobeacon bands is recommended for coastal and harbor marine coverage. However, before these recommendations can be considered as final, it will be necessary to conduct field tests of a data link in this band. Such a program is now in progress, being conducted by the U.S. Department of Transportation's Coast Guard and Research & Special Programs Administration[1].

The Pseudolite Working Group has developed a concept which holds promise for interference-free operation, using the GPS frequency band for transmission of corrections. However, the concept must be proved in the field before the recommendations can be considered final. This concept will not be described here in detail. It is described in a companion paper written by the working group chairman, Thomas A. Stansell, Jr.[2]

Special Committee 104 will soon issue two documents: the first will be a set of firm recommendations on the Data Message and Format, and the second will be a set of provisional recommendations on the frequency bands and the pseudolite design concept.

The Committee has attempted to accomodate the widest possible user community, including not only marine users, but land-based and airborne users. Both stationary positioning and navigation applications are supported. Provision is made for a standard data link interface, which will enable a receiver vendor to utilize different data links to receive corrections. In addition to a number of standard message types, provision is made for agencies or companies to define their own special messages as well. We are confident that the recommended format will support the most stringent and unique applications of this new high-accuracy positioning technique.

## DIFFERENTIAL GPS DESCRIPTION

Differential operation of the NAVSTAR/GPS offers the possibility of accuracies of 5-20 meters for dynamic navigation applications, and better than 3 meters for stationary applications using code-phase measurements. The basic concept of differential GPS is similar to that employed in differential LORAN-C, differential OMEGA, and the translocation mode using TRANSIT. A reference receiver is placed at a known, surveyed-in point. By comparing the known location with that predicted by the GPS, corrections can be determined. These corrections can then be broadcast to nearby users, who can use them to improve their position solutions. One purpose of this Committee is to determine the content and format of these broadcasts.

The differential technique works if the preponderant errors are bias errors common to all users. This is the case for GPS. The major sources of error are the following:

1. *Selective Availability errors*—artificial errors introduced at the satellites for security reasons. Pseudorange errors of this type are about 30 meters, 1-sigma. Users having access to the GPS Precise Positioning Service (PPS) have the key to these artificially induced errors and can eliminate them entirely.
2. *Ionospheric delays*—signal propagation group delay, which can be as much as 20–30 meters during the day to 3–6 meters at night. In the two-frequency mode of operation of the PPS, this effect is largely removed by applying the inverse square-law dependence of delay on frequency.
3. *Tropospheric delays*—signal propagation delays caused by the lower atmosphere. While the delays are as much as 30 meters at low satellite elevation angles, they are quite consistent and modellable. Variations in the index of refraction can cause signal delay differences between reference station and user of 1–3 meters for low-lying satellites.
4. *Ephemeris errors*—differences between the actual satellite location and the location predicted by the satellite orbital data. Normally these will be quite small (less than 3 meters) in the operational system.
5. *Satellite clock errors*—differences between the actual satellite clock time and that predicted by the satellite data. The oscillator that times the satellite signal is free-running; the GPS ground control station monitors it, and establishes corrections, which are sent up to the satellite to set the data message. The user reads the data and adjusts the signal timing accordingly.

Satellite clock errors are completely compensated by differential operation, as long as both reference and user receivers employ the same satellite data. Ephemeris errors, unless they are quite large, say 30 meters or more, are similarly compensated by differential operation. Selective Availability errors affecting the timing of the signals are also compensated by differential operation, except that the corrections lose their validity after a period of time. For users near the reference station, the respective signal paths to the satellites are close enough together that the compensation is almost complete. As the user-reference station separation is increased, the different ionospheric and tropospheric paths to the satellites can be far enough apart that the atmospheric inhomogeneities cause the delays to differ somewhat. To the extent they differ, they constitute an error in the differential GPS measurement. This type of error will be greater at larger user-reference station separations.

It is apparent that differential GPS is a promising high-accuracy navigation and positioning technique that will find numerous important applications in the civil community.

## DATA MESSAGE AND FORMAT RECOMMENDATIONS

### *General*

The general message format was patterned after the GPS Navigation Message Format as specified in the system document ICD-GPS-200[3]. The recommended format differs somewhat from that of the GPS message, however, because different requirements surfaced during its development. The GPS word size, word format, parity algorithm and other features survived. The major difference between the formats is that the differential messages will use a variable length format, whereas the GPS format has fixed length subframes.

The reasons for maintaining commonality with the GPS satellite data format are the following:

1. To provide some degree of commonality in the user software.
2. To avoid precluding the use of pseudolites for data transmission.
3. To provide a strong parity algorithm to detect errors, in order to prevent the use of erroneous corrections that could affect user safety.
4. Most users do not want to rely on some data link parity scheme over which they have no control. In fact, data link parity might not be required in many applications.
5. The GPS parity algorithm is a known and proven technique.
6. The parity algorithm overlaps word boundaries and resolves sign ambiguities encountered in biphase modulation data transmissions.
7. The GPS format provides a convenient timing capability.

The format does not dictate the data rate, but the recommendations are premised on a minimum 50 bits per second, the same as GPS (highly desirable if pseudolites are employed). The variations introduced by Selective Availability drive this data rate requirement, which supports better than 5 meter accuracy even with the loss of two successive messages. If the policy on Selective Availability is ever changed to eliminate the artificial variations, the data rate requirement will be reduced by more than an order of magnitude.

The general message format is illustrated in Figure 1. Each message frame is made up of several 30-bit words, always headed by two standard words which provide the following:

1. Fixed preamble
2. Message type identifier
3. Station identification
4. Timing data
5. Message frame length
6. Station health information.

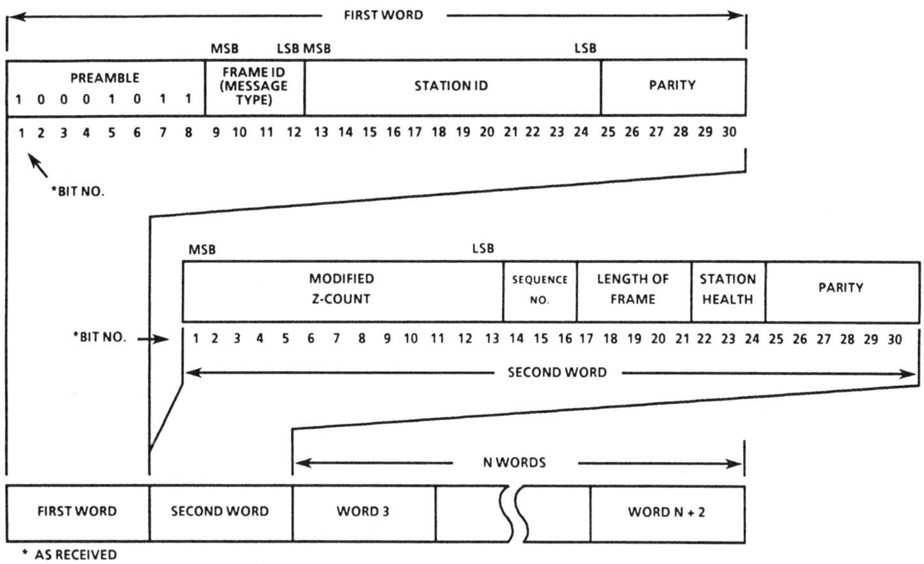


*Fig. 1—General Message Format*

Table 1 shows the number and meaning of the bits in the first two words. The number of other words can vary from zero to 31 (rarely more than twelve in practice), depending on the message type, and the number of satellites in view of the reference station. The Z-count is modified from its GPS definition in that its scale factor is 0.6 seconds, instead of the 6-second scale factor of the satellite message. The range is only one hour.

Table 1.—Content of First and Second Words

| Word | Content | Number of Bits | Scale Factor and Units | Range |
|---|---|---|---|---|
| First | Preamble | 8 | See Ref. 2 | |
| | Message Type | 4 | 1 | 1–16* |
| | Station Ident | 12 | 1 | 0–4095 |
| | Parity | 6 | See Ref. 2 | |
| Second | Modified Z-Count | 13 | 0.6 Seconds | 0–3599.4 |
| | Sequence Number | 3 | 1 | 0–7 |
| | Length of Frame: N+2 | 5 | 1 Word | 2–33 |
| | Station Health | 3 | — | 8 States |
| | Parity | 6 | See Ref. 2 | |

*16 is indicated with all zeros

Instead of verifying frame synchronization with an incrementing Z-count, as is done with the satellite messages, the receiver can use the incrementing sequence number. The location of the start of the next frame is indicated with the length-of-frame number.

The reference station health element is a 3-bit code which indicates how recently the reference station measurements were taken.

Allowance is made for sixteen message types; eleven of these have been defined, and two others reserved for particular applications. Table 2 lists each of them.

Table 2.—Message Types

| Type No. | Message Type | Type No. | Message Type |
|---|---|---|---|
| 1 | Differential Corrections | 9 | High Rate Differential Corrections |
| 2 | Delta Differential Corrections | 10 | P-Code Differential Corrections (Reserved) |
| 3 | Station Parameters | 11 | C/A-Code L1, L2 Delta Corrections (Reserved) |
| 4 | Surveying (Carrier Phase) | 12 | Health Message (ASCII String) |
| 5 | Constellation Health | 13 | Undefined |
| 6 | Null Frame | 14 | Undefined |
| 7 | Beacon Almanacs | 15 | Undefined |
| 8 | Pseudolite Almanacs | 16 | Special Message (ASCII String) |

### *Message Type 1*

Message Type 1 is the primary message; it contains the pseudorange and range-rate corrections to be used for navigation (see Figure 2 and Table 3). That is, each differential GPS service user applies these corrections to the GPS measurements to achieve the high service accuracy available with this technique.

Pseudorange corrections are used rather than latitude/longitude corrections because there is no way of assuring that both reference station and users employ

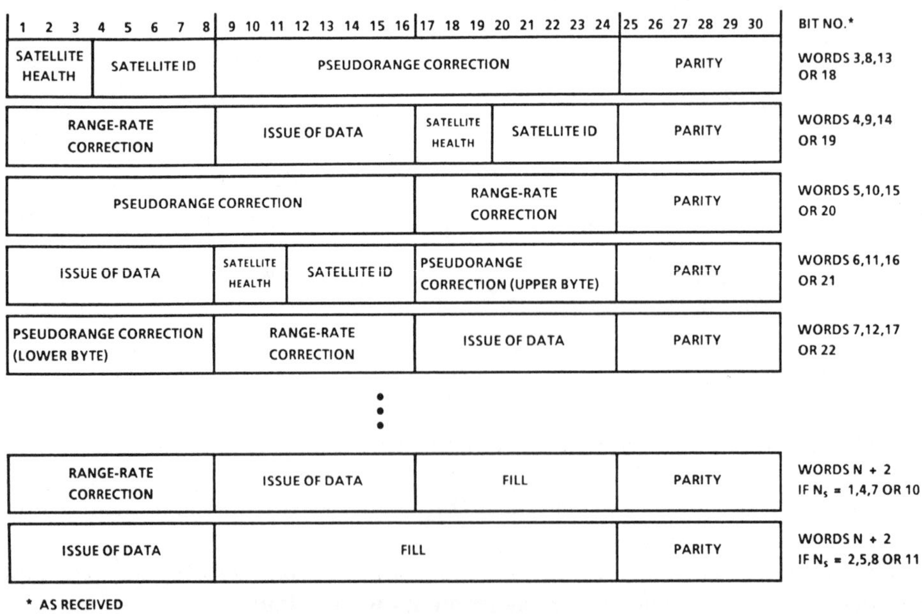


*Fig. 2—Type 1 Message Format*

Table 3.—Contents of the Type 1 Message

| Parameter | No. of Bits | Scale Factor and Units | Range |
|---|---|---|---|
| Pseudorange Correction* | 16 | 0.02–0.64 meters | ±655.36 – ±20971.52** |
| Range-rate Correction* | 8 | 0.002–0.064 m/s | ±0.256 – ±8.192** |
| Satellite Identification | 5 | $1 \times N_s$ | 1–32*** |
| Satellite Health Summary and UDRE | 3 | — | 8 states |
| Issue of Data | 8 | See Ref. 2 | |
| Fill (Alternate 1's & 0's) | 8 or 16 | — | |
| Parity | $6 \times N_s$ | See Ref. 2 | |

* 2's complement
** Scale factor varies with UDRE
***32 is indicated with all zeroes

the same satellites. In applying the corrections, the user receiver algorithms must apply the usual ephemeris and satellite clock correction terms in the satellite data, but must not apply any ionospheric or tropospheric models. In this manner the problem of potentially different models being used by user and reference station is avoided.

The satellite pseudorange measured by the receiver is corrected as follows, using the Message Type 1 parameters:

$$PR(t) = PR_m(t) + PR_o + (dPR_o/dt)(t - t_o)$$

where $PR_m(t)$ is the measured pseudorange at time t in meters, $PR_o$ is the 16-bit pseudorange correction at $t_o$, $dPR_o/dt$ is the 8-bit range-rate correction, and $t_o$ is the 13-bit modified Z-count of the Second Word, applied to that satellite indicated by the 5-bit Satellite ID, which indicates its PRN number. Also provided is a 3-bit satellite health indication (see Table 4), which includes an estimate of the differential error in terms of User Differential Range Error (UDRE) as estimated by the reference station. The satellite health is not just an echo of the satellite data, but is determined by the reference station, based on a comparison of true and measured pseudorange, as well as the quality of the data and the measurement.

Table 4.—Satellite Health and UDRE

| Code | No. | Indication |
|---|---|---|
| 111 | 7 | Don't use |
| 110 | 6 | See Message Type 5 for details |
| 101 | 5 | 1-sigma differential error > 8 meters |
| 100 | 4 | 1-sigma differential error > 4 meters |
| 011 | 3 | 1-sigma differential error > 2 meters |
| 010 | 2 | 1-sigma differential error > 1 meter |
| 001 | 1 | 1-sigma differential error > 0.5 meters |
| 000 | 0 | 1-sigma differential error < 0.5 meters |

The scale factors of the pseudorange and range-rate corrections vary with the value of UDRE. With a UDRE of 000, or zero, the scale factors are 0.02 meters and 0.002 meters/second, respectively. With UDRE values of 1 to 5, the scale factors are multiplied by 2 raised to the power of the UDRE code. Thus a UDRE of 5 means the scale factors are 0.64 meters and 0.064 meters/second, respectively. The rationale for the variable scale factor is to allow for occasional high values of pseudorange and range-rate, while still maintaining the high precision for the normal situation.

"Issue of Data" (previously called "Age of Data") tells the user whether the reference station is basing the corrections on the most recent satellite data. Most of the time the user and reference station will base the position estimates on the same satellite data.

If the user's data is older than that of the reference station, the information contained in the Type 2 Message should be used to further correct the measured pseudorange. It will rarely occur, but if the user's data is newer than that of the reference station, the new data should not be used until the reference station is also basing the corrections on the new data.

It is important to note that the pseudorange correction described above is a predicted correction that will have an error buildup over time. Thus, it will be updated and transmitted as often as required to maintain high accuracy. The user should update the solutions using all the corrections as they are received. The user should avoid using the range-rate correction to alter the carrier Doppler measurement, as it may well degrade it. Message Type 4 is intended for accurate Doppler corrections.

To facilitate the possibility of rapidly changing pseudoranges or range-rates, a short Type 9 Message will be transmitted as soon as possible, repeating the parameters of the Type 1 Message for that (those) satellite(s) only. The contents will be identical except that the number of satellites will rarely exceed one. it would be during the condition that the UDRE might take on non-zero values.

### *Message Type 2*

Message Type 2 contains a second set of pseudorange corrections, termed "Delta Differential Corrections", which enable a user to modify the received corrections when the satellite ephemeris and clock data set with which it is operating is older than the data set being used by the reference station. The reference station will transmit a Type 2 Message as soon as it starts using new GPS Navigation Message data. It will indicate this by changing the Issue of Data parameter in Message Types 1 and 4. It is likely to be transmitted every two or three minutes or so. The parameters of the Type 2 Message will be updated with each transmission, since the corrections will change slowly with time. This is a result of the fact that the satellite ephemeris data parameters are related to orbital parameters, while the Delta Differential Corrections are satellite positional parameters.

Table 5.—Contents of the Type 2 Message

| Parameter | No. of Bits | Scale Factor and Units | Range |
|---|---|---|---|
| Delta Diff. Correction* | $16 \times N_s$ | 0.02–0.64 meters | $\pm 655.36$ – $\pm 20971.52$ |
| Fill (Alternate 1's & 0's) | 8 or 16 | — | |
| Parity | $6 \times N_s$ | See Ref. 2 | |

*2's complement

The Message Type 2 content is given in Table 5. Figure 3 shows the format of the transmissions. Fill words are used when the number of satellites is not an even mulitple of three. The order of transmission of the Delta Differential Corrections will be the same as that of the previously transmitted Type 1 Message.

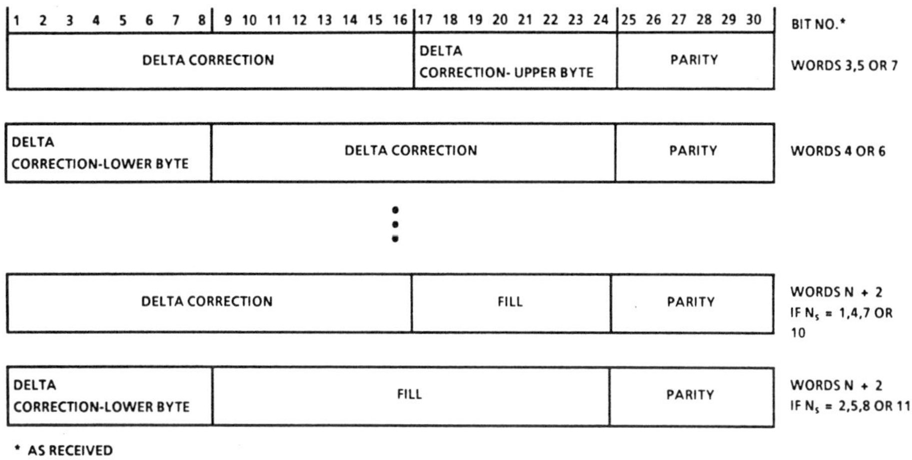


*Fig. 3—Type 2 Message Format*

## *Message Type 3*

Message Type 3 is a fixed-length message which contains essential auxiliary information about the reference station (see Figure 4):

1. GPS station coordinates (earth-centered-earth-fixed)
2. Tropospheric delay at the zenith of the reference station
3. Station health (expanded)
4. Estimated station clock parameters
5. Average carrier-to-noise density figure of merit

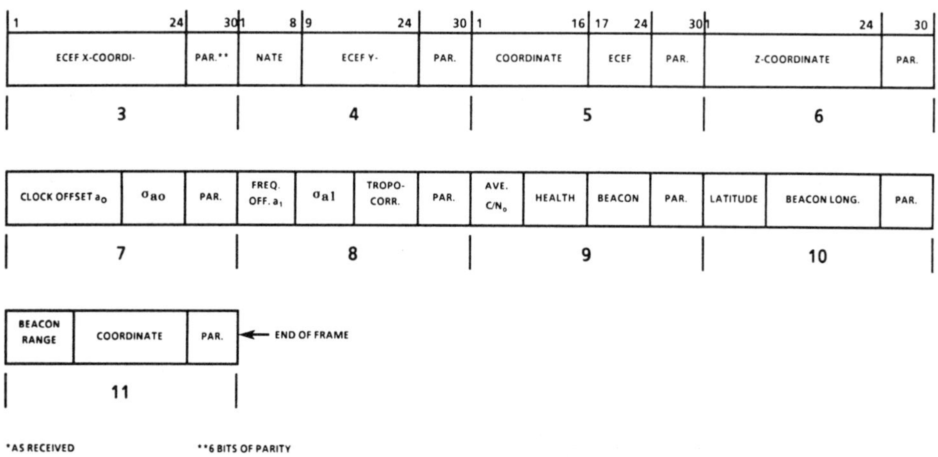


*Fig. 4—Type 3 Message Format*

The tropospheric delay is provided for those users at an altitude different from that of the reference station. The parameter provided is in the form of a vertical correction in meters, representing the delay of the signal from a satellite at the zenith over the reference station. Any of several tropospheric models defined by the user can be used.

The reference station clock time and frequency offsets are provided for users who wish to determine and transfer time accurately. To provide this service, however, the reference station must have some means of synchronizing its time to GPS time, such as via a common-mode, common-view procedure with the U.S. Naval Observatory.

The average $C/N_0$ is an indication of station health. It is the average (in decibels) over all satellites being tracked.

If radiobeacon stations are used to broadcast corrections, it would be useful to have information which aids the receiver/data link set to select which radiobeacon is to be used. Some information is provided in the Message Type 3, as well as in Message Type 8, described below. Here reference station and beacon health indications are provided, although not all have been determined. Beacon latitude, longitude and transmission range are provided, if radiobeacons are used for broadcasting corrections.

### *Message Type 4*

Message Type 4 has been designed with surveying applications in mind. It was decided to design this message to include pseudorange corrections, in order to make the Type 4 Message stand by itself. That is, a survey receiver can operate without Type 1 Messages being interposed between the Type 4 Messages. The Type 4 Message thus contains the pseudorange corrections, the Issue of Data information, and the instantaneous cumulative fractional Doppler count as obtained from the carrier phase measurement at the ground station. By knowing the precise location of the differential station (from the Type 3 Message), a comparison of the Doppler counts at the user and reference station over a period of time can provide relative accuracies at the centimeter level.

Surveyors typically determine relative positions using continuously integrated carrier Doppler measurements, where the change in geometry of the satellites over time enable them to remove system biases. Normally, the ionosphere does not change in an entirely predictable manner, so that the ionospheric delays limit the achieveable accuracy, especially when only one signal frequency is available. However, the relative effects of the ionosphere can be cancelled, at least over short baselines of a few tens of miles (much longer, if meter accuracy is adequate). Differential GPS surveying will also reduce dramatically the time required to take the measurements, compared to alternative techniques.

Because the measurement of interest to the surveyor is the integrated Doppler count accumulated over a period of time, the information transmitted by the reference station must be transparent to any event that might have occurred during that period of time—such as loss of lock, a change in the GPS satellite navigation message, etc. To realize this, a raw measurement that is not corrected for anything is transmitted, that being the reference station's complete instantaneous carrier phase. This measurement is defined as the beat phase between the received carrier phase and that of the reconstructed carrier derived from the station's local clock, including whole cycles which have accumulated since the initial measurement was made. It is the same as the continuously integrated Doppler count biased by the initial phase measurement. Using the location of the reference station provided in the Type 3 Message, the user will then determine his own differential Doppler correction. The Type 2 Message is still applicable to accommodate changes in the Issue of Data, which is also provided in the Type 4 Message.

Table 6.—Contents of the Type 4 Message

| Parameter | No. of Bits | Scale Factor and Units | | Range |
|---|---|---|---|---|
| Pseudorange Correction* | 16 | 0.02–0.64 meters | | ±655.36 – ±20971.52** |
| Complete Instantaneous Phase | 32 | 1/256 cycle | | ±8,388,608** |
| Satellite Identification | 5 | 1 | | 1–32*** |
| Satellite Health Summary and UIPE | 3 | — | × $N_s$ | 8 states, 6 error levels |
| L1/L2 Indicator | 1 | — | | 2 states |
| Cumulative Loss of Lock Count | 7 | 1 | | 0–127 |
| Issue of Data | 8 | See Ref. 2 | | |
| Parity | 6 | See Ref. 2 | | |

* 2's complement
** Scale factor varies with UIPE levels
***32 is indicated with zeros

A reference station loss of carrier phase lock will be indicated by setting the integer portion of the complete instantaneous phase to zero and by incrementing a cumulative loss-of-lock counter. If in the future the L2 signal becomes available to civil users, the use of L2 instead of L1 would be indicated by changing L1 to "1" from its normal "0" state. Table 6 gives the details of the Type 4 Message, and Figure 5 shows the format. The satellite identification and Health Summary are as described for the Type 1 Message. However, the User Instantaneous Phase Error (UIPE) replaces the UDRE as the measure of expected error. These error levels are shown in Table 7. The scale factor of the pseudorange corrections are variable and handled in the same way as in the Type 1 Message, except that the UIPE code will be used as the power of 2 multiplier. The scale factor of the instantaneous phase is constant, since it is not affected by large changes in carrier frequency.

The Type 4 Message will be transmitted only by a reference station set up for specific service to surveying users, because the transmission of that message would reduce the update rate of the Type 1 Message to navigation users. For the surveying users, the Type 4 Message would be the primary message, along with the periodic transmission of the Type 2, 3, and 5 messages.

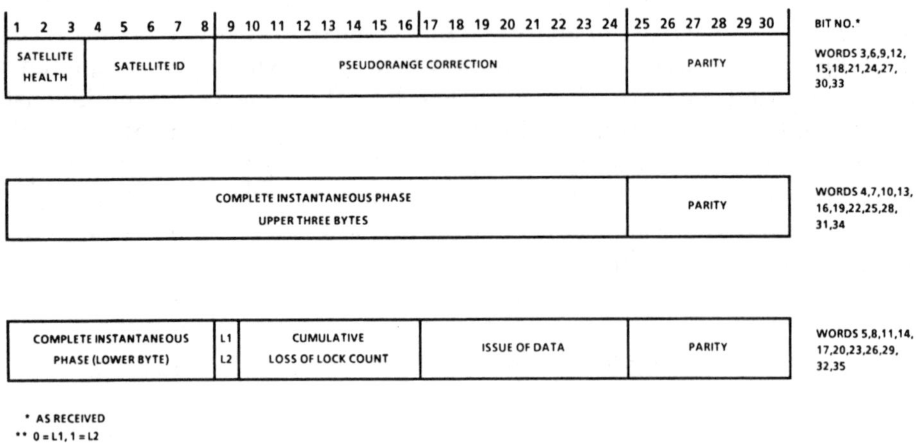


*Fig. 5—Type 4 Message Format*

Table 7.—User Instantaneous Phase Error (UIPE)

| Code | Indication |
|---|---|
| 101 | 1-sigma instantaneous phase error > 0.16 cycles |
| 100 | 1-sigma instantaneous phase error > 0.08 cycles |
| 011 | 1-sigma instantaneous phase error > 0.04 cycles |
| 010 | 1-sigma instantaneous phase error > 0.02 cycles |
| 001 | 1-sigma instantaneous phase error > 0.01 cycles |
| 000 | 1-sigma instantaneous phase error < 0.01 cycles |

### *Message Type 5—Constellation Health*

Message Type 5 provides the user with the health status of all satellites in view of the reference station, as well as an estimate of the health of the next satellite coming into view. This message will be transmitted periodically. It contains the following:

1. Satellite identification
2. Data health, i.e., the three most significant bits of the 8-bit almanac health and status word provided by in the GPS satellite message.
3. $C/N_o$ as currently estimated by the reference station for the satellites in view, and for the next satellite.

### *Message Type 6—Null Frame*

Message Type 6 contains no parameters, only the information contained in the first two words. It is used for maintaining frame synchronization when there are no messages to send (if such a situation were to occur). In addition to the first two words, it can contain one data word of alternating "1"s and "0"s.

### *Message Type 7—Beacon Almanac*

Since maritime operators are anticipated to be major users of the differential GPS service, the working group developed a message to assist coastal users to automatically select the optimum radiobeacon transmitter to obtain the differential corrections, and to automatically switch from station to station as the ship travels along the coast. Message Type 7 provides the location, frequency, service range, and health information for a number of networked radiobeacon transmitters broadcasting differential GPS corrections obtained from a single reference station. This operation is described by Quill[1].

### *Message Type 8—Pseudolite Almanac*

Message Type 8 provides pseudolite almanac information. It is similar to the radiobeacon almanac message (Type 7), and is only provided in pseudolite transmissions. It provides the location, code, and health of a number of pseudolites.

### *Message Type 9—High Rate Differential Corrections*

Message Type 9 provides extra differential corrections (following the Type 1 message format) in the rare event that pseudorange or range-rate excursions for a particular satellite are excessive. This increases its average rate of transmission. Only the corrections for the affected satellite would be involved. It is not certain that this message is needed.

### *Message Type 10—P-Code Differential Corrections*

This message type is reserved for applications involving authorized users who can receive the P-Code signals; it has not yet been defined.

### *Message Type 11—C/A-Code L1, L2 Delta Corrections*

This message type is reserved for possible future applications if the C/A Code is ever transmitted on the L2 frequency; it has not yet been defined.

### *Message Type 12—Health Message*

This type enables a free-form health message of variable length, containing a string of 8-bit ASCII printable characters.

### *Message Types 13–15—Spares*

These messages have not been defined, and are available for special uses.

### *Message Type 16—Special Message*

This message enables a free-form special message using the same ASCII format as that of Message Type 12.

## MESSAGE RATE OF TRANSMISSION CONSIDERATIONS

The purpose of the variable length format is to maximize the rate of transmission of the differential corrections. It is up to the reference station designer to determine which messages are sent, and how often. A typical scenario would be to transmit the Type 1 Message as often as possible, interleaved with Message Types 2, 3, 5, and 7. Suppose the Type 2 Messages were transmitted at once per minute, the Type 3 and 5 Messages alternated once per minute, and the Type 7 Message (nine words long) once every 5 minutes. At a rate of 50 bits per second the average transmission rate of differential corrections can be computed for different numbers of satellites in view. Table 8 shows the results for 4, 7, and 11 satellites in view. In the planned constellation, there will be an average of between 6 and 7 satellites in view.

Table 8.—Average Transmission Rate of Type 1 Messages

| No. of Satellites | Average Transmission Rate* (Type 1 Message) | Average Transmission Rate* Using Type 9 Messages Once Per Type 1 Message: Non-Repeated Satellites | Repeated Satellites |
|---|---|---|---|
| 4 | 1/6.23 sec | 1/9.15 sec | 1/4.58 sec |
| 7 | 1/10.22 sec | 1/13.14 sec | 1/6.57 sec |
| 11 | 1/16.11 sec | 1/19.18 sec | 1/9.59 sec |

*Over a 10-minute period

The significance of the transmission rates is illustrated in Figure 6, which shows the error growth after a correction is made. The error growth is dominated by the Selective Availability variations. The results are based on data supplied to the Department of Transportation by the GPS Joint Program Office[4] Also provided are the errors anticipated if the Selective Availability errors were doubled. For the errors to remain under 5 meters (2drms), the transmission rate must exceed 1/26 seconds. It can be seen that the transmission rate would be better than this for 4, 7 and 11 satellites in view.

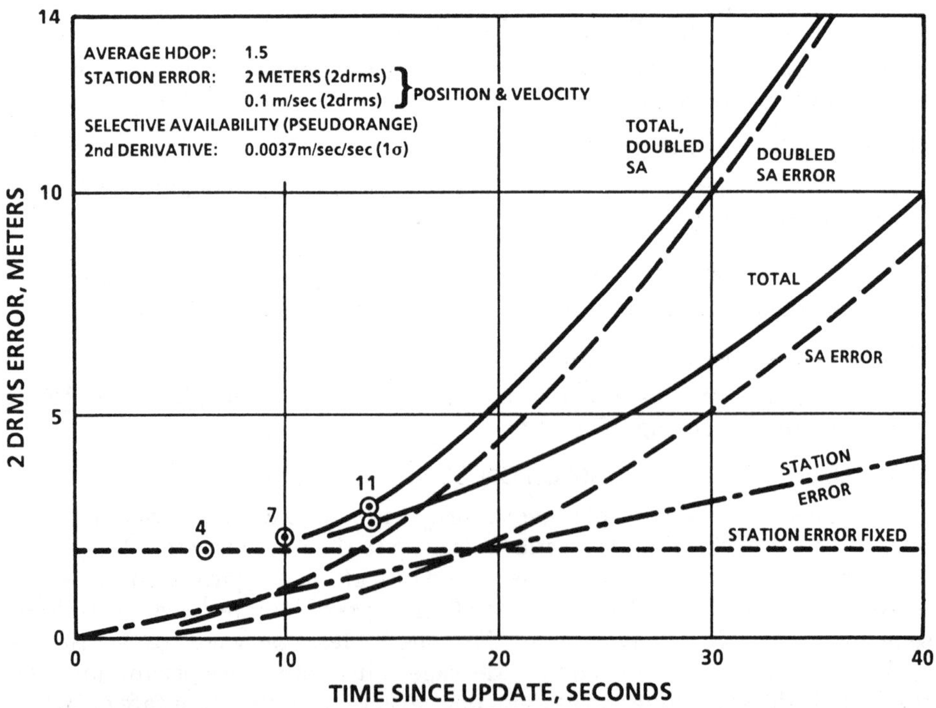


*Fig. 6—Differential GPS Error Growth Due to Selective Availability*

## COMMUNICATIONS RECOMMENDATIONS

A major goal of SC-104 was to identify a number of frequency bands which could be used for broadcasting the differential GPS corrections. There are both political and practical considerations which make certain bands more attractive than others.

There are a number of bands allocated to radionavigation, and a number of others allocated to radiolocation. Radionavigation differs from radiolocation in that in the former, position information is intended for the guidance of a vessel (including hazard avoidance), while in the latter the information is intended to provide the position of a stationary or slowly moving platform. In the US only the government is currently allowed to provide radionavigation service, since it involves the safety of life.

Radiolocation services are usually provided by private concerns that petition a local authority, such as the Federal Communications Commission in the US to obtain permission to use radiolocation bands. A number of such bands were identified by the Communications Working Group, ranging from HF to microwave frequencies. It appears that differential GPS broadcasts for radiolocation purposes should be able to operate within the existing procedures.

Identifying frequency bands for radionavigation is more problematical. A very attractive choice for coastal networks is to operate in the current radiobeacon band (285–335 kHz). This might be accomplished by modulating existing radiobeacons, if such an approach can be shown to not interfere with direction-finders that use the radiobeacons. If they would interfere, it might still be possible to use beacon frequencies for differential GPS broadcasts. If, however, it does not prove to be feasible to provide a data link of high enough integrity in this frequency band, there could be a problem. The reason is that there are no available frequency bands allocated for navigation at VHF or UHF. The case would then have to be made that maritime communications bands at VHF or UHF should be available for this use. There are good precedents for making such a case, but it would be preferable if it weren't necessary. Table 9 summarizes the frequencies allocated to radionavigation, and provides some useful information on each band of interest to marine users. The column headings have the following interpretations:

**FREQUENCY BAND**—a listing of frequency bands that are allocated for radionavigation in at least one of the regions. Reference must be made to the ITU Radio Regulations for limitations of the allocation in each region.

**REGIONAL ALLOCATION**—indicates which region has a radionavigation allocation in the specified frequency band. It also indicates if the frequency band, or a portion of the band, is allocated for radionavigation in the United States.

**MEET NEEDS**—indicates if a transmission in this band would meet the needs of a mariner using differential GPS for radionavigation.

**TECHNICALLY FEASIBLE**—indicates if it is technically feasible to transmit differential GPS corrections to a mariner with current state-of-the-art technology in this band. A question mark indicates that the committee could not make a determination.

**MARINE USE**—indicates if marine radionavigation is an allocation in this band. A "no entry" indicates that either there is no allocation or the allocations were not specifically for that purpose.

**COMMENTS**—comments the committee has on the suitability of the frequency band or constraints that may be placed on the use of the frequency band.

In summary, there appear to be a number of bands that could be used for a wide variety of applications of differential GPS. The low channel bandwidth and data rate requirements would make it possible in some applications for the corrections to be transmitted using existing communications services without adding a significant burden. Also, if the Selective Availability variations were ever removed, dedicated differential GPS data links could be shared with other services.

## ACKNOWLEDGEMENTS

The authors would like to thank the members of RTCM Special Committee 104, who provided their unique insights and expertise in developing the recommendations described here. They represented a considerable cross-section of industry, military and civil governments, both domestic and international. Also essential to the development of the recommendations were the offices of RTCM, in particular the efforts of William T. Adams, president of the commission. The funding of the government portion of the effort was provided by the Office of Budget and Programs of the U.S. DOT Research and Special Programs Administration.

## REFERENCES

1. Quill, LCDR John, USCG, *U.S. Coast Guard Differential GPS System Development*, Proceedings of the RTCM Annual Meeting, April 29–May 1, 1986
2. Stansell, T. A., Jr., *RTCM SC-104 Recommended Pseudolite Signal Specification*, Navigation, Vol. 33, No. 1, Spring 1986, pp 42–59
3. Anon., *Interface Control Document, NAVSTAR GPS Space Segment/Navigation User Interfaces*, ICD-GPS-200, Rockwell International Corporation, November, 1982, revised 1985
4. Kalafus, R. M.; Knable, N.; Kraemer, J. and Vilcans, J.; *NAVSTAR GPS Simulation and Analysis Program*, Technical Report No. DOT-TSC-RSPA-83-11, December 1983

# RTCM SC-104 Recommended Pseudolite Signal Specification

THOMAS A. STANSELL, JR.
Magnavox Advanced Products and Systems Company, Torrance, CA
*Received June 1986*

## ABSTRACT

Anyone involved in the design of GPS user equipment should be concerned with the potential impact of pseudolite signals. It is quite possible that pseudolite transmitters will be located near the majority of all airports and heliports, both rural and urban, in the United States and around the world. Therefore, even if a receiver is not intended for aircraft navigation or will not use the pseudolite signals, it may be affected by these signals.

Differential GPS can improve the reliability, integrity, and accuracy of GPS navigation, but it cannot overcome the inherent geometric limitations of an 18 satellite constellation, especially when one of the satellites is out of service. Pseudolites transmit the differential correction message at the GPS L1 frequency, thus also providing an additional pseudorange measurement which overcomes poor satellite geometry. The key problem was to develop a GPS-like pseudolite signal structure which does not interfere with reception of GPS satellite signals even when a receiver is very close to a pseudolite transmitter. A simple and effective signal has been defined, but receiver manufacturers must assure that their designs are immune to pseudolite interference. This paper defines the proposed signal and describes the appropriate receiver design considerations.

## BACKGROUND

This paper is a companion to [1] which describes the recommendations of RTCM Special Committee 104 (SC-104) for the data message and the communication methods to implement various differential GPS services. SC-104 also developed specifications for a pseudo-satellite (pseudolite) form of differential GPS, which is the subject of this paper.

The Radio Technical Commission for Maritime Services (RTCM) established SC-104 in November 1983 to develop recommendations on standards for the broadcast of differential corrections to users of the Global Positioning System (GPS) and to define the data link between the reference station and the users. Recognizing that differential GPS has potential for other than marine use, the committee's terms of reference call for maintaining liason with the Radio Technical Commission for Aeronautics (RTCA) and other organizations and for ensuring that the standards and recommendations do not restrict differential operation to marine usage. This is particularly appropriate for the pseudolite form of differential GPS because its primary application would be for terminal area (non-precision approach) guidance of aircraft and helicopters.

The conventional form of differential GPS consists of a fixed monitor station at a known location which determines the range measurement error in each received GPS signal and communicates these as corrections to be applied by local users. As a result, the local users are able to obtain more accurate navigation results relative to the monitor station location. With proper equipment a relative accuracy of 5 meters 2drms should be possible at distances of a few hundred kilometers from the monitor station.

In addition to better accuracy, differential GPS also improves navigational reliability by assuring the user that the GPS signals being used have been checked and corrected. This may even permit use of signals from satellites otherwise deemed unhealthy. Furthermore, system integrity is improved because the monitor station will immediately warn the user if there is an uncorrectable satellite signal error.

In spite of the improvements in navigational accuracy, reliability, and integrity offered by differential GPS, it cannot overcome the geometric deficiencies inherent in an 18 satellite GPS constellation. The original satellite deployment plan called for 24 satellites, but budget restraint has trimmed the number to 18. As a result, many locations on Earth will suffer 30 minute periods of poor satellite geometry once or twice per day, and these problem areas will spread whenever an operational satellite is out of service. Pseudolites can overcome this problem as well as provide other advantages, which is why a subcommittee of SC-104 was formed to find a practical pseudolite signal structure. This paper details the proposed solution.

Before starting the pseudolite signal description it is important to note that a pseudolite is simply one specific form of differential GPS. Therefore, the pseudolite message structure is the same as for any other form of differential GPS, as recommended by the committee.[1] Also, it is important to note with appreciation the work donated to the pseudolite subcommittee by, among others, representatives of:

FAA-TC
GPS Joint Program Office
Intermetrics
Litton Aero Products
Magnavox
Mitre
Rockwell-Collins
Texas Instruments
The Aerospace Corporation
Trimble Navigation

## PSEUDOLITE ADVANTAGES

Figure 1 is a pseudolite system block diagram which shows that the monitor receiver processes all available GPS satellite signals as well as the pseudolite signal and compares them with each other and with the known monitor and pseudolite antenna positions. The monitor controls the pseudolite transmission and provides the differential GPS message. Because the pseudolite transmits its signal precisely at the GPS L1 frequency of 1575.42 MHz, the GPS user equipment receives both the satellite signals and the pseudolite signal.

It is useful to note that both international and U.S. radio regulations allocate the 1559–1610 MHz frequency band both for aeronautical navigation and for radio navigation-satellite (space to Earth) applications. Therefore, pseudolite signals at the 1575.42 MHz L1 frequency are in an appropriate band for aircraft navigation. Offshore navigation applications may require a modification of the frequency allocation regulations.

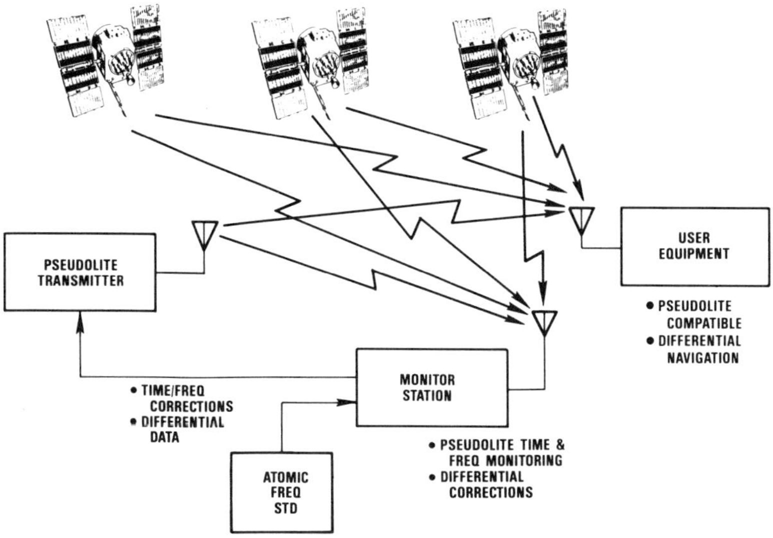


*Fig. 1—Pseudolite System Configuration*

Because the pseudolite signal epochs are precisely defined in GPS time, the user equipment can obtain an additional pseudorange measurement from the pseudolite, thus augmenting the satellite coverage and strengthening the navigation geometry. As a result, excellent navigation can be obtained even during periods of poor satellite coverage. This capability is graphically illustrated by Figures 2 and 3, obtained by permission.[2] Figure 3 emphasizes that the additional line of position from a pseudolite can be of great value in minimizing the harmful impact of a satellite signal failure. This characteristic is especially

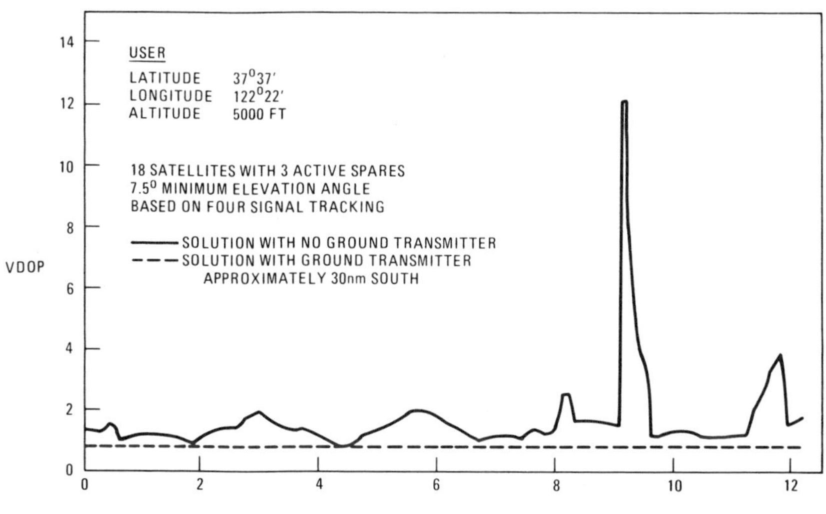


*Fig. 2—VDOP vs* Time at San Francisco Airport

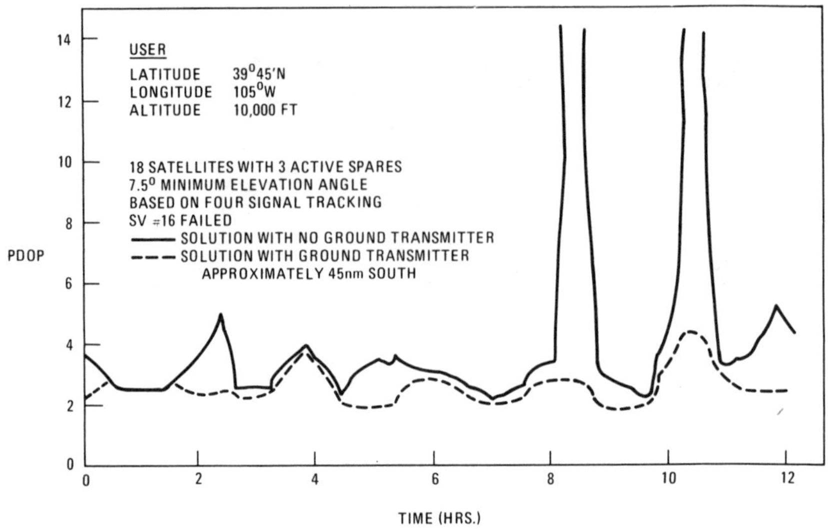


*Fig. 3—PDOP vs* Time in the Denver Area with SV-16 Failed

significant for the presently-planned, 18-satellite GPS constellation, which does not have a large amount of redundancy.

The pseudolite form of differential GPS also may be less costly than the conventional form because there is no need for a separate receiver. The antenna, RF electronics, etc., are common with the GPS receiver. Furthermore, system coordination problems of frequency selection and channel sharing are eliminated.

Operationally, these characteristics imply that within their sphere of influence, pseudolites can support a number of more demanding navigation tasks than can be accomplished with conventional differential GPS or with unaugmented GPS. Specifically, the reduction or limitation of altitude errors obtainable by use of pseudolites can support non-precision approaches of aircraft to airports or heliports down to lower decision heights than would be the case otherwise. This could be useful to DoD as well as civilian aircraft.

Significantly, non-precision approaches with GPS are not limited to straight-line segments. The aircraft is not following a narrow signal beam. It's position can be computed and displayed with respect to any selected approach pattern, which can be useful in areas with difficult terrain. This also permits the approach pattern to be changed whenever necessary without affecting the installation or the operation of the ground equipment. In fact, it may be desirable for the pseudolite to transmit a special message which defines the approach path in digital form for automatic insertion into the aircraft navigation equipment.

Pseudolites also can yield substantial incremental benefits over conventional GPS for surface vehicles operating in difficult (mountainous) terrain by filling in for satellites that may be occulted temporarily by the terrain. This is also true for certain types of range applications.

## SOLUTIONS CONSIDERED

The most significant problem in developing a pseudolite signal structure is that of interference with the GPS satellite signals. If the pseudolite signal strength equals that of the GPS satellites when the user is, say, 50 kilometers from the pseudolite, the same pseudolite signal will be 60 dB stronger than the satellite signals when the user is 50 meters from the pseudolite. The signal structure must be designed to cope with this near-far problem (i.e., how to have sufficient pseudolite signal strength at the maximum usable range while not jamming the desired GPS signals when the user is near the pseudolite transmitter). We recommend that receivers be designed to operate properly in the presence of a pseudolite signal at least 60 dB stronger than the nominal GPS satellite signals.

Three basic methods of achieving this objective were evaluated. The first is to use a more powerful pseudorandom code (CDMA). However, to obtain 60 dB of protection would require not only a much longer code but one which was clocked at a rate many times that of the GPS C/A codes. In addition to other practical difficulties, this would reduce compatibility with the GPS signal structure and significantly increase receiver costs. Therefore, the use of a new code to resolve the near-far problem was deemed impractical.

The second approach is that of frequency offset (FDMA). By moving the pseudolite signal away from 1575.42 MHz, filters in the receiver could prevent the pseudolite signal from interfering with the desired satellite signals. However, the pseudolite and satellite signals would then follow separate paths in the receiver, so that very careful control of relative time delays between the two paths would be needed to avoid combinational errors. This approach also would increase receiver costs and reduce compatibility with the GPS signal structure. Therefore, the frequency offset technique has been set aside pending evaluation of the more desirable time division technique described below.

The third and preferred approach is to employ a time division signal structure (TDMA) in which the pseudolite signal is transmitted in short, low duty cycle pulses which interfere with the satellite signals for only a small fraction of the time. The TDMA scheme is attractive because it appears to offer the lowest implementation cost for both the pseudolites and user equipment. This is because the additional equipment required is primarily digital and of a relatively simple, low cost nature.

## SIGNAL SPECIFICATION

### *General Concept*

The pseudolite signal is time division multiplexed so as to avoid the near-far problem. However, the underlying signal structure must be as nearly like a GPS satellite signal as possible in order to simplify the receiver design. Therefore, the pseudolite message is transmitted at 50 bits per second, which matches the satellite message rate exactly. The pseudolite code and bit epochs are carefully aligned with GPS time in order to permit pseudorange measurements to the pseudolite antenna location.

Besides a difference in message content, there are several other key differences between the pseudolite signal and that from a GPS satellite. The pseudolite employs one of 51 non-GPS, 1023-bit, one msec Gold codes. Furthermore, within each half-bit period (10 msec), the signal is gated on (transmitted) only during eleven 90.91 microsecond intervals (one-eleventh of a one-msec, 1023 chip code, or 93 chips per transmission), thus forming a pulsed waveform with a 10% duty cycle. In this way the pseudolite could interfere with the desired GPS signals no more than 10% of the time, which represents no more than 1 dB loss in the received GPS average signal power, even when the receiver is close to a pseudolite. The pulse position is changed from millisecond to millisecond, both to randomize the spectrum and to assure that one complete code sequence is transmitted every 10 milliseconds. All pulse positions change from one bit to the next over a 10-bit, 200 msec interval so that all possible pulse positions will have been transmitted after ten bits. This is important to assure compatibility with multiplexed receivers which only integrate received signals for a fraction of each bit interval. Details of this concept are provided in the following sections.

### *Pulse Position*

Figures 4 and 5 illustrate how the transmitted pulse position is modulated from msec to msec and from bit to bit over a 10-bit, 200 msec interval synchronized and aligned with GPS time. There are 11 possible pulse positions in each msec, numbered 0 through 10. For example, note that during the first msec of the first bit of the 10-bit sequence the pulse is transmitted in position five. In bit number two the pulse is transmitted in position nine during the first msec. This progression continues until all eleven pulse positions in the first msec have been filled after 10 bits. The sequence is repeated every 200 msec.

The pulse position differs from msec to msec. This randomizes the pulse positions and also assures that each of the 11 portions of the Gold code have been transmitted within each 10 msec interval. The pattern is fixed and is identical for all pseudolites. Note that the pseudolite Gold Code runs continuously; the pulses gate the signal on at the designated time to transmit 93 chips of the continuously running code.

An algorithm which defines the pulse positions is given in the RTCM specification. However, it is probably best simply to use 2200 bits of memory to map the pulse positions in the repeating 200 msec pattern.

### *Pulse Width*

Each transmitted pulse contains exactly 93 code chips, or one-eleventh of the entire, continuously-running, 1023-chip Gold code. Therefore, the pulse width is 1000/11 or approximately 90.91 microseconds.

### *Gold Code Selection*

The satellite C/A codes are 1023-chip Gold codes generated by combining the outputs of two 10-stage code generators termed G1 and G2. The individual codes are obtained by delaying the G2 code by a specified number of chips before combining it with the G1 code. The delay is normally implemented by combining the output of two specified taps on the ten-stage G2 coder. The initial condition of both the G1 and G2 coders is all binary ones, (i.e., 1777 octal).

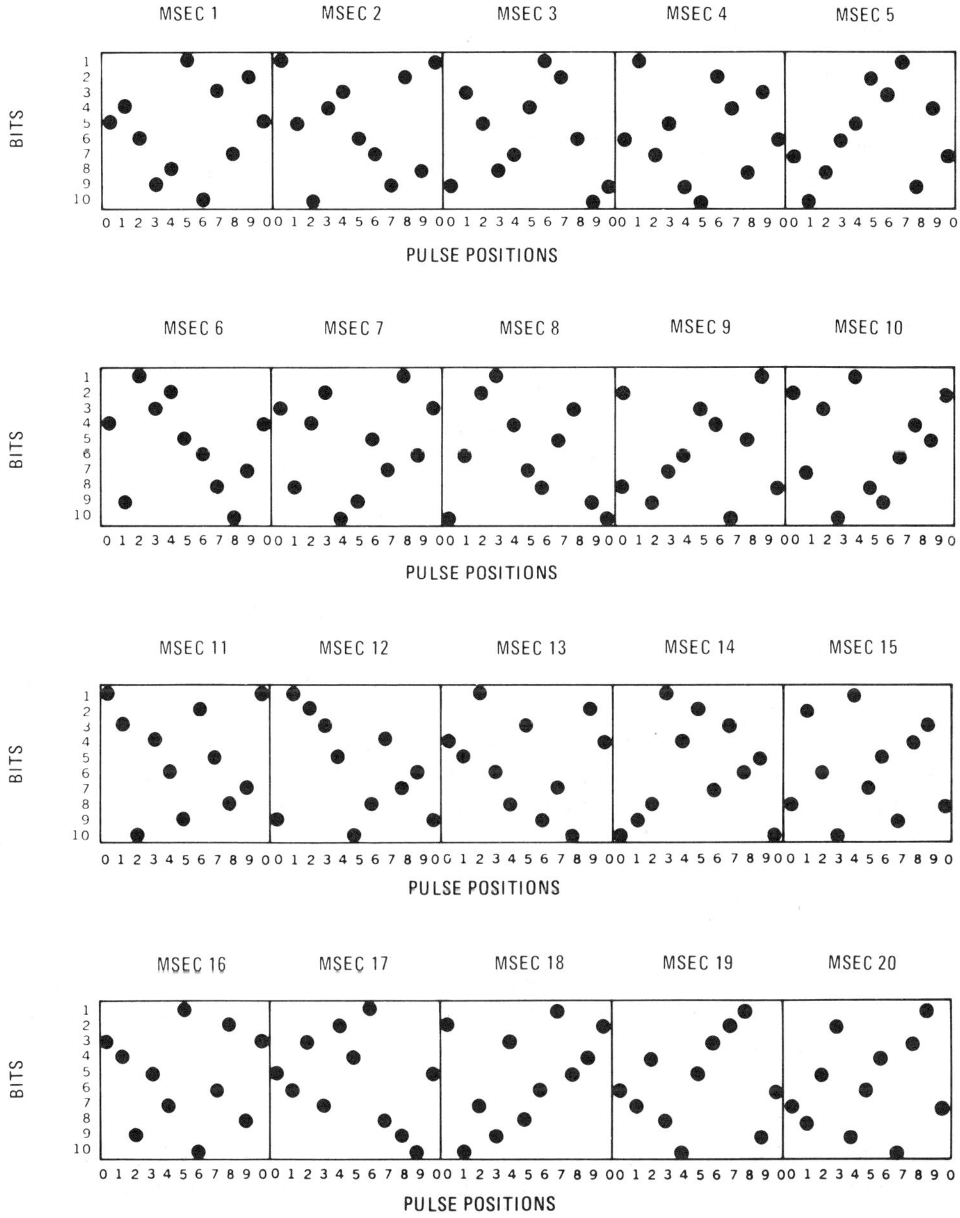


*Fig. 4—Pseudolite Pulse Positions within the 10 Data Bit, 200 msec Pattern*

The objectives in defining the set of pseudolite Gold codes were: (1) to achieve maximum commonality in the code generation hardware, and (2) to obtain as many balanced, non-GPS codes as possible. The selected approach is to use the same coder hardware, but to have a different initial condition for the G2 coder, namely 0722 octal rather than 1777. This initial condition can be achieved either by presetting the G2 coder to 0722 at the G1 epoch or by delaying it by 701 chips from the G1 epoch. The available codes using the G2 coder taps one

| | MILLISECOND | | | | | | | | | | | | | | | | | | | |
|---|---|---|---|---|---|---|---|---|---|---|---|---|---|---|---|---|---|---|---|---|
| DATA BIT | 1 | 2 | 3 | 4 | 5 | 6 | 7 | 8 | 9 | 10 | 11 | 12 | 13 | 14 | 15 | 16 | 17 | 18 | 19 | 20 |
| 1 | 5 | 0 | 6 | 1 | 7 | 2 | 8 | 3 | 9 | 4 | 0 | 1 | 2 | 3 | 4 | 5 | 6 | 7 | 8 | 9 |
| 2 | 9 | 8 | 7 | 6 | 5 | 4 | 3 | 2 | 1 | 0 | 6 | 2 | 9 | 5 | 1 | 8 | 4 | 0 | 7 | 3 |
| 3 | 7 | 4 | 1 | 9 | 6 | 3 | 0 | 8 | 5 | 2 | 1 | 3 | 5 | 7 | 9 | 0 | 2 | 4 | 6 | 8 |
| 4 | 1 | 3 | 5 | 7 | 9 | 0 | 2 | 4 | 6 | 8 | 3 | 7 | 0 | 4 | 8 | 1 | 5 | 9 | 2 | 6 |
| 5 | 0 | 1 | 2 | 3 | 4 | 5 | 6 | 7 | 8 | 9 | 7 | 4 | 1 | 9 | 6 | 3 | 0 | 8 | 5 | 2 |
| 6 | 2 | 5 | 8 | 0 | 3 | 6 | 9 | 1 | 4 | 7 | 4 | 9 | 3 | 8 | 2 | 7 | 1 | 6 | 0 | 5 |
| 7 | 8 | 6 | 4 | 2 | 0 | 9 | 7 | 5 | 3 | 1 | 9 | 8 | 7 | 6 | 5 | 4 | 3 | 2 | 1 | 0 |
| 8 | 4 | 9 | 3 | 8 | 2 | 7 | 1 | 6 | 0 | 5 | 8 | 6 | 4 | 2 | 0 | 9 | 7 | 5 | 3 | 1 |
| 9 | 3 | 7 | 0 | 4 | 8 | 1 | 5 | 9 | 2 | 6 | 5 | 0 | 6 | 1 | 7 | 2 | 8 | 3 | 9 | 4 |
| 10 | 6 | 2 | 9 | 5 | 1 | 8 | 4 | 0 | 7 | 3 | 2 | 5 | 8 | 0 | 3 | 6 | 9 | 1 | 4 | 7 |

NOTE: THERE ARE 11 PULSE POSITIONS IN EACH MILLISECOND, NUMBERED 0 THROUGH 10. WHENEVER A PULSE IS TRANSMITTED IN POSITION 0, IT ALSO IS TRANSMITTED IN POSITION 10.

*Fig. 5—Pseudolite Pulse Position in Each Millisecond of the 10 Data Bit, 200 msec Pattern*

or two at a time are listed in Figure 6. The 35 primary codes are obtained by using identically the same taps defined by ICD-GPS-200[3] for GPS and "other" C/A codes. The 16 auxiliary codes use other tap combinations, including the use of only one at a time. The auxiliary codes will be used only if required to eliminate interference between a cluster of pseudolites, so existing GPS coders are sufficient to provide the most often used pseudolite codes. However, it is recommended that new equipment be designed to generate all 51 pseudolite codes.

### *Pseudolite Address*

Each pseudolite in a local (line-of-sight) area is assigned a unique Gold code from the 51 listed in Figure 6. In addition to a unique Gold code, the message also defines the pseudolite position. Together these provide a unique address for each pseudolite.

### *Pseudolite Clock Offset*

The transmitted pseudolite signal epochs are defined to be precisely correct in GPS time. Thus, the pseudorange and pseudorange rate correction parameters transmitted in the pseudolite message correct the GPS satellite signals with respect to the monitor antenna location and the time epochs being transmitted by the pseudolite. This avoids the necessity of transmitting a pseudolite clock offset term. Even so, the pseudolite epochs will be controlled to be in close agreement with GPS time in order to minimize the magnitude of the satellite pseudorange correction values. An agreement to within ± 100 nanoseconds should be achievable in practice.

### *Signal Level*

The average pseudolite signal power should approximately equal that of the C/A code GPS signal at the maximum usable range from the pseudolite. The

| PSEUDOLITE CODE NO. | G2 TAPS | G2 DELAY (CHIPS) | FIRST 10 C/A CHIPS (OCTAL) | |
|---|---|---|---|---|
| 1 | 2 6 | 706 | 1426 | |
| 2 | 3 7 | 707 | 0613 | |
| 3 | 4 8 | 708 | 0305 | 35 PRIMARY PSEUDOLITE CODES |
| 4 | 5 9 | 709 | 0142 | |
| 5 | 1 9 | 718 | 0660 | |
| 6 | 2 10 | 719 | 0330 | |
| 7 | 1 8 | 840 | 1524 | |
| 8 | 2 9 | 841 | 1652 | |
| 9 | 3 10 | 842 | 0725 | |
| 10 | 2 3 | 952 | 1362 | |
| 11 | 3 4 | 953 | 1571 | |
| 12 | 5 6 | 955 | 0336 | |
| 13 | 6 7 | 956 | 0157 | |
| 14 | 7 8 | 957 | 1067 | |
| 15 | 8 9 | 958 | 0433 | |
| 16 | 9 10 | 959 | 0215 | |
| 17 | 1 4 | 147 | 0156 | |
| 18 | 2 5 | 148 | 0067 | |
| 19 | 3 6 | 149 | 1033 | |
| 20 | 4 7 | 150 | 0415 | |
| 21 | 5 8 | 151 | 1206 | |
| 22 | 6 9 | 152 | 1503 | |
| 23 | 1 3 | 187 | 0350 | |
| 24 | 4 6 | 190 | 1235 | |
| 25 | 5 7 | 191 | 1516 | |
| 26 | 6 8 | 192 | 0647 | |
| 27 | 7 9 | 193 | 0323 | |
| 28 | 8 10 | 194 | 1151 | |
| 29 | 1 6 | 537 | 0414 | |
| 30 | 2 7 | 538 | 0206 | |
| 31 | 3 8 | 539 | 0103 | |
| 32 | 4 9 | 540 | 1041 | |
| 33 | 5 10 | 541 | 1420 | |
| 34 | 4 10 | 628 | 0523 | |
| 35 | 1 7 | 625 | 1234 | |
| 38 | 1 5 | 705 | 1055 | |
| 39 | 6 10 | 710 | 0061 | |
| 40 | 1 2 | 951 | 0745 | |
| 41 | 7 10 | 153 | 1641 | |
| 42 | 2 4 | 188 | 1164 | |
| 43 | 3 6 | 109 | 0472 | |
| 44 | 3 9 | 627 | 1247 | |
| 45 | 1 10 | 688 | 1302 | |
| 46 | 1 | 692 | 1754 | |
| 47 | 2 | 693 | 0766 | 16 AUXILIARY PSEUDOLITE CODES |
| 48 | 4 | 695 | 0175 | |
| 49 | 5 | 696 | 1076 | |
| 50 | 6 | 697 | 0437 | |
| 51 | 7 | 698 | 1217 | |
| 52 | 8 | 699 | 1507 | |
| 53 | 10 | 701 | 1321 | |

NOTES:

(1) PSEUDOLITE GOLD CODES ARE GENERATED IN THE SAME WAY AS GPS C/A CODES, EXCEPT THAT THE G2 CODER INITIAL STATE IS SET TO 0722 OCTAL, WHICH INTRODUCES AN ADDITIONAL DELAY BETWEEN THE G1 AND G2 CODERS OF 701 CHIPS.

(2) THE 35 PRIMARY CODES EMPLOY THE G2 CODER TAP COMBINATIONS ALSO USED TO GENERATE THE FIRST 35 GPS C/A CODES. THE 16 AUXILIARY CODES EMPLOY ONE OR TWO TAPS, BUT THE COMBINATIONS ARE NOT EQUIVALENT TO THOSE FOR THE GPS C/A CODES.

(3) OF THE 55 CODES WHICH CAN BE GENERATED WITH ONE OR TWO TAP COMBINATIONS, FOUR HAVE BEEN DELETED FROM THE LIST BECAUSE THEY ARE NOT BALANCED, i.e., THEY DO NOT YIELD 511 ZEROS AND 512 ONES.

(4) CODE NUMBERS 36 AND 37 HAVE NOT BEEN USED BECAUSE THE PSEUDOLITE EQUIVALENT OF C/A CODE 36 IS NOT BALANCED, AND BECAUSE ICD-GPS-200 LISTS CODE 37 AS BEING THE SAME AS CODE 34.

*Fig. 6—The 35 Primary and 16 Auxiliary Pseudolite Gold Codes Defined by the G2 Coder Taps and Equivalent Code Delays*

typical C/A satellite signal level is −126 dBm. Therefore, because of its 10% duty cycle, the received power during each pseudolite pulse should be −116 dBm at maximum range.

It is expected that pseudolite signals will be used only for aircraft final approach control or perhaps for limited area marine coverage. Therefore, the maximum desired pseudolite range typically will be 50 km. Less than 30 milliwatts of power radiated from a circularly polarized hemispheric pattern antenna is required to provide −116 dBm at 50 km into a similar receiving antenna.

Figure 7 is a diagram showing the received pseudolite pulse signal power as a function of distance from the pseudolite to the receiver for the pseudolite power recommended above. At 0.1 km (100 meters distance) the pseudolite pulse power is −62 dBm, or 54 dB above typical received GPS signal levels. Therefore, the dynamic range is great, but the maximum received power is quite modest.

### *Transmitter Signal Leakage*

Because it is vital for the pseudolite signal to be a pulsed waveform and not continuous, it is extremely important that when the pulse is "off" that it be truly off, i.e., below GPS tracking thresholds. At 10 meters distance the pulse power will be −42 dBm. Assuming that −142 dBm is "off" at this distance, the signal leakage should be at least 100 dB below the transmitted pulse power.

## ANTENNA LOCATION AND PATTERN SHAPING

### *Installation Criteria*

The pseudolite antenna should be positioned and its pattern shaped in order to optimize its benefit while minimizing potential interference to non-users. For example, by placing the pseudolite antenna on a tower some distance from the airport, the reception by top-mounted GPS aircraft antennas will be improved. Also, by mounting the antenna on an extensive ground plane, the signal at ground level near the tower will be attenuated, thus minimizing possible interference with GPS receivers in local vehicles. If aircraft coverage is not needed in certain zones, then the signal should be attenuated in those directions. Proper

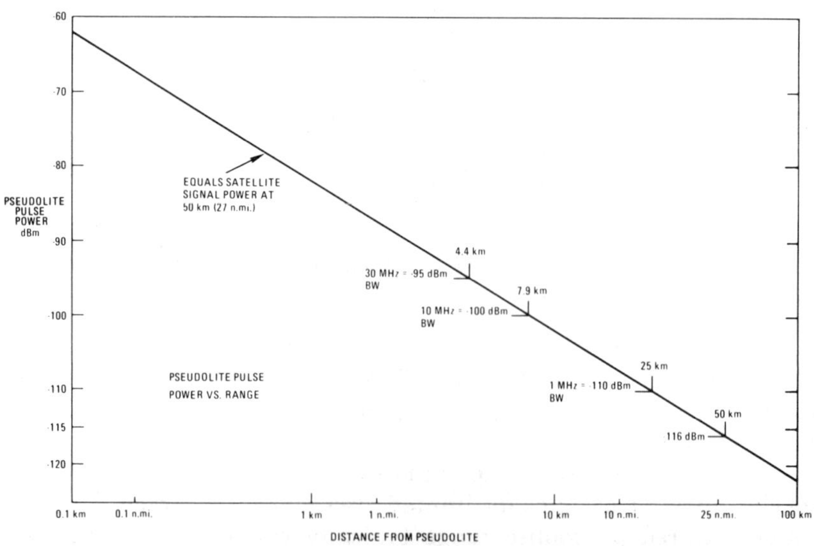


*Fig. 7—Pseudolite Pulse Power vs.* Range

antenna location and pattern shaping can improve performance and minimize the potential for interference.

### *Minimum Spacing of Pseudolites With the Same Code*

Two other issues affect the placement of pseudolites. First is the question of how far apart pseudolites must be in order to use the same Gold code. Our assumption is that all pseudolites are designed to serve a radius of 50 km and that the transmitted power is adjusted so the average signal power level at 50 km is −126 dBm, which matches the received GPS signal power. The question is whether at 50 km from a pseudolite the receiver might acquire a more distant pseudolite which was using the same code.

There are two ways to answer this question. Assuming that essentially all pseudolite acquisitions are aided by knowledge of GPS time, user location, and pseudolite location, the only requirement is to assure no pulse overlap from the two transmissions. This requires that the pseudolites be separated by at least 130 km, which assures that when the user is 50 km from the desired pseudolite A, he will be at least 80 km from undesired pseudolite B. Thus, the pulses from B must travel at least 30 km further than those from A, causing them to arrive 100 microseconds later and thus avoiding overlap.

A more conservative approach is to require that at 50 km from pseudolite A the signal strength from pseudolite B will be too weak to interfere. This occurs with a spacing of approximately 300 km, so that the minimum distance to B is 250 km when 50 km from A. The signal from B at 250 km will be 14 dB below that at 50 km, or approximately −140 dBm.

In summary, two pseudolites using the same code may be placed as close as 130 km without causing a problem. However, a spacing of 300 km is more conservative and should be used if possible.

### *Minimum Spacing of Pseudolites Using Different Codes*

The second issue involves the minimum spacing of two pseudolites which use different codes. Figure 8 makes the issue clear. There are two types of zones. In the central zone the pulses from the two pseudolites will overlap, whereas they will not overlap in the two outer zones. The hyperbolic curves which separate the zones are the locus of points at which the difference in distance to the two pseudolites is 30 km, which is approximately the distance a radio signal will travel during the time of one pseudolite pulse. (The precise value is 27.25 km, which has been rounded up both for convenience and to avoid interference if pulses are spread due to receiver response times.) Therefore, the minimum width of the central zone is 30 km.

The pseudolites must be moved back from the central zone a sufficient distance that when their pulses do overlap they will not interfere with each other. Let us assume that we wish to track the signal from A when we are closer to B. To assure that the signal from B does not interfere with that from A, it is first important that the receiver not clip or limit on the signal from B. Here we must assume a well-designed receiver. As will be seen later, such a receiver will clip the incoming signal at ± 2 sigma of the precorrelation noise power. The precorrelation bandwidth will be at least 1 MHz, so the incoming signal will be clipped at a power level of −104 dBm, assuming a 4 dB receiver noise

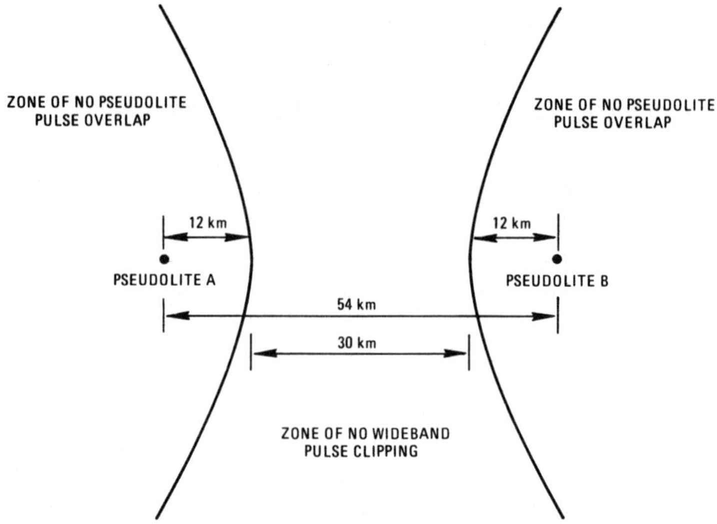


*Fig. 8—Minimum Spacing of Pseudolites*

figure. If the pseudolite pulse power is set at −116 dBm at 50 km, it will be −104 dBm at 12.5 km. Because most receivers will have a wider bandwidth, thus clipping at a higher level, and because the central zone is actually only 27.25 km wide, it is reasonable to place the pseudolites 12 km outside the central zone. Note that the worst case signal power ratio on the hyperbolic boundary occurs when the user is 42 km from A and 12 km from B, a signal power ratio of 11 dB. Because different codes provide at least 23 dB of separation, the stronger signal will be at least 12 dB below the desired, correlated signal.

In summary, Figure 8 shows the recommended minimum spacing for intervisible pseudolites. In the central zone there is no wideband pulse clipping, so code separation assures little or no interference from the undesired pseudolite. In the outer zones, the pulses will not overlap, so there is no possibility of interference no matter how close the user comes to the undesired transmitter. Therefore, the recommended minimum pseudolite spacing is 54 km.

## RECEIVER DESIGN CONSIDERATIONS

### *General*

With regard to the effect and/or use of pseudolite signals, there are two classes of user equipment receivers. A "participating" receiver is one which is currently tracking the specified pseudolite signal, and it should be designed to take maximum advantage of that signal. A "non-participating" receiver is being affected by but is not tracking a pseudolite signal. This can be a receiver which is not designed to track pseudolite signals, or it can be a receiver designed to track pseudolite signals but in the presence of one or more such signals which it is not presently tracking. A non-participating receiver should be designed to minimize the effect of such signals on its GPS tracking performance.

For a non-participating receiver which is well designed to minimize the impact of a pulsed signal, the effect will be a reduction in GPS signal-to-noise ratio (S/N) of .92 dB (20 LOG .9). In such a receiver the pseudolite pulse simply overpowers and therefore eliminates the satellite signal 10% of the time.

Receivers also may be designed to blank the input to the GPS tracking channels whenever the pseudolite signal is being received. In this case the received noise power also is reduced by 10%, so the net effect on S/N is only .46 dB.

If there are two or more pseudolites within line-of-sight, even a receiver participating with one will be receiving the other(s) as well, so it is best to design every receiver to cope well with pulsed interfering signals. The following sections discuss the signal and receiver design factors which influence the ability both to cope with and to obtain benefit from the pseudolite correction signals.

### *Non-Participating Receiver Design*

These paragraphs discuss the receiver design parameters which affect performance in the presence of a pulsed signal such as the proposed pseudolite. All GPS receivers should be designed with such interference in mind, with pulse power levels as defined by Figure 7.

### *Wide Band Circuits*

Figure 9 is a generic GPS receiver block diagram with notes relating to design factors which affect its ability to cope with a pulsed jammer. There are two requirements for the wideband circuits before the correlator. First is that the amplifiers must not latch up when receiving a strong input signal. If they saturate, they must recover very quickly when the pulse ends. Otherwise the pulse will have an effect for more than 10% of the time.

Second, it is very desirable to clip the wideband signals so as to limit the

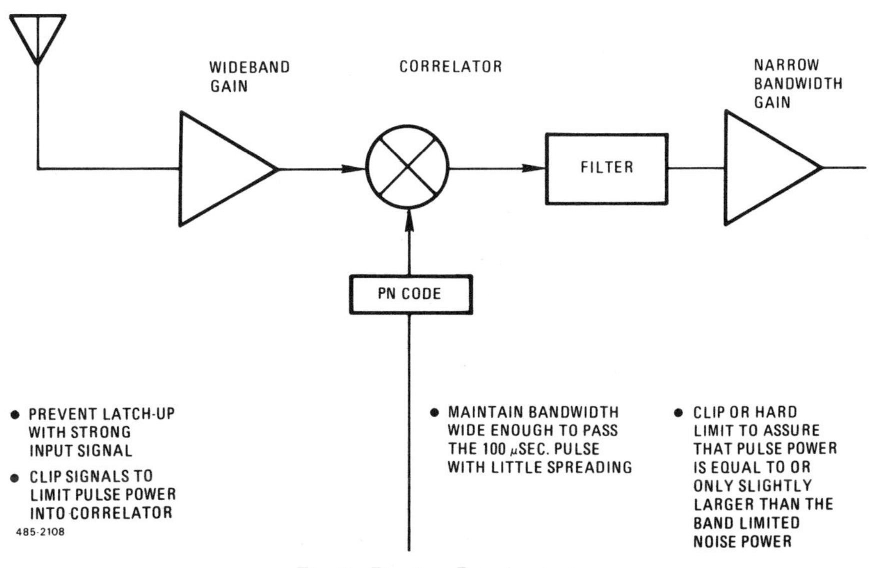


*Fig. 9—Receiver Requirements*

maximum pulse amplitude before it reaches the correlator. There are several design trade-offs which must be considered when implementing such a limiter. The primary issue is where to set the clipping level with respect to the wideband noise amplitude. The noise amplitude is determined by the bandwidth. For example, a C/A code receiver must have at least 1 MHz bandwidth, which yields −110 dBm of equivalent noise input power if the noise figure is 4 dB. Wider bandwidths are required for a P-code receiver, and C/A code receivers usually have wider bandwidths, depending on cost and performance goals.

An inexpensive limiter approach would heavily clip the wideband noise, thus assuring a clipped signal regardless of wideband amplifier gain variations. The disadvantages would be 1.05 dB of bandpass limiter loss in the GPS S/N and the increased risk of signal suppression if a CW jammer were to capture the limiter. On the other hand, this inexpensive approach minimizes the wideband pulsed jammer amplitude.

To avoid the 1.05 dB bandpass limiter loss it is necessary to clip above most of the wideband noise (e.g., at the ± 2 sigma noise amplitude). To maintain this clipping level might require a noise AGC circuit. If CW Jamming is of concern, the AGC might have to be adaptive. Another excellent approach would be to suppress the input only when a pulse was detected. This would be most effective for very wideband receivers where the clipping level would otherwise be many dB higher, but using a pulse detector having the optimum 1 MHz bandwidth.

Whatever the approach, the objective is to minimize the pseudolite pulse amplitude as it passes through the correlator and reaches the narrowband filter.

### *Post Correlation Filter*

If the limiter is set deep into the frontend noise, and if the frontend bandwidth is nearly the same as the pseudolite code clock bandwidth of ± 1 MHz to first nulls, there will be no additional wideband energy striking the narrowband filter during the pulse. On the other hand, if the input filter is, say, 30 MHz wide, and if the limiter is set at ± 2 sigma, the limited pulse power *density* (noise power within ± 500 KHz of the center frequency) will be about 20 dB greater than the wideband noise power density. The filter must be able to handle this 20 dB noise pulse and recover quickly so as not to spread the effect. This suggests that for wideband receivers with ± 2 sigma clipping the post-correlation filter bandwidth should be at least 25 KHz (i.e., the pseudolite pulse envelope two-sided bandwidth plus a Doppler window). (If a narrow bandwidth filter is used, the 20 dB noise pulse with 10% duty cycle would cause a continuous 10 dB increase in filter output noise power).

### *Cross Correlation Effect*

The next concern is how much of the pulse power cross-correlates with the satellite Gold code and appears as an interfering signal at the filter output. The pulse power level depends on the precorrelation bandwidth and on the limiting level. Examples are listed in Figure 10. The precorrelation bandwidth determines the wideband noise power. The type of limiter determines how large the pulse power can be relative to the wideband noise power. A deep limiter

| PRE-CORRELATION BANDWITH (MHz) | WIDEBAND NOISE POWER (dBm) | WITH DEEP LIMITER | | WITH ±2σ LIMITER | |
|---|---|---|---|---|---|
| | | PRE-CORRELATOR PULSE POWER (dBm) | FILTERED INTERFERENCE LEVEL (dBm) | PRE-CORRELATOR PULSE POWER (dBm) | FILTERED INTERFERENCE LEVEL (dBm) |
| 1 | -110 | -110 | -154 | -104 | -148 |
| 3 | -105 | -105 | -149 | -99 | -143 |
| 10 | -100 | -100 | -144 | -94 | -138 |
| 30 | -95 | -95 | -139 | -89 | -133 |

*Fig. 10—Equivalent Interference Level Near a Pseudolite with Non-participating GPS Receivers Having 4 dB Noise Figure, Different Precorrelation Bandwidths, and Two Types of Limiters*

allows no increase, and a ± 2 sigma limiter allows the pulse to be up to 6 dB larger than the wideband noise power. If the receiver tracking loops average the interference over at least 10 msec, which is the minimum time required for an entire 1 msec pseudolite Gold code to be transmitted because of its 10% duty cycle, the precorrelation pulse power will be attenuated by at least 24 dB due to cross-correlation with a GPS Gold code. Also, the 10% duty cycle causes a 20 dB reduction in equivalent S/N of the interfering signal. The resulting interference levels as given in Figure 10 range from −154 dBm to −133 dBm. A GPS signal at −135 dBm is considered weak and marginal. Therefore, these interference levels range from 19 dB below a weak GPS signal to 2 dB stronger. Thus, it is obvious that the receiver design parameters must be considered very carefully.

### P-Code Receivers

Receivers with a very wide bandwidth are likely to be using the P-code, in which case the pulse power will be suppressed by at least another 30 dB. Therefore, the problem when using such a receiver will occur only during a C/A code search for initial acquisition.

### Tracking Loop S/N

Again with reference to Figure 9, it is important to minimize the impact of any noise increase during the pulse which might affect the tracking loop S/N. Many receivers hard limit the output of the narrow filter, thus maintaining a constant total power input to the tracking loops. For this type of receiver there will be no increase in noise power during a strong pseudolite pulse. The desired GPS signal simply will disappear 10% of the time. If hard limiting is not employed, some form of limiting may be needed to control the maximum noise power increase, depending on the input filter and limiter characteristics.

### Multiplexed Receivers

The interference situation with multiplexed receivers is not much different than with slow sequenced or multi-channel receivers. Although the multiplexed receiver typically loses 7 dB of S/N because it tracks each satellite signal for only 4 msec every 20 msec, the same reduction applies to the cross-correlated pseudolite signal. Therefore, the only significant difference is evident from

Figure 4. It requires 10 msec for all 1023 chips of a pseudolite Gold code to be transmitted. Because a multiplexed receiver samples a satellite signal only 4 msec every 20 msec, that channel will not experience all 1023 pseudolite Gold code chips until 20 successive samples have been taken. In other words, signal averaging requires 200 msec with the multiplexed receiver rather than 10 msec with other types.

### Summary

Actual interference from or false lock to a pseudolite signal will depend on receiver design, distance from the pseudolite, GPS signal strength, the relative Doppler shift of the two signals, and whether the worst case (24 dB) cross-correlation exists between the two codes. In most cases the interference will be marginal or negligible and certainly temporary, but worst case situations do occur. The substantial advantage of detecting the leading edge of the pulse in a 1 MHz bandwidth and blanking the input for the 91 microsecond duration is evident.

These paragraphs have discussed the receiver design parameters which affect performance in the presence of a pulsed signal such as the proposed pseudolite signal. All GPS receivers should be designed with such interference in mind. It is recommended that each receiver design be tested by the manufacturer so as to specify both the radius of potential false receiver lock on a 30 milliwatt pseudolite signal and the degradation in S/N when tracking a −135 dBm GPS signal due to a 30 milliwatt pseudolite only 10 meters from the receiving antenna.

### Participating Receiver Design

A participating receiver must obtain the pseudolite message continuously. Therefore, the receiver must have a channel which can be assigned on a continuous basis. For example, a one-channel slow sequencing receiver cannot be used.

The pseudolite receiver must be designed to generate the pseudolite C/A codes of Figure 6. Although it would be possible to track the pseudolite signal with no other modification, a 10 dB S/N advantage is obtained by blanking or ignoring the noise between the pseudolite pulses. Only in this way will the full signal strength and maximum range be obtained.

Because the pseudolite signal is pulsed interference to those channels tracking the GPS satellite signals, it is very desirable for the participating receiver to blank all such channels whenever the pseudolite signal pulse is being received. The advantage would be to eliminate all interference effects from that pseudolite signal, except for .46 dB of satellite signal loss.

By making the pseudolite signal structure very similar to the satellite signal, a channel capable of tracking a pseudolite also can be used to track a satellite. Therefore, maximum use may be made of all receiver hardware, and interchannel bias can be measured and eliminated by occasionally tracking common satellite signals.

### *Pseudolite Signal Acquisition*

There are four modes of pseudolite signal acquisition defined by two parameters. These parameters are: (1) whether or not the receiver already is tracking GPS satellites and therefore knows GPS time, and (2) whether or not a specific pseudolite code has been designated for acquisition or whether the receiver is simply searching for any available pseudolite signal.

If GPS time is known, the receiver can employ a pulsed waveform when searching for the pseudolite signal. By blanking the noise between the pseudolite pulses, a 10 dB S/N advantage is obtained. Assuming that any useful pseudolite will be within 50 km, the search window is less than 170 microseconds, i.e., no more than 170 C/A code chips. Even with a code search rate of only 25 chips per second, acquisition to a specific pseudolite code should occur within 7 seconds maximum. However, to search all possible pseudolite codes would require up to six minutes. A frequency search also may be required if moving at a high speed and if the pseudolite location is not already known so that a relative velocity can be computed.

Although not recommended, if GPS time is not known it is possible to perform a blind search for a pseudolite signal. In this case it is not possible to blank the noise between pulses, thus losing 10 dB of S/N. The receiver would search for C/A code acquisition as if for a satellite signal, but perhaps at a slower search rate. It would be necessary to search the entire 1023 chip code sequence. Once the code was acquired, bit synchronization would be achieved in the normal way, and message demodulation would establish GPS time. As soon as GPS time is known, the pseudolite pulse waveform can be applied by the receiver to recover the 10 dB of S/N. Knowing GPS time and approximate position could then be useful in acquiring the GPS satellite signals.

## PSEUDOLITE ALMANAC MESSAGE

Figure 11 shows the proposed pseudolite almanac message, which is Message Type 8 (see Table 2 in Reference 1). Latitude and longitude are two's complement binary numbers defining the approximate pseudolite position with a resolution of 180/65,536 degree for latitude and 360/65,536 degree for longitude. The appropriate Gold code number as defined by Figure 6 is given by the code number in binary form. The health bits are defined by the figure. Four 7-bit ASCII alphanumeric characters are provided to name the pseudolite (e.g., LAX1 or JFK3). Four spare bits are included for future use and to round the message to three whole 30-bit words per pseudolite. The entire almanac message describes 3 pseudolites, requires 330 bits, and takes 6.6 seconds to transmit.

## SUMMARY

This paper has defined a pseudolite signal structure which can provide a differential GPS correction message plus another line of position for local users of GPS. The signal structure minimizes interference with all well designed, non-participating receivers, even when they are very close to the pseudolite. Participating receivers can be designed to have negligible interference while obtaining substantial reliability, integrity, and accuracy benefits.

| WORD | PARAMETER | NUMBER OF BITS | RANGE |
|---|---|---|---|
| FIRST | SAME AS TYPE 1 | 30 | |
| SECOND | SAME AS TYPE 1 | 30 | |
| REMAINING | BEACON ALMANAC DATA X3 | | |
| | LATITUDE | 16 | ±90 DEGREES |
| | LONGITUDE | 16 | ±180 DEGREES |
| | CODE NO. | 6 | 0 - 63 |
| | HEALTH | 2 | SEE NOTE |
| | 4 ALPHANUMERIC | 28 | 7 - BIT ASCII |
| | SPARE | 4 | |
| | PARITY | 18 | |
| | | 90 x 3=270 | |

NOTE: PSEUDOLITE HEALTH STATES ARE:
00 STATION NORMAL
01 STATUS UNKNOWN
10 GOING DOWN FOR MAINTENANCE
11 STATION DOWN OR DO NOT USE

*Fig. 11—Sub-Frame Message Type 8 Pseudolite Almanac Message*

## REFERENCES

1. Kalafus, R. M.; Van Dierendonck, A. J. and Pealer, N. A.; *Special Committee 104 Recommendations for Differential GPS Service,* Navigation, Vol. 33, No. 1, Spring 1986, pp 26–41.
2. Klein, D. and Parkinson, B. W.; *The Use of Pseudo-Satellites for Improving GPS Performance,* Proceedings of the 40th Annual Meeting of the Institute of Navigation, Cambridge, Massachusetts, June 1984.
3. Anon.; *Interface Control Document, NAVSTAR GPS Space Segment/Navigation User Interfaces,* ICD-GPS-200, Rockwell International Corporation, November 1982, revised in 1985.

# The Use of Pseudo-Satellites for Improving GPS Performance

DALE KLEIN
Intermetrics, Inc., Cambridge, MA

BRADFORD W. PARKINSON
Stanford University, Stanford, CA
*Received June 1984*

## ABSTRACT

The NAVSTAR Global Positioning System has proven itself to be extremely accurate under tests of normal operations conducted at the Yuma proving ground. During the same tests of several years ago, a special technique called Differential GPS was also evaluated and shown to have even more startling accuracies. The basic concept of Differential GPS is to have a local receiving station continuously calibrate the remaining bias in the GPS signal and transmit it to the local users. This allows aircraft, ships or land vehicles to refine their estimate of position to the last meter or two.

One technique for transmitting such a correction is to use a GPS-like transmitter itself. The exact form of the signal used may be quite significant to operations and accuracy enhancement. The "pseudo-lite" or GPS ground transmitter also provides powerful geometric leverage for the total navigation solution. A pseudo-lite will be called a PL in this paper.

This paper will address the geometric problems encountered by the current GPS satellite constellation. Simulation results will show the value of PLs to deal with both poor satellite geometry and potential satellite outages (in the event of failure). Quantitative examples of the performance improvements are included.

This paper will also outline potential design alternatives for the PLs.

## INTRODUCTION

In the last few years, the civil aviation and maritime communities have shown increasing interest in the GPS system to satisfy their need for accurate positioning. Because the military has not provided access to the higher accuracies of the P-code, alternative means of insuring accuracy have led to "differential GPS," which will be called DGPS in this paper. In its most fundamental form, DGPS consists of (1) a ground calibration receiver which measures the time-biases in the GPS satellite signals and (2) a transmitter (not necessarily at GPS frequencies) which relays the corrections to an appropriately equipped user.

For the C/A code receiver, the fundamental form of DGPS provides accuracies which approach or exceed that of P code receivers in the local area of the DGPS transmitter. While this is a very significant improvement, there may be circumstances when this is not enough. For example, if elevation angles are constrained to exceed 7½ degrees the planned constellation of 18 satellites and 3 spares still has some periods of very poor geometry, for elevation angles which exceed the 7½ degree unit. In addition, local conditions may have terrain or man made features which further reduce the availability of satellites. Another important case is that of satellite outages. While a DGPS can recognize and flag this problem, the inherently poor geometry has not been corrected. Therefore, the user may be in a situation where a navigation solution does not exist.

One solution is to combine the DGPS transmitter with a GPS-like ranging signal, synchronized to GPS. This has been called a pseudo-satellite or pseudolite. It will be abbreviated as PL in this paper.

The RTCM Special Committee on Differential NAVSTAR/GPS Service has defined a PL in the following way:

(1) Receives GPS satellite signals and computes pseudo-range and range rate corrections, etc., as needed to satisfy the requirements of any other differential GPS monitor station.
(2) Transmits the correction information at 50 bits per second on an L-band frequency, preferably but not necessarily $L_1$.
(3) The transmitted signal is GPS-like, in that a PN code is used to permit local user equipment to obtain an additional pseudorange measurement to the transmitting antenna.
(4) The transmitted signal is designed to prevent interference with other equipment.

This paper will analyze two typical cases where the existing constellation would provide poor navigation results and show that PLs can overcome these difficulties by restoring fundamental accuracy. This is due to adding another ranging signal which can improve the geometry significantly. Note that the added benefit of differential corrections is not included in these results.

## CONSTELLATION

The constellation currently chosen for GPS satellite deployment consists of 18 satellites in 6 orbital planes with 3 active spares, one in every other plane. Table 1 lists the parameters for the planned constellation.

While this constellation offers four or more satellites in view (for a 7.5 mask angle) at any point on the earth or in time, very poor geometries do occur at certain locations. Figure 1 is a composite of all locations that experience poor geometries. The measure of poor geometry is a three dimensional Position Dilution of Precision (PDOP) of greater than 6.0 with a 7.5 elevation mask angle. Also highlighted on this figure is one typical poor coverage period.

Each of the periods of poor geometry appears for 20 to 30 minutes and then disappears, only reappearing 24 hours later. A new set appears later in time at new locations. At any one time a maximum of four of the poor geometry locations may be occurring.

Examination of the figure shows that with the 7.5 degree elevation mask angle some locations on the earth actually experience periods of poor coverage twice a day. It should be noted that previous results of a similar nature were reported in Ref. 1. However the difference in those results and these can be categorized as occurring in three areas: first, this paper assumed a 7.5° elevation mask angle instead of a 5° angle; second, the three active spares were included

Table 1—Symmetric 18 Satellite Constellation (18/6/2)

| Satellite Number | Longitude of Ascending Node* | Argument of Latitude* |
|---|---|---|
| 1 | 0 deg | 0 deg |
| 2 | 0 | 120 |
| 3 | 0 | 240 |
| 4 | 60 | 40 |
| 5 | 60 | 160 |
| 6 | 60 | 280 |
| 7 | 120 | 80 |
| 8 | 120 | 200 |
| 9 | 120 | 320 |
| 10 | 180 | 120 |
| 11 | 180 | 240 |
| 12 | 180 | 360 |
| 13 | 240 | 160 |
| 14 | 240 | 280 |
| 15 | 240 | 40 |
| 16 | 300 | 200 |
| 17 | 300 | 320 |
| 18 | 300 | 80 |
| Spare #1 | 0 | 30 |
| Spare #2 | 120 | 170 |
| Spare #3 | 240 | 310 |

*At Epoch
Period = 11 hr. 58 min.
Inclination = 55 deg.
Eccentricity = 0.

in this study which were not in the previous work; finally, these results are presented on a different map projection than was used in Ref. 1. Note that the Mercator's Projection used in Figure 1 has a tendency to distort the results at high latitudes by enlarging their perceived area.

## PROBLEM DEFINITION

Based on the coverage chart, Figure 1, the original system goals of GPS for 24 hour global coverage have not been achieved with the planned constellation. An accurate receiver clock or baro-altimeter can be used to maintain the solution during the periods of poor geometry; however without external aides the accuracy degrades and solution initialization or reinitialization is extremely difficult. While differential GPS without PLs improves upon the basic GPS accuracy, it suffers from the same geometry problems as basic GPS during periods of poor coverage.

The use of an extra GPS type transmitter in the form of PLS would be one possible solution, especially in the terminal area of a mission. The PLs may also be used to communicate the differential correction terms to a user in the vicinity.

This paper will examine:

(1) The geometrical improvement obtained with a PL for the nominal constellation.

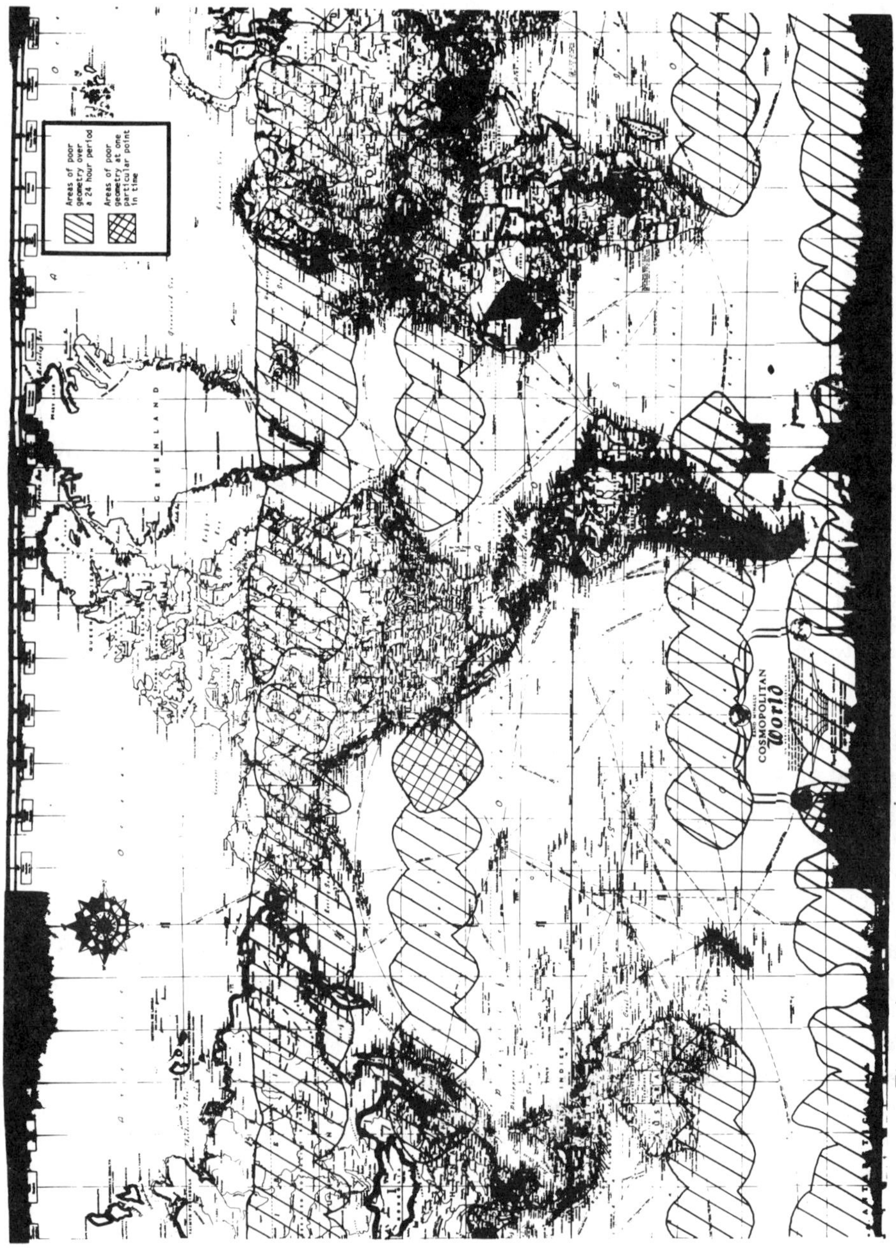


*Fig. 1—Approximate location of areas of poor geometry (PDOP > 6 at a 7.5° minimum elevation angle) over a 24-hour period.*

(2) The area or extent of the improvement.
(3) The geometric improvement for the case of satellite outage.
(4) Some of the other advantages of PLs.

For this study the PL was assumed to replace one of the satellites in the receiver tracking sequence. Only four signals were used in the dilution of precision calculation at any point in time. These four signals can be all satellites, or three satellites and one PL. This approach was taken so that increased complexity (the use of a fifth channel on a receiver) would not be mandated. Clearly a receiver which already has five channels can potentially improve accuracy at no cost.

## THE CALCULATIONS

The approach used in preparing the results that follow was to calculate the DOP values at each point in time using all combinations of available satellites (over a 7.5° elevation mask angle) or PLs taken four at a time. The combination which then provided the minimum DOP value at a particular point and time was the value used for plotting the results.

## PERFORMANCE DURING PERIODS OF POOR OBSERVABILITY

For illustrative purposes the San Francisco area was chosen. This area lies on the fringe of a region of poor observability. The results are therefore not as dramatic as a point in the center of a region chosen or a point where two periods of poor geometry occur. However it is one of the areas of poor geometry existing within the continental United States and therefore is of key interest.

Using a 7.5° mask angle, a time history of the DOP values were calculated. Only a 12 hour period when the major discontinuity in performance occurred is presented. A plot of PDOP versus time at the San Francisco airport is shown in Figure 2. As can be seen from the plot, the PDOP values get extremely large at approximately 9 hours. There are also other points in time when the PDOP attains a value of four or above.

Next a ground transmitter was introduced to evaluate the effect of the additional transmitter on the poor geometry. Not only was the poor geometry problem solved but in general PDOP was reduced for other points in time also. The amount of improvement was shown to be a function of the ground transmitter placement. To illustrate this point a series of time history and contour plots have been used.

First, if we place the ground transmitter approximately 30 nm south of the airport, we get a time history of PDOP as presented in Figure 3. As can be seen by comparison with Figure 2, we get an improvement not only during the period of poor coverage but at other times also.

The above results apply at only one point with respect to the ground transmitter. Due to the great distances of the satellites from the earth, all points in a small area exhibit similar DOP values when observing satellites only. Therefore, one method of determining the effect of the geometric relationship between a user and the PL is to fix the PL and move the user around it, calculating DOP values. A contour plot of the results of this analysis is presented in Figure 4 for the time 9 hours 4 minutes which is a point of poor PDOP. As can be seen from this plot there is generally a preferred location of the PL south of the user

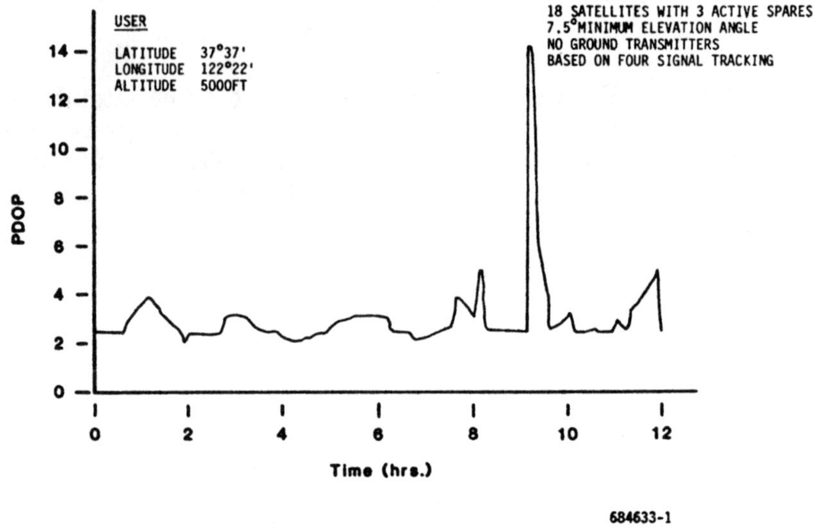


*Fig. 2—PDOP vs. time at San Francisco Airport.*

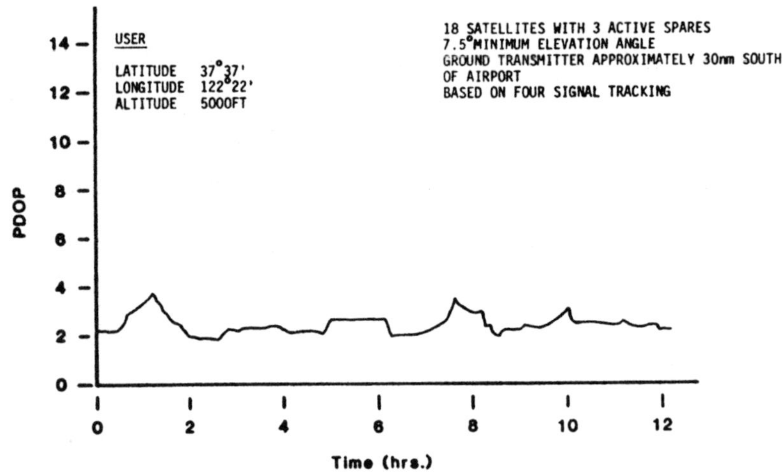


*Fig. 3—PDOP vs. time at San Francisco airport (with ground transmitter).*

even though the PDOP is dramatically improved at all locations. (The upper left hand portion of the plot is the area which had poor PDOP without the PL.)

Since this plot is for only one point in time, the preferred PL location may not be a time invariant process. To further examine this point, a second contour plot was generated at a time of 9 hrs. and 8 mins. (Figure 5). As with the first, the preferred transmitter location is still south of the users. To evaluate this effect over all time, a series of time histories were run for various points about the transmitter. A user north of the transmitter performed much better than users located south of it for all times. This indicates there is definitely a time invariant preferred ground transmitter location. This information would be

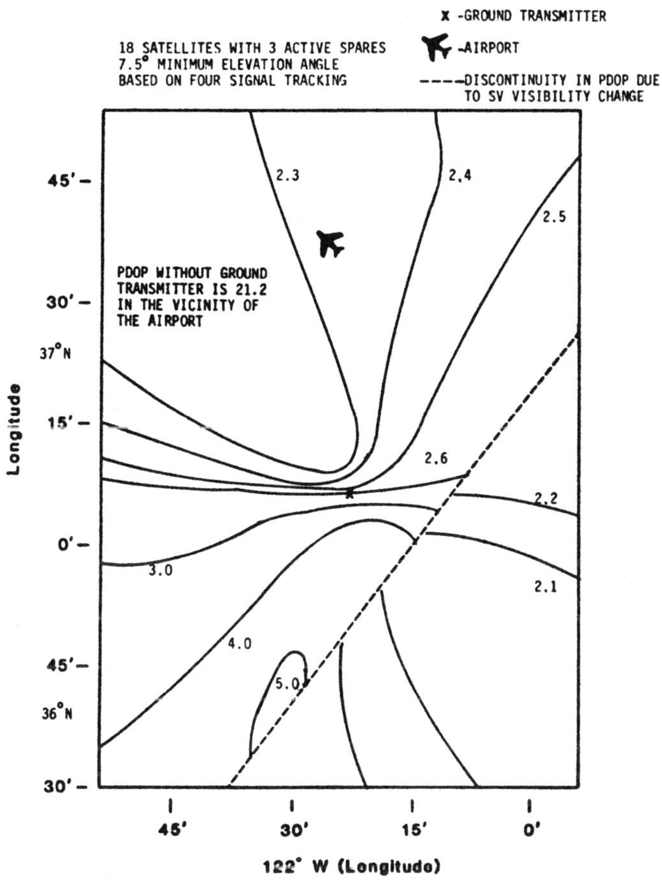


*Fig. 4—Contours of PDOP in the vicinity of the San Francisco Airport (time 9 hrs. - 4 minutes).*

most useful when considering the positioning the ground transmitter for a particular application such as an airport, oil rig, etc.

In some applications it is desired to minimize the vertical error instead of the total position error. To demonstrate that a ground transmitter is also useful for this application, a time history of VDOP at the San Francisco Airport with and without a ground transmitter is presented in Figure 6. (Note that a single degree of freedom DOP (i.e., VDOP) can be less than one with GPS since the time variable is included in the observation matrix and the minimum set of measurements is four allowing multiple observations of a single dimension.) As can be seen from this figure, a dramatic improvement in VDOP occurs even with the transmitter 30 nm south of the user. This occurs since it is not necessary for the ground transmitter to provide the vertical component (a satellite could be used to provide information in this dimension), but to provide information in a direction which is poorly observed from the satellites.

## PERFORMANCE WITH A FAILED SATELLITE

Denver is a location at which good coverage exists continuously. Again using a 7.5° visibility angle, a time history plot of PDOP was produced for a location

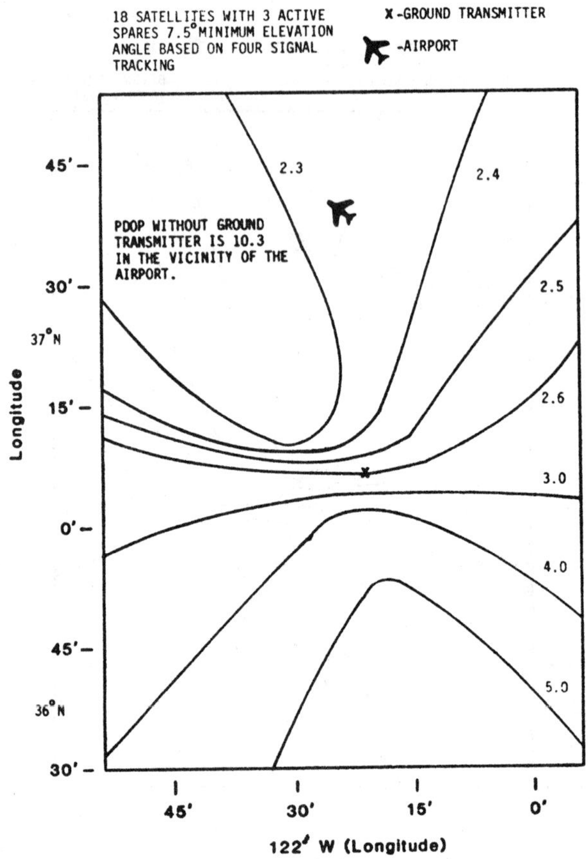


*Fig. 5—Contours of PDOP in the vicinity of the San Francisco Airport (time 9 hrs. - 8 minutes).*

near Denver. The solution with and without an external transmitter is presented in Figure 7. Examination of this figure shows that, unlike the San Francisco area, no large increases in PDOP occurs due to poor geometry. The use of one ground transmitter does improve the PDOP in general, reducing the PDOP during some periods by a factor of two.

The most dramatic improvement available from a ground transmitter occurs in the case of a failed or temporarily inoperative satellite. If we perform the same calculations as illustrated in Figure 7 but assume that satellite #16 is inoperative, the time history PDOP plot is that illustrated in Figure 8. As can be seen, the loss of just one satellite can cause a dramatic deterioration of the available satellite geometry, thus significantly increasing PDOP. However, the introduction of one ground transmitter not only eliminates the deterioration due to the satellite loss but also reduces the PDOP to near the level available with no failed satellites on a ground transmitter.

The subject of preferred ground transmitter placement becomes very complex when considering the question of failed SV's. If a ground transmitter were located after SV failure, a preferred location could definitely be determined. However, if a site is to be determined prior to SV failure as a safety factor, the

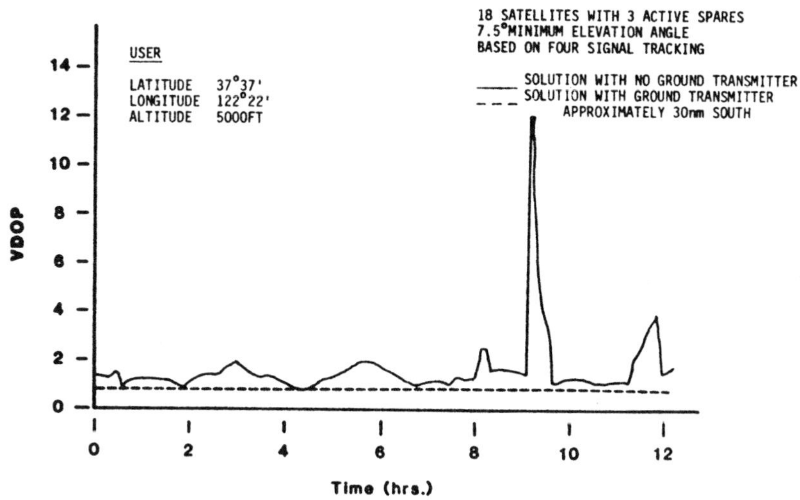


*Fig. 6—VDOP vs. time at San Francisco airport.*

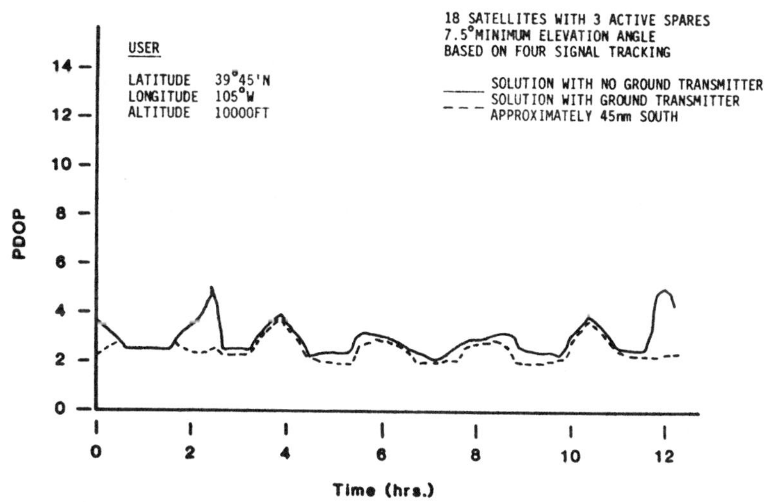


*Fig. 7—PDOP vs. time in the Denver area.*

choice becomes very complex since the particular SV that will fail is unknown. While it is beyond the scope of this paper to address this issue, we feel that there are preferred placements which will maximize ground transmitter usefulness prior to SV failure and provide good compensation in the event of any SV failure visible at that location. This subject is probably best addressed separately for each location and application.

## DESIGN ALTERNATIVES

Over the last several months, these have been a series of meetings under the sponsorship of RTCM to discuss design issues associated with DGPS. Included

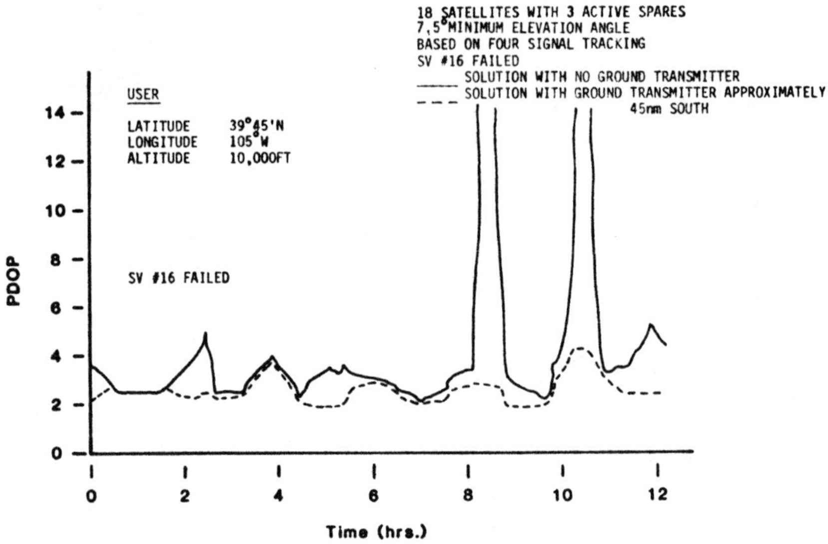


*Fig. 8—PDOP vs. time in the Denver area with SV #16 failed.*

has been a separate effort called the Pseudo Lite Working Group. This section of the paper will summarize the progress that has been made in defining a cost effective and user-acceptable PL design.

User Community. There seem to be numerous potential applications in the near term the applications for both PLs and DGPS include:

(a) Airports of all sizes
(b) Heliports
(c) Harbor and harbor control
(d) Rapid Survey
(e) Oil and Gas Drilling

Signal Alternatives. A major problem facing PL signal designers is potential interference with satellite signals, particularly the C/A code. This is driven by the ratio of maximum to minimum range to the PL, since the power variations are proportioned to the inverse square of distance. Thus, a user will face 40db of power difference as his range to the PL changes by a factor of 100, if antenna gains are constant.

To avoid this "near-far" problem PL signals must be designed for low-cross-correlation with the satellite signals. For example the C/A codes have roughly −20db of peak cross correlation. This appears to be the greatest acceptable level of cross-correlation for conservative system operation. The usual methods of reducing cross-correlation are:

(a) Pulsed signals with random or fixed cycle rates. These are variations of TDMA.
(b) Frequency offset. This could be at a frequency within the currently allocated band, or altogether separated.
(c) Alternative codes. These codes would have to be longer than the current C/A code to improve on the −20db level of cross-correlation.

All of these alternatives can be considered for the PL, including combinations of methods. The two currently favored alternatives are:

(1) Pulsed signal at $L_1$ (1575 MHZ) with 100 usec pulses at quasi-random spacing of about 1 ms. The chipping rate would be the same as the C/A code. The code length could be quite long. Since the signal is narrow and pulsed, the near far problem would be helped by clipping or limiting which is assumed a part of all receivers. The signal could pass data at a rate of 50 pbs or more.

(2) A continuous signal on the fringe of the $L_1$ allocated band. This could be a longer PN code with a band width of 1 mhz. It could be at 1560 or 1590 without obtaining a special frequency allocation waiver.

At this point, the Differential GPS working group favors alternative (1), but have agreed to be open to other suggestions.

Method of Comparison. To choose the acceptable alternatives, a comparison must be made on all significant aspects of the question. As a minimum, the following should be considered.

(1) The ability to handle the near-far problem.
  (a) PL designed receiver
  (b) non-PL designed receiver
(2) Frequency allocation and FCC licensing
(3) Impact on PL designed receivers
  (a) Performance
  (b) Cost
(4) Impact on non-PL designed receivers
  (a) Performance
  (b) Cost
(5) Cost of PL
(6) Ability to handle DGPS data rates
(7) Operational aspects
  (a) Safety & graceful degradation
  (b) Calibration
  (c) Location
  (d) Acquisition

Detailed comparisons are not complete, but should be developed in the next six months.

## UTILITY OF PLS BEYOND FUNDAMENTAL GPS

During the development of GPS, it became obvious that the system had many potential civilian uses. The major representatives of the civilian communities (FAA, DOT, Coast Guard) were reluctant to endorse the system until its capabilities and reliability had been more fully tested. Included in their concerns were the issues of error detection and user notification. If PLs offer significant advantages in these areas, these would be of value above that of accuracy improvements alone.

The following summarizes the utility of PL's:

(I) Improves Geometry. This is the point of the paper. The improvements can be dramatic, since poor geometry can lead to no navigation solution at all. For example, it has been shown that PLs produce VDOPs of less than 1 during a period that would otherwise support no solution at all. This example was for a fully functioning constellation.

(II) Improves Availability. In the event of a *known* outage PLs can insure basic system accuracy in the area of their coverage. The user can work with less than four satellite solutions.

(III) Self Contained Failure Notification. Assuming the PL is implemented in a way that can be received by a GPS User Set without external connections, it allows the user to perform his own failure assessment.

With 1 PL, he can detect a problem, since he has 5 pseudo ranges to measure. If there are 2 PLs in view, he can both detect a single failure and isolate the cause, *without additional external help.* This would meet the FAA 10 second notification requirement.

(IV) Communication of Corrections. The use of PLs allows the user to receive DGPS corrections without resort to additional external receiving equipment. This potentially will save money and reduce integration complexity.

(V) Solves Local Problems. Shading of the line-of-sight satellite signal can be caused by terrain, buildings, or the receiving platform (e.g., aircraft) itself. The additional PL line of signal will help avoid these difficulties.

(VI) Earlier System Operation. Using PLs, the system can be made locally operational at an earlier time. When the deployment has reached a guaranteed "three in view," one or two PLs will probably give full solutions with guaranteed accuracy.

Thus the PLs can answer many of the concerns regarding reliability and guaranteed accuracy.

## CONCLUSIONS

This paper has shown that PLs produce PDOPs of about 3, even during degraded operation. If differential corrections are being supplied it is likely that accuracies in the 1 to 2 meter area can be realized. However the utility of PLS goes far beyond that of accuracy alone. As listed above there are 5 additional operational improvements which enhance reliability and usefulness of the system. We believe that the early development of DGPS with PLs is critical and should be undertaken immediately.

## REFERENCES

1. Kruh, Pierre, The Navstar Global Positioning System Six Plane, 18 Satellite Constellation.
2. Denaro, R. P., "Simulation and Analysis of Differential GPS," Proceedings of ION National Technical Meeting, San Diego, January 1984.
3. Kalafus, R., et al., "Navstar GPS Simulation and Analysis Program," DOT-TSC-RSPA-83-2, May 1983.
4. Kalafus, R., "Synopsis and Recommendations of the TSC Workshop on Differential Operation of Navstar GPS," DOT-TSC-RSPA-83-10, October 1983.
5. "Global Position System," The Institute of Navigation, Washington, D.C., (1980) reprint of the summer, 1978, issue of Navigation, *Journal of the Institute of Navigation,* Vol. 25, No. 2 (together with additional articles concerning the system).
6. Beser, J., and B. W. Parkinson, "The Application of NAVSTAR Differential GPS in the Civilian Community," *Journal of the Institute of Navigation,* Vol. 29, No. 2, 107–136 (1982).
7. Teasley, S. P., W. M. Hoover, and C. R. Johnson, "Differential GPS Navigation," Texas Instruments, Inc., PLANS Symposium (December 1980).

# Codeless GPS Systems for Positioning of Offshore Platforms and 3D Seismic Surveys

P. F. MacDORAN, R. B. MILLER, L. A. BUENNAGEL,
H. F. FLIEGEL and L. TANIDA

ISTAC, Inc., Pasadena, Calif.

*Received January 1984*

## ABSTRACT

A technique known as SERIES (Satellite Emission Range Inferred Earth Surveying), is a method by which radio signals from artificial satellites already in earth orbit can be exploited for positioning and navigation. The existing satellites of choice are those of the Global Positioning System (GPS). The SERIES technique makes it possible to perform simultaneous pseudo ranging to multiple NAVSTAR-GPS satellites without knowledge of the codes usually required to make use of these satellites. The accuracy achievable with the SERIES system in a dynamic environment is at the meter level when operated in a differential mode with a single base station which may be several hundred km away. To achieve real time operation, it is necessary to cross-link the base station receiver output to the dynamic vessel which determines its position and velocity relative to the geodetic position of the base. The demands upon the cross-link communications are very modest at an average rate of only one byte per second. Since the technique has no requirement for a knowledge of the GPS codes, it is possible to use both of the P-code channels at L1 and L2 to accomplish the ionospheric calibrations which are essential for achieving one meter accuracy over distances of several hundred kilometers. The conditions under which the SERIES technique will function are compatible with all civilian needs but are not useful in military operations because of electronics-countermeasures already developed.

By the use of a pair of receivers on a single vessel, it is possible to configure an angular orientation measurement system using the NAVSTAR satellites as the frame of reference and with an accuracy of 0.1 degree or better without any moving parts.

## INTRODUCTION

The SERIES invention was originally intended for sub-decimeter accuracy measurements of the crust of the earth in search of tell-tale patterns that could be exploited for research into earthquake prediction. The SERIES innovation was conceived of and feasibility demonstrated under the leadership of MacDoran while he was employed by the California Institute of Technology, Jet Propulsion Laboratory under contract to the National Aeronautics and Space Administration. MacDoran was granted waivers of commercial rights to his invention by Caltech and NASA and ISTAC was formed in July 1982 to commercialize the SERIES techniques.

The specific application of the SERIES technology being addressed in this paper is in meeting the challenge of high accuracy positioning related to exploration for oil and gas reserves in the offshore environment. One of the most advanced methods of exploration for hydrocarbon resources is known as 3D seismic surveying, a detailed discussion of which is given by Morgan.[1] In Morgan's paper, the possible benefits of using GPS are discussed. However, reliable access to the NAVSTAR Precise Positioning Service (PPS), allowing 15m accuracy, continues to be a major inhibiting factor in making GPS a broadly used navigational method. This paper will present the SERIES-GPS method and how wide civil use of the NAVSTAR is possible to levels of accuracy well beyond PPS and without dependence upon the Department of Defense for NAVSTAR codes and orbits. Because of electronic countermeasures already devised, the SERIES-GPS is not useful in military environments. However, the SERIES methods are compatible with virtually every civil application. For a complete discussion of the NAVSTAR Global Positioning System, the reader is referred to Parkinson and Gilbert.[2]

The original SERIES stations developed at JPL used steered 1.5 meter antennas sequentially pointed at each of the NAVSTAR satellites such that 10 minutes were needed to acquire measurements sufficient to determine the baseline between the antennas. The antenna controller and data acquisition equipment necessary was moderately expensive and housed in a truck camper enclosure. (See MacDoran, Spitzmesser and Buennagel.[3]) Where sub-decimeter accuracy is required in a few minutes, a pointed antenna implementation of the SERIES technique remains a good approach. However, the majority of civil precision positioning needs are not of the sub-decimeter type. Specifically, the needs of most geophysical exploration are in the regime of 1 to 10 meters while in a dynamic state or with short periods available on-site. To be commercially attractive, the SERIES technique had to be further developed.

## PRINCIPLE OF OPERATION

The SERIES receivers are fundamentally codeless spectral compressors where the spread spectra modulations of several NAVSTAR satellites can be simultaneously received at L-band and compressed by seven orders of magnitude into audio or sub-audio bandpasses. A computer extracts the frequency and phase of each satellite in view. The data processing proceeds by a combination of Doppler positioning and phase ranging to levels of precision as may be required even to the millimeter level by using the L-band carriers. The possibility of exploiting this high precision as accuracy remains substantially a problem of the calibration of the system for instrumental, ionospheric and tropospheric delay effects. Because SERIES devices are pseudo ranging receivers, it is possible for a single receiver to measure the ionosphere along each line of sight and perform point positioning given the satellite ephemerides.

The major advantage of the SERIES technique is the ability to use the NAVSTAR satellite without code knowledge (either C/A or P) or the requirement to know the receiver position and velocity and the satellite positions to begin the data acquisition process. The compressed spectral output can be processed into positional data in the receiver or the compressed data can be buffered and transferred over a modest telephone-type channel. For example, a 300 baud channel can transfer 1.5 hours of receiver observations in 3 minutes of telephone time. The option to buffer and dump data to another location greatly reduces the data acquisition receiver complexity and affords differential cancellation of the NAVSTAR signal biases for these civilian applications in addition to attenuating the ephemeris error contribution to the baseline computation.

Because several NAVSTAR satellites can be received simultaneously, it is possible to achieve sub-decimeter accuracy using a crystal reference oscillator with stability requirement of only $10^{-9}$ at averaging times of 10 seconds.

Although the SERIES-GPS technique is fundamentally a methodology for codeless pseudo ranging, the highest accuracy is obtained by operating pairs of receivers in a differential mode so as to cancel or attenuate certain error sources. For example, when the NAVSTAR-GPS is fully implemented with its 18 satellite constellation, the DoD may at its discretion incorporate into the transmitted signal structure what are known as methods of accuracy denial. Such denial will be achieved in several ways including altering the frequency and phase of the transmitted signals (dither) and changing the broadcast ephemeris message. Then only those receivers with the Captain Midnight secret decoder rings will be able to acquire the P-code channel and know the real clock offsets and satellite orbital elements. However, by operating the receivers in a differential mode, it is possible to correct for these conditions. (See Kalafus, Vilcans and Knable.[4]) A particular type of differential operation is preferred in the SERIES method which is insensitive to the dither and allows the NAVSTAR orbits to be determined independently by DoD activity.

## EQUIPMENT DESCRIPTION

For ease of operation and flexibility in application, it was necessary to develop an L-band radio and data acquisition system capable of using an unsteered antenna. For reasons of desiring the enhanced signal reception and avoiding multi-path error effects near the horizon, an antenna with gain in the zenith direction was selected. A 13 cm helix over a 20 cm diameter ground plane provides the desired properties of dual L1 (1575.42 MHz) and L2 (1227.6 MHz) reception with approximately 10 dB on axis gain. Data are typically acquired at elevation angles above 25 degrees. An ambient temperature field effect transistor preamplifier is used at the antenna. Following a hetrodyne down conversion to an intermediate frequency, the 20 MHz wide spread spectra from multiple NAVSTAR satellites are compressed into audio bandpasses and spectrally analyzed to extract their various amplitudes, frequencies and phases. A microcomputer then differences these values, removing any dither effects and attenuating satellite orbital errors by the ratio of the distance to the satellite divided by the separation between the receivers.

The SERIES system does not incorporate any telemetry extraction. Thus, the satellite ephemerides must be provided separately. However, the independent determination of the NAVSTAR satellite orbit is possible by exploiting the codeless pseudo ranging capabilities of the SERIES method. For example, the reference station in differential observations, usually operated at a known geodetic position can be exploited as a satellite orbit monitoring station. Because of the nature of the non-gravitational disturbing forces on the satellites, the orbital element most affected is the along track position. By observing the closest approach event from a single station, it is possible to estimate the range to the satellites with an accuracy of approximately 50 meters which is approximately equivalent to the broadcast ephemeris accuracy on the C/A telemetry channel. With inclusion of additional ephemeris monitoring stations, it is possible to develop the other orbital elements and finally with four regional sites a completely geometric approach to deriving the satellite positions can be done which avoids orbital kinematics. (See MacDoran.[5])

## EXPERIMENTAL RESULTS

Tests on the generic prototype SERIES receivers have shown an instrument accuracy of 20 cm with 6 minutes of data acquisition on a short baseline of 20 meters. Over such a short baseline, the satellite positional errors are substantially attenuated. For example, a 50 km on-orbit error contributes a 5 cm baseline error over a 20 meter distance.

SERIES measurements have also been conducted on a 1.9 km baseline. The standard for comparison came from 22 passes of Transit Doppler translocation which were taken by GSI using Magnavox MX 1502 Geoceivers. The Doppler positioning required 48 hours of data acquisition to achieve its sub-meter accuracy. The differential SERIES measurements using one hour of data demonstrated 60 cm or better accuracy relative to the Transit satellite Doppler positioning, Table 1.

The fact that a consistently negative difference exists between the SERIES and Doppler solutions is probably due to the rather limited accuracy of the "almanac" NAVSTAR satellite ephemerides to which ISTAC has access. However, the sub-meter agreement with the Transit results are essentially all that can be expected in the way of accuracy from a 22 pass Doppler translocation solution.

A more complete accuracy demonstration has been started on 50 km baselines. Data have been acquired from two locations at Santa Monica Bay in Southern California relative to ISTAC in Pasadena. However, data reduction must await the availability of NAVSTAR satellite positions with an accuracy of 400 meters on-orbit in order to demonstrate meter level differential positioning accuracy between the two National Geodetic Survey marks 42 km apart across the bay. It is estimated that only a few minutes of data will be required from each site to demonstrate accuracy at the few meter level.

## AIR-MOBILE POSITIONING

The above discussed positioning demonstrations offer some interesting applications where it is important to achieve greater efficiency in the use of personnel

Table 1—Summary of 1.9 km baseline results
(in meters)

| | East | North | Vertical | Length |
|---|---|---|---|---|
| SERIES | 164.7 | 1871.3 | 35.6 | 1878.9 |
| DOPPLER | 164.9 | 1871.9 | 36.2 | 1879.5 |
| Difference | −0.2 | −0.6 | −0.6 | −0.6 |

and equipment resources. For example, by timing the arrival of the SERIES equipment at the remote site to coincide with the NAVSTAR passage over a particular area, it is possibile to accomplish a geodetic positioning mission in a small fraction of the time usually required for Transit satellite Doppler positioning. In the above discussed results, sub-meter accuracy is obtainable in a few minutes to an hour which is at least a factor of 50 improvement over Transit techniques. Of course, as the NAVSTAR constellation of 18 satellites is deployed, no special scheduling of observations will be necessary since four or more satellites will always be above every horizon.

Consider now the task of position verification for an offshore platform. A multi-purpose helicopter might be configured to carry a SERIES receiver which could even deliver positions in real time with meter level accuracy, Figure 1. This helicopter-borne operation could provide the platform operations staff with accurate positioning during the few minutes that a helicopter spends at a platform during routine flight operations.

Such real time operation assumes that a base station is available in the region, typically within a few hundred kilometers of the air-mobile operation. This base station transmits over a narrow band radio link its SERIES receiver compressed spectral L1 and L2 observations of the NAVSTAR satellites as they arrive at the reference site. The radio cross link needs to be capable of moving perhaps only a one byte per second data rate to achieve a positioning fix every 30 seconds. At the helicopter, the SERIES receiver phase and frequency obser-

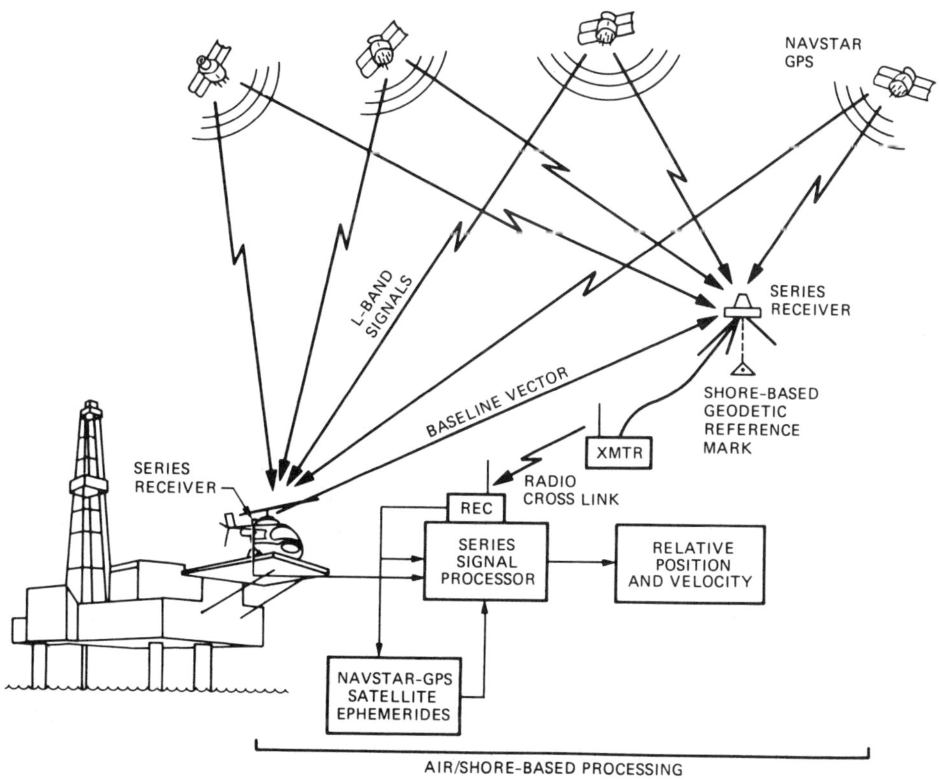


*Fig. 1—SERIES-GPS air mobile positioning.*

vations are differenced with those of the shore based reference site. The ionospheric delay differences are determined and applied to the differential phase measurements and then solved in three dimensions to derive the baseline vector.

Helicopter operations in the Arctic are commonly employed to establish geodetic control for oil and gas exploration. Again, the efficiency of SERIES operations will be attractive.

## ATTITUDE OPERATION

By the use of two or more antennas on a single platform, it is possible to configure a 3D radio compass which functions as an attitude orientation system with a wide range of accuracies, depending upon need. Consider three L-band antennas, each simultaneously receiving multiple NAVSTAR signals as the signals illuminate each of the antennas. The wide band (20 MHz) signals from each of the antennas are brought to a SERIES Spectral Compression and Translation (SCaT) unit whose outputs are at audio. This audio spectrum is analyzed for its phase and frequency content. The distinction being drawn here between a SERIES receiver and a SCaT unit is one of completeness. Every SERIES receiver contains a SCaT unit in addition to a positional computation processor. The corresponding spectral response from each satellite is differenced between the three antennas to produce essentially an interferometic-like response with high phase measurement precision. The angular measurement precision is governed by the ratio of the phase precision divided by the separation between the antennas.

Three fundamental data types are available, C/A channel, P-code channel and L-band carriers from both L1 and L2. Angular measurement precision increases with radio frequency used; being the least with the C/A channel (1.023 MHz) and the most with the L1 channel (1575.42 MHz). However, the increase in precision comes at the expense of ambiguity. Consider the ship illustrated in Figure 2 with the athwart mounted antennas 20 meters apart and the fore and aft antennas separated by 100 meters. The fore and aft pair are sensitive to the pitch and yaw orientation of the vessel while the athwart pair sense the roll and yaw attitude with respect to the directions of the distant satellites. Using the C/A channel, an electrical phase measurement precision of one degree can be done which results in an unambiguous angular measurement with a precision of 0.5 degrees for pitch and yaw and 2.5 degrees for roll. A desire for greater precision is met by going to the P-code channel where the 10.23 MHz modulation gives ten times the precision of the C/A channel but with an 8 degree ambiguity in pitch which is resolved by the 0.5 degree precision from the unambiguous C/A channel. The P-code channel phase is measureable with a precision of about 3 degrees which provides angular measurements equivalent to 0.1 degrees in pitch and yaw and 0.5 degrees in roll. This approach of using higher frequencies and resolving ambiguities is summarized in Table 2.

The data types most likely to find applications in an at-sea environment are those that effectively replace a magnetic compass or a gyro stabilized platform, precision of 0.1 degree. However, performance at the 4 arc second precision level, equivalent to measuring the L1 channel phase with 2 millimeter preci-

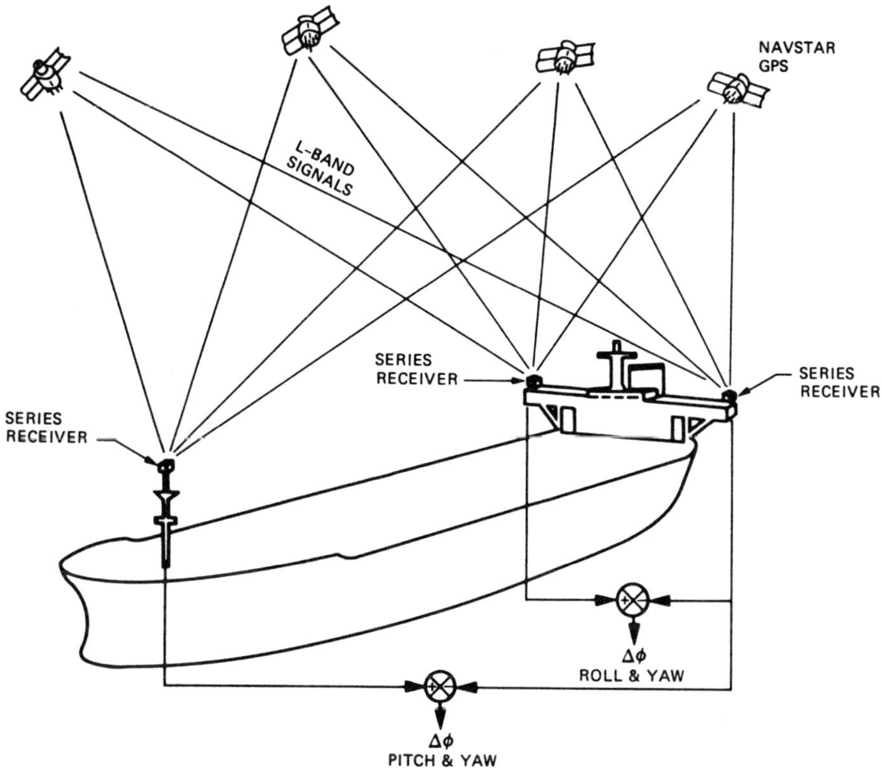


*Fig. 2—SERIES-GPS attitude orientation measurement system.*

Table 2—Attitude measurements for 100 meter antenna separation

| Data Type | Angular Precision | Ambiguity | Resolved by |
|---|---|---|---|
| C/A | 0.5 deg. | none | — |
| P | 0.1 deg. | 8.0 deg. | C/A |
| (L1–L2) | 0.004 deg. | 0.25 deg. | P |
| L2 | 1.4 mdeg. (5 arc sec) | 0.07 deg. | (L1–L2) |
| L1 | 1.1 mdeg. (4 arc sec) | 0.05 deg. | (L1–L2) |

sion, might find application as a strain gauge to monitor large structures such as offshore platforms and oil tankers.

## POSITIONING FOR 3D SEISMIC SURVEYING

As discussed by Morgan, it is an appealing idea to use the GPS in order to track both the position of the seismic boat and tail buoy at the end of the seismic streamer. The actual implementation of SERIES to accomplish this task is the next topic to be discussed. Consider Figure 3. A SERIES SCaT unit is located at a shore based geodetic reference site and a radio cross link provides the frequency and phase values of the arriving NAVSTAR signals to the remote user which performs the data reduction. The seismic boat is such a remote user

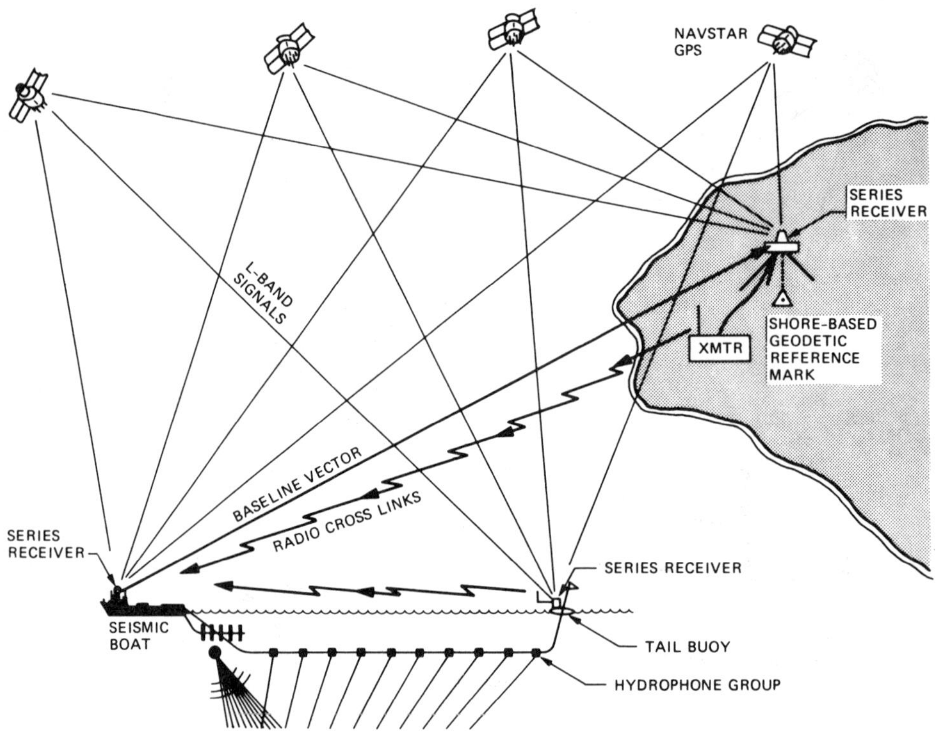


*Fig. 3—SERIES-GPS positioning for 3D seismic surveying.*

and it performs a differencing of the shore base SCaT information with its own SCaT output at each time a relative position is required. In this manner the SERIES receiver aboard the seismic boat positions itself with respect to the shore based SCaT. The tail buoy of the seismic streamer also carries a SCaT unit and links its output to the seismic boat and the boat again computes the vector between the boat mounted SERIES receiver and the SCaT unit on the buoy. The demand upon the radio cross link varies according to requirements for independent vector computation. For example, if the SCaT performs the audio spectral analysis and outputs only time, amplitude, frequency and phase then the cross link requirements are a minimum at an average rate of one byte per second. That one byte/sec rate will allow an independent position computation every 30 seconds. A cross link rate of 10 bytes/sec will allow a position computation every three seconds and 100 bytes per second will allow three positional computations per second.

The SCaT unit need not contain the digital spectral analysis hardware and associated timing circuitry. For example, for the buoy mounted SCaT, a voice grade (few KHz band width) channel can easily be established over the roughly 5 km separation between the boat and tail buoy. In that way the analog output of the SCaT can be directly linked to the boat which performs the spectral analysis to extract frequency and phase of the NAVSTAR satellites as they arrive at the buoy location. Again, by differencing the boat and buoy phase observations, the relative vector separation of boat and buoy is determined with meter level accuracy. By the previous discussion this boat to buoy vector is also positioned relative to a shore based geodetic reference.

As discussed by Morgan, the seismic streamers contain magnetic compasses which aid in the real issue of 3D seismic surveying which is to locate the hydrophone groups relative to the sea sub-floor geology. Since the SCaT antenna on the seismic boat can be located relative to the first streamer hydrophone group and it is assumed that the tail buoy mounted SCaT can be located relative to the last hydrophone group it follows that the end points of the seismic streamer are constrained. By using the magnetic compass information as the first derivative of the streamer figure in the water and the streamer ends constrained by the SERIES SCaT units, the hydrophone groups can be positioned with high precision relative to a distant shore based reference. That shore based reference site is of course assumed to be rigidly attached to the subsurface geology lying below the ocean floor.

Additional methods of determining the seismic streamer shape in the water might be accomplished by exploiting the SERIES positioning information to measure the amount by which the seismic boat is being "set" by currents. The velocity of the boat can be measured with a precision of 1 cm/sec (equivalent to 0.02 knot). With onboard meteorological information, the wind imparted set can be modelled to separate out the water current. Having a measurement constraint on the seismic streamer ends the possible transverse current forcing functions should aid in the modelling of the streamer figure in the water.

## NAVSTAR SATELLITE EPHEMERIDES

The desire for a truly independent NAVSTAR-GPS positioning service must contend with the necessity of information on the NAVSTAR satellite positions in their orbits. In the conventional GPS positioning implementation, the correlation receiver extracts the 50 bit/sec telemetry message which contains the information that enables the receiver to compute the geocentric position of each of the received satellites for the instant at which the receiver desires to compute its position. The satellite orbit mechanics of which the receiver is aware are quite rudimentry and it is valid for about one hour. The NAVSTAR system compensates the lack of sophisticated orbit software in the user receiver by changing the telemetry message. The principal complications in the NAVSTAR orbits result from the gravitational effects of the sun and the moon. In addition to the Keplerian parameters of the ephemeris message, correction terms for the lunar and solar gravity effects valid for about one hour are also part of the telemetry message.

An important ingredient in matching or exceeding the PPS accuracy of the conventional GPS is to operate in a differential mode so as to attenuate the orbital uncertainties. For example, differential observing attenuates the on-orbit error contributions to baseline estimation by the ratio of the baseline/satellite height. Thus, for a baseline of 200 km and a satellite height of 20,000 km, on-orbit error effects are reduced by a factor of 100. Therefore, a 50 m on-orbit error contributes 50 cm of baseline equivalent error to a 200 km distance.

To achieve an independent NAVSTAR-GPS positioning service, ISTAC has developed microcomputer orbit software which makes it possible to exploit the single station codeless pseudo ranging capabilities of the SERIES system to determine the NAVSTAR satellite positions as they pass over one or more reference locations. Thus, as the seismic boat operations are proceeding, the reference station SCaT output can be separately analyzed aboard the boat for its satellite orbital positional information content. The operations would be accomplished by having the boat use the previous day observations from the base station to predict NAVSTAR orbits for the present day. As the data from the base station arrives at the seismic boat, the boat compares the base station observations of frequency and phase for each of the satellites with the predictions based upon the previous day's data. In this manner, real time validation can be performed to guarantee that orbit data of requisite quality is in use (i.e., 50 m on-orbit to support 1 m accuracy on 400 m baselines). Real time re-estimation of the present orbital positions might also be desirable. The physics of the situation greatly contributes to the accomplishment of this task because the principal causes of orbital errors are from non-gravitational forces. The main non-gravitational force is unmodelled solar radiation pressure whose effect is to move the orbit's true anomaly. A change in true anomaly manifests itself in the time of closest approach. When viewed from the reference station of known position, the time of closest approach can be measured with good precision allowing the satellite's along track position to be estimated.

The key to this capability is in the ISTAC microcomputer software which models the solar and lunar perturbations, earth's gravitational field (through the sixth zonal harmonic) and a nominal solar radiation pressure. This software can be executed in a microcomputer whose sophistication is not more than that of an IBM PC. Using these programs it will be possible to derive satellite state vectors with 50 meters of on-orbit accuracy in real time. By post processing multiple revolutions of the orbit, 20 meter on orbit accuracy will be possible with a 1 or 2 day delay. To illustrate the point, the solar and lunar gravitational forces amount to 2 km on-orbit effects upon inclination over a three day interval which maps into a 20 meter positioning error on a 200 km baseline. Figure 4

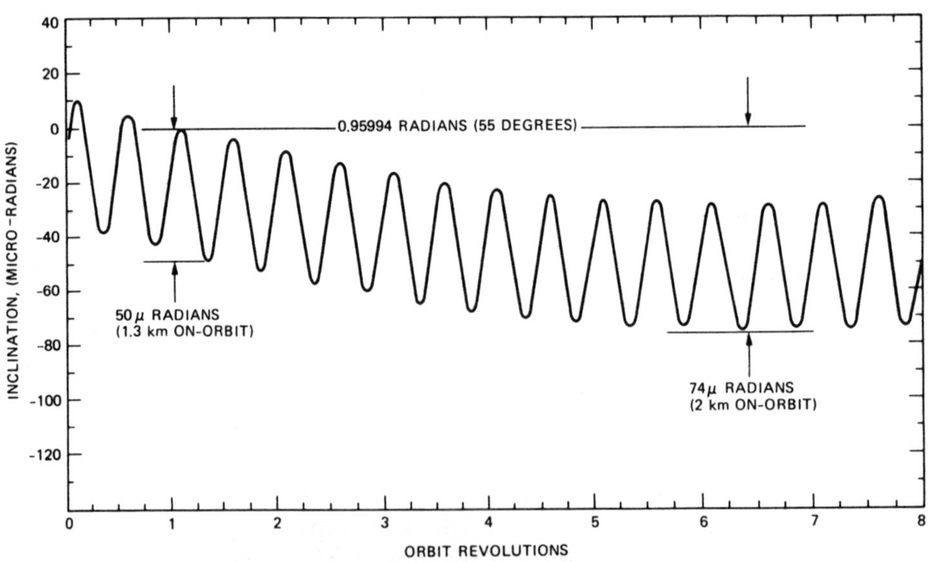


*Fig. 4—Perturbations to NAVSTAR orbital inclination.*

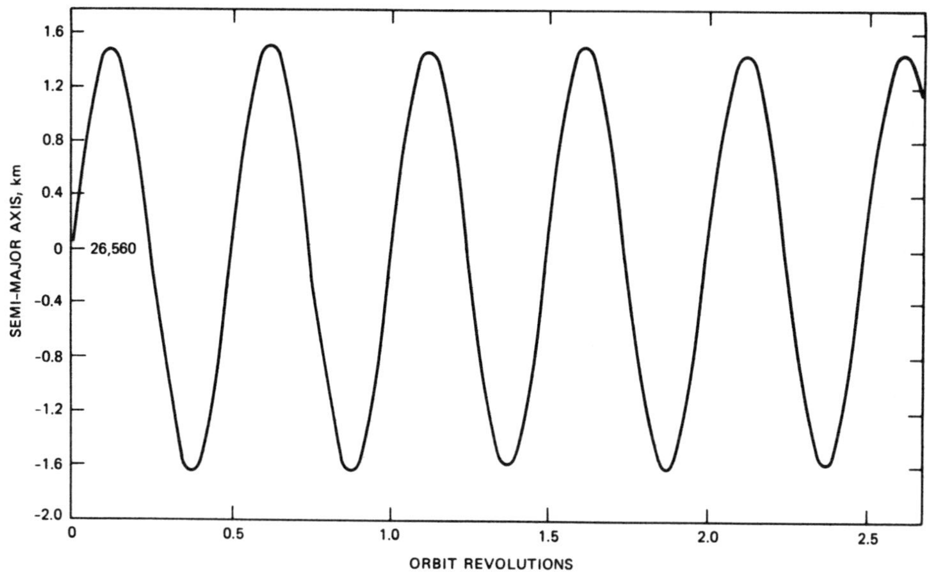


*Fig. 5—Perturbations to NAVSTAR orbital semi-major axis.*

Table 3—Orbit error limitations to positioning accuracy

| BASELINE LENGTHS: | 20 km | 200 km | 2,000 km |
|---|---|---|---|
| ACCURACY | | | |
| REAL TIME: | 5 cm | 50 cm | 5 m |
| POST-PROCESSED: | 2 cm | 20 cm | 2 m |

illustrates the perturbational effects upon the orbital inclination and Figure 5 shows perturbations on the semi-major axis of the orbit. Table 3 summarizes positioning accuracy attainable versus baseline length.

Thus, private positioning services can develop their own NAVSTAR satellite ephemerides with an accuracy sufficient to support one part per million marine geodesy in a mode completely independent of conventional methods.

## IONOSPHERIC CALIBRATION

Where meter level positioning accuracy is required, the delay effects of the ionosphere must be calibrated. Modelling the delay can be done with some success but where missions must be performed while underway or with only a few minutes at a remote site, modelling may not be an adequate method for calibration. A far more desirable approach is to exploit the existence of the coherent dual L-band transmissions from the NAVSTAR to explicitly measure the total electron content of the ionosphere at the specific instant that the pseudo ranging is performed. In the conventional GPS mode, such an option is not readily available because the L2 channel normally contains only the P-code modulation to which there is no open access being planned. Thus, the motivation to have the C/A modulation also available on the L2 channel so that conventional civil GPS equipment can derive ionospheric calibration. The absence of a C/A code of L2 is, of course, intentional and is part of the NAVSTAR accuracy limitations upon non-DoD users.

Such a change in the NAVSTAR-GPS transmission implementation is, however, totally unnecessary since the SERIES-GPS technology makes use of the NAVSTAR signals without code knowledge. Thus, SERIES receivers exploit the dual L1 and L2 channel P-code modulations to extract the ionospheric delay from a single station along the line of sight to each of the satellites. This subsystem of a SERIES receiver is known as SLIC (Satellite L-band Ionospheric Calibration) and has been described in Reference 3. Briefly, SLIC operates by using the fact that the P-code modulations in L1 and L2 are nominally in-phase and coherent in the space vehicle at the time of transmission. As these two wide band (20.46 MHz) signals pass through the ionosphere, the L2 channel experiences 1.65 times the delay effects that occur at L1. By measuring the phase shift of L2 relative to L1, it is possible to directly deduce the columnar electron content along the individual lines of sights.

Actual results with the ISTAC SERIES system using unsteered antennas have demonstrated multiple satellite simultaneous SLIC measurements with 5 degree or better differential P-code phase precision in 10 seconds. That precision corresponds to 0.4 TEC where 1 TEC unit corresponds to $10^{17}$ electrons per square meter which is equivalent to 0.6 m of L1 pseudo ranging. The NAVSTAR space vehicles (SV) appear to agree among themselves in the parameter of ionospheric zenith electron content (ZTEC) to within 0.2 TEC except for SV 6 which has consistently low values by 0.6 TEC. A bias of 0.6 TEC implies that the L1 and L2 P-code phases are offset with the spacecraft, NAVSTAR 3, by 3 nano seconds. However, in the differential mode, any biases originating in the spacecraft will undergo common-mode cancellations.

The most complete accuracy assessments of the SLIC method have been performed at JPL by Royden, Miller and Buennagel.[6] In that analysis, using the original SERIES system with steered 1.5 m antennas, agreement with Faraday Rotation ionospheric measurement methods has been demonstrated to 0.5 TEC or better with the Faraday Rotation accuracy estimated at 0.3 TEC.

## SUMMARY AND CONCLUSIONS

The frequently held notion that the high accuracy positioning needs of the civil user should somehow be solved by the NAVSTAR Global Positioning System (GPS) has led to the perception that the only way to achieve that accuracy is to access the GPS in the same way the DoD users do it, through the Precise Positioning Service (PPS). Such a perception is not strictly true. An alternative exists in the form of the SERIES-GPS technology which does not require knowledge of the classified P-code pseudo ranging and restricted orbit information. The SERIES system innovations make it possible to perform simultaneous pseudo ranging to all the NAVSTAR satellites above the horizon from a single station at both L1 and L2, thus enabling ionospheric calibrations to support a wide variety of high precision civil user applications.

ISTAC SERIES equipment now operates with submeter three dimensional accuracy with a requirement of only a few minutes of on-site data acquisition. This level of demonstration is believed applicable to a helicopter-borne rapid surveying system suitable for offshore platform position verification. In 1984, experiments will be conducted in a low dynamic mode to demonstrate the applicability of the technology to the needs of 3D seismic surveying with meter level accuracy at distances of hundreds of km. The SERIES techniques can also solve the problem of tracking the tail buoy of the seismic streamer to aid in the solution of the real problem of 3D seismic surveying, locating the hydrophones relative to the sub-sea floor geology. The use of multiple SERIES receivers aboard a platform such as an oil tanker can be configured into a three dimensional radio compass of unprecedented precision perhaps to 2 arc seconds.

The quest by the civil user community for access to the NAVSTAR PPS will have the effect of degrading the national security interests. The likelihood of diversions and/or reverse engineering of civilian P-code receivers into smart weapons by an enemy is significant and must be of serious concern. P-code receivers, by their nature, possess high levels of electronic warfare resistance. By contrast, the SERIES-GPS technology is very vulnerable to electronic countermeasures but is capable of high accuracy in virtually all civil applications. For these reasons, the release of NAVSTAR classified codes and telemetry formats is considered inadvisable and counter to U.S. national interests. As for the implementation of user fees to underwrite the GPS operations, such an approach is unlikely to generate any offset over its own costs. Furthermore, the use of SERIES techniques, which requires no knowledge of codes and telemetry messages, makes the concept of a user fee imposition entirely moot.

In the event that the U.S.S.R. implements its Global Navigational Satellite System (GLONASS), SERIES technology will operate just as well with their microwave signals. SERIES is an equal opportunity radio signal consumer.

## ACKNOWLEDGEMENTS

The authors wish to acknowledge the assistance of Geophysical Services, Inc. (GSI), especially to Jack Lane for the many discussions on the needs of the marine exploration community and for the loan of a pair of cesium frequency systems, a field instrument van and a navigation engineer, Tom Games, who assisted in the SERIES data acquisition. GSI also performed the Transit Doppler measurements at four sites in Southern California (thanks due to Steve Williams) and GSI made available NAVSTAR "almanac" ephemerides. Thanks also to Kinemetrics, Inc. for the loan of a GOES satellite time synchronization receiver.

## REFERENCES

1. Morgan, James G., "The Challenge of Precisely Positioning a 3D Seismic Survey," *Navigation,* Vol. 30, No. 3, Fall 1983.
2. Parkinson, Bradford W., Gilbert, Stephen W., "NAVSTAR: Global Positioning System—Ten Years Later," *Proceedings of the IEEE,* Vol. 71, No. 10, October 1983.
3. MacDoran, P. F., Spitzmesser, D. J., Buennagel, L. A., "SERIES: Satellite Emission Range Inferred Earth Surveying," Proceedings of the Third International Geodetic Symposium on Satellite Doppler Positioning, New Mexico State University, February 1982.
4. Kalafus, R. M., Vilcans, J., Knable, N., "Differential Operations of NAVSTAR-GPS," *Navigation,* Vol. 30, No. 3, Fall 1983.
5. MacDoran, Peter F., "Satellite Emission Radio Interferometric Earth Surveying, SERIES—GPS Geodetic System," *Bulletin Geodesique,* 53, 1979.
6. Royden, H. N., Miller, R. B., Buennagel, L. A., "Comparison of NAVSTAR Satellite L-band Ionospheric Calibrations with Faraday Rotation Measurements," submitted to *Radio Science,* to be published in 1984.

# Global Positioning System Differential Navigation Tests at the Yuma Proving Ground

L. R. KRUCZYNSKI, W. W. PORTER
United States Air Force

D. G. ABBY, E. T. WESTON
General Dynamics Service Corporation
*Received January 1985*

## ABSTRACT

The Global Positioning System (GPS) Joint Program Office has conducted the majority of user equipment testing at the U.S. Army Yuma Proving Ground (YPG), Yuma, Arizona. This paper briefly reviews the history of GPS testing at YPG, describes the major assets of the test range and presents the results of dynamic and static differential GPS tests conducted at YPG.

## INTRODUCTION

In 1976, the Global Positioning System (GPS) Joint Program Office (JPO) opened an operating location at the U.S. Army Yuma Proving Ground (YPG), Yuma, Arizona. After an extensive survey of available ranges, the JPO had determined that YPG was the only range that had the combination of high accuracy trajectory determination over a large airspace, real-time monitoring of testing and ready availability of range time.

By March 1977, four ground transmitters (GTs) had been installed at YPG and the Phase I test program began. The GTs simulated satellite signals so that tests could be conducted prior to the availability of GPS satellites. The four GTs comprise the Inverted Range (IR). Operation of the IR is controlled by the Inverted Range Control Center (IRCC) which contains a Phase I Magnavox four-channel receiver, a cesium frequency standard, computers and communication equipment.

The first GPS satellite was launched on 22 February 1978. The fourth satellite was launched on 10 December 1978 and was available for navigation on 3 January 1979. The orbits of the first four satellites and all launches to date have been designed to optimize coverage at YPG. With the exception of SV 12 (Nav 10) and a planned change to the orbital inclination of the Block II satellites, all satellites are in slots that are a part of the fully operational constellation. We currently have over five hours of horizon to horizon satellite coverage with good geometry at YPG.

When fewer than four satellites were available, we used combinations of satellites and GTs in a "hybrid" mode. We continue to use GTs for selective availability/anti-spoofing (SA/A-S) tests and for differential navigation tests. Almanac data for the GTs is transmitted by the satellites but the health words are set to indicate that the GTs are unavailable.

## YUMA PROVING GROUND RANGE DESCRIPTION

### *Laser Trackers*

The laser trackers are the primary source of trajectory data. The laser trackers are manufactured by Sylvania and are known as the Precision Automated Tracking System (PATS). Six laser trackers are available for operation at the locations shown in Figure 1. Each laser tracker determines the position of an aircraft or ground vehicle fitted with a retroreflector. The PATS is capable of automatically tracking the vehicle up to a range of approximately 30 km. Advertised accuracy of the PATS is 0.1 mrad elevation, 0.1 mrad azimuth and 1 meter range at 10 km. This accuracy is obtained through rigorous premission and post-mission calibrations. The laser data is transmitted to the YPG Range Operations Center (ROC) where it is processed in real time for display. Laser data is also recorded at the ROC for post-processing.

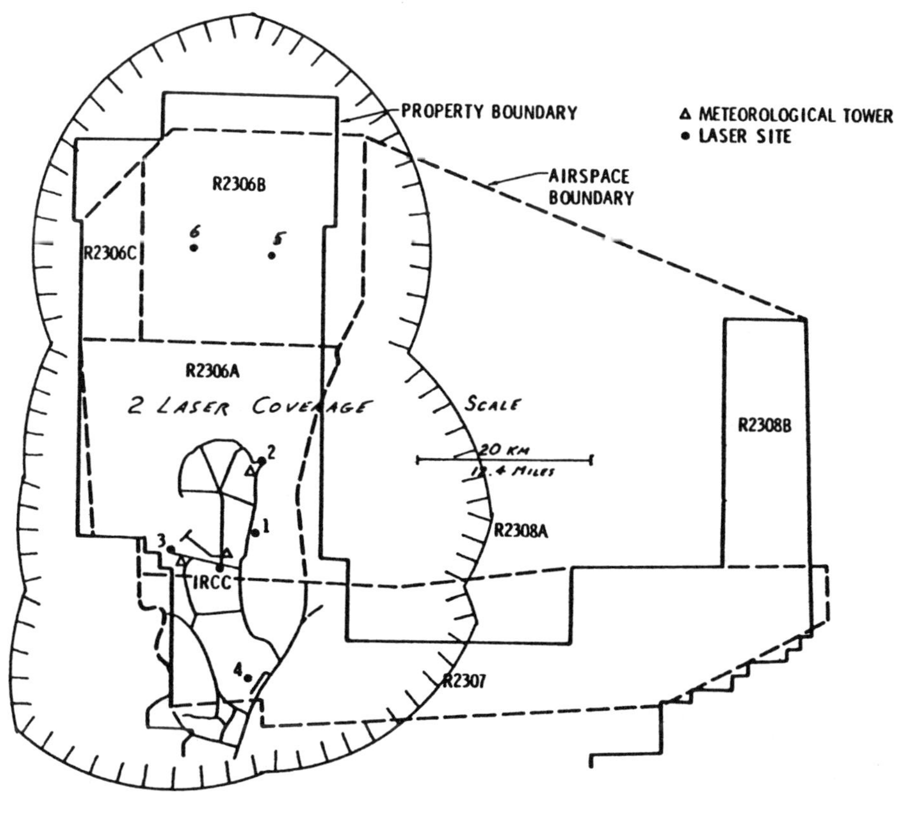


*Fig. 1—U.S. Army Yuma Proving Ground, Arizona*

### Position Locating System

The primary function of the Position Locating System (PLS) is to provide acquisition data for the laser trackers. The PLS includes transponder units onboard the target and ground interrogators. The PLS determines a target's location using multi-lateration techniques at a rate of 10 times per second. Up to 64 targets may be tracked by the PLS.

### Telemetry

Each GPS Host Vehicle (HV) except the M-60 tank is equipped with a telemetry system to transmit the complete GPS data stream to the ROC for real time display and analysis and for recording for post-mission analysis. Typically the HV uses two transmitters and two antennas to ensure continuous data reception. YPG receives the data at the Range Telemetry Ground Station and transmits it to the ROC. Two separate systems are available to support two simultaneous missions.

### Meteorology Equipment

Three meteorological towers are located on the YPG Cibola range as shown in Figure 1. These 300-ft. towers have pressure, humidity and temperature sensors at the 50-ft. level and temperature sensors at the 100-ft., 200-ft. and 300-ft. levels. Data are automatically transmitted to the ROC for performing laser refraction corrections. A rawinsonde sounding system located on the Cibola range monitors atmospheric parameters up to 30,000 ft. prior to and after each test. Data collected include temperature, humidity, and wind direction and velocity at selected altitude intervals.

### Range Time

The YPG range timing system consists of a central clock with a cesium standard for frequency control and a 150 mhz. radio system that transmits time in Inter-Range Instrumentation Group (IRIG) format to all range instrumentation sites. Time synchronization is maintained by synchronizing all systems to GPS coordinated universal time (CUT). The GPS satellites will transmit GPS time and all data needed for the user to compute CUT. The IRCC tracks the GPS satellites, obtains GPS CUT and then synchronizes the YPG clock to GPS CUT.

## GPS ERRORS

### Satellite Ephemeris and Clock (Common Terms)

Space and Control Segment errors in estimation of satellite orbit and clock behavior cause a large portion of the user's navigation error. Inaccuracies in satellite orbit and clock parameters are independent of the user and vary slowly with time. At any instant, all users should have identical predictions of a satellite's time and position. Within a fairly large geographical area, the orbit and clock estimation errors contribute common ranging errors.

For colocated users with identical user clock solutions, the pseudo range measurements should also be identical. Two colocated users should maintain identical navigation solutions. The error in their relative position (zero in this case of colocated users) is a function only of receiver differences. This effectively means the Space and Control Segment errors can be eliminated since they are common errors.

### *Geographically Separated Users*

When the users are geographically separated, the pseudo range measurements will probably be different. The measurement difference is a complex function of the actual satellite position error (along track, cross track and radial), the actual satellite clock error and the locations of the two receivers. See Ref. 1 for some approximations.

### *Estimation of Ephemeris and Clock Errors*

The prime source of the satellite ephemeris and clock errors is the inability of the Control Segment to precisely predict the satellite orbit and clock behavior. Two sources of information exist to describe the performance of the Space and Control Segments. The first is the User Range Error (URE) plots produced by the Control Segment at Vandenberg AFB California. To calculate a URE, the Control Segment first calculates the satellite location and clock state using the data contained in the navigation message. The Control Segment then calculates satellite location and clock state using all available monitor station data, including new data not used in calculating the navigation message data. Using a known user location, typically the IRCC at YPG, the URE plots depict the range differences using the old (currently transmitted) data and the new data. Note on Figure 2 that, at upload, the URE drops to near zero indicating that, in the opinion of the Control Segment, the satellite's navigation data is almost exact.

The second source of Space and Control Segment performance data is the ground truth data generated by the IRCC. Two basic plots are provided. The first plot contains the "Corrected Pseudo Ranges" and the second depicts the "Ground Truths."

A Corrected Pseudo Range is the actual error in the pseudo range measurement plus any uncorrected error in the IRCC clock. Because the IRCC clock error is included in the Corrected Pseudo Range, a perfect Space and Control Segment may have non-zero, time-varying Corrected Pseudo Ranges but all satellites will have identical Corrected Pseudo Ranges.

Figure 3 depicts Corrected Pseudo Ranges for the same day as the UREs in Figure 2. Note that, at upload, the total URE change for SV 11 is about five meters. This agrees with the change in the Corrected Pseudo Range. Since all Corrected Pseudo Ranges are roughly equal (about 10 to 12 meters) after all uploads, the user's navigation solution should be excellent.

After applying the Corrected Pseudo Ranges to the navigation filter, the IRCC clock error is separated. Using a least squares navigation filter, the IRCC then calculates a second form of Space and Control Segment performance data, the "Ground Truth." Figure 4 shows the Ground Truth using SVs 6, 8, 9 and 11. After uploads, total position error is about ten meters until satellite geom-

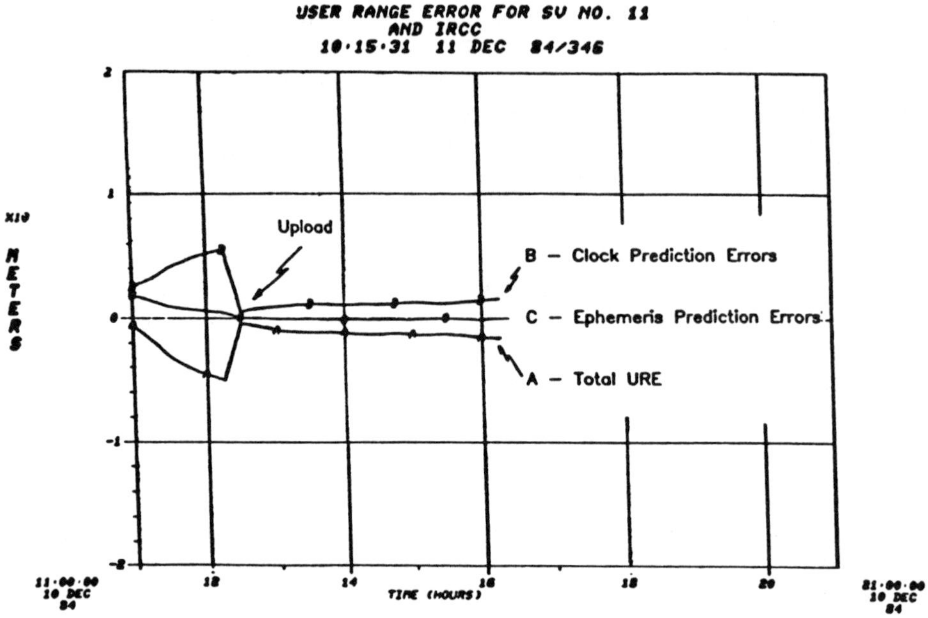


*Fig. 2—10 December 1984 User Range Error (URE)*

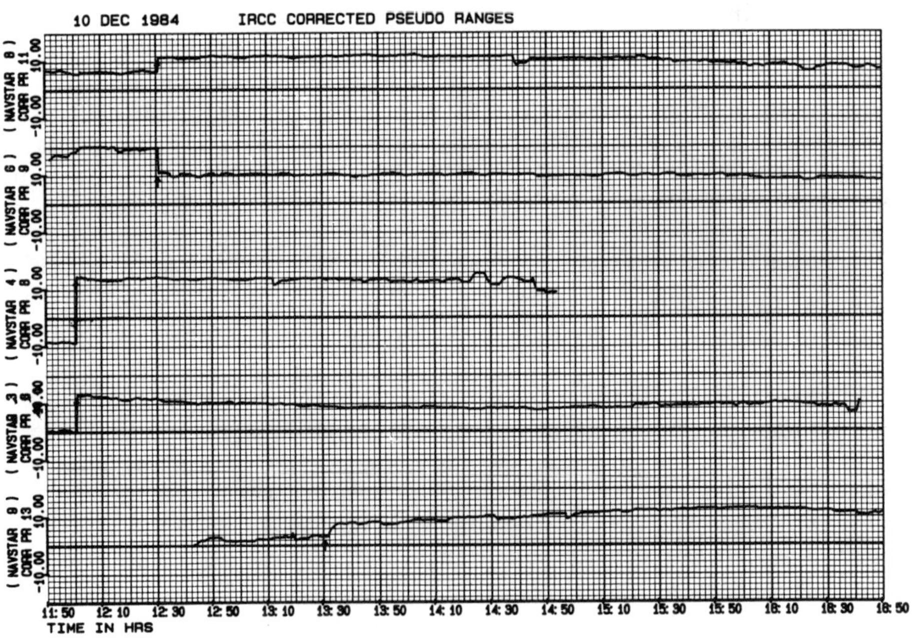


*Fig. 3—10 December 1984 corrected pseudo-ranges*

etry worsens. Although the individual satellites are performing within expectations, the poor geometry contributes to filter instability for the SV 6, 8, 9 and 11 combination. Properly working User Equipment would change to a different

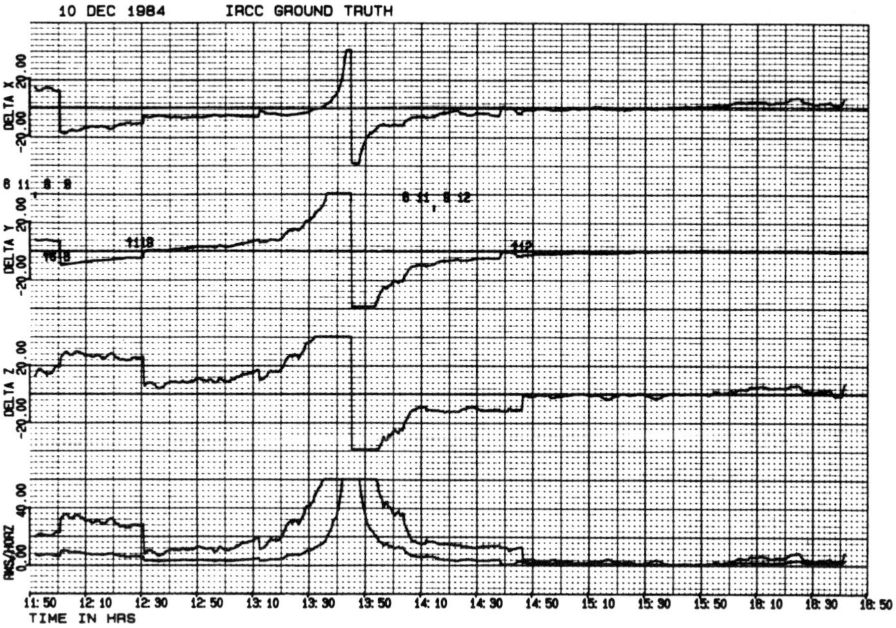


*Fig. 4—10 December 1984 IRCC ground truth*

combination around 1300Z. Figure 4 Ground Truth data changes to the SV 6, 9, 11 and 12 combination about the time of SV 8 set.

The Ground Truth represents the navigation error in East, North and vertical components. Ground Truths are computed for all combinations of four satellites if four or five satellites are visible. When more than five satellites are visible, Ground Truths are explicitly computed for five combinations of four satellites. A user with a constellation other than one with an explicitly computed Ground Truth must apply the Corrected Pseudo Ranges to his own navigation filter.

The IRCC data is more valuable to testers at YPG because it is available in real time and because the form of the data applies directly to evaluation and user equipment navigation. In addition, use of the IRCC data avoids any concerns about incestuous relationships in the Control Segment URE data. Several organizations have indicated the need for an independent ground station to perform the quality assurance (QA) function and operational status evaluation of the Space and Control Segments. The IRCC currently performs this independent QA function.

### *Ionospheric Delays (Almost Common Terms)*

For colocated users, ionospheric and tropospheric delay errors should be identical. Ref. 2 showed that using two frequency measurements to compute ionospheric delays will contribute noisier results with no increase in differential navigation accuracy. Ref. 2 recommended that, in differential mode, no ionospheric delays need be calculated. Although this is acceptable for colocated users, geographically separated users may have to consider the effects of varying ionospheric delays. The best solution would be to provide the differential GPS corrections without ionospheric calculations but to include the results of two frequency ionospheric delay calculations on the data link. Our tests at YPG have not indicated any significant difference in navigation errors due to a difference in ionospheric delays experienced by separated users.

### *Tropospheric Delays (Not Necessarily Common Terms)*

Tractable models of tropospheric delay typically are functions of user altitude, atmospheric conditions (wet and dry refractivities) and elevation angle to the satellite. For colocated users, no differential corrections, the use of identical models or transfer of tropospheric delay calculations from the reference receiver to the user all result in good differential navigation. As the reference receiver and the user separate, however, tropospheric delays will differ. This is particularly apparent with altitude differences and low elevation angle satellites. An example in Ref. 4 shows a 20 meter delay difference assuming a 15 km altitude separation and a five degree elevation angle. If we remember that the function of differential navigation is to remove the **common** Space and Control Segment error terms, then we should conclude that differential navigation will do nothing to remove the non-common errors caused by tropospheric delays. Since pseudo range calculations are significantly affected by tropospheric delays, it is incumbent on the user to use a model that satisfies the user's accuracy requirements. The reference must also use an accurate tropospheric delay model.

## DIFFERENTIAL TECHNIQUES

All differential techniques require a reference receiver which is a well calibrated GPS receiver located at a surveyed site. The difference between the reference receiver's navigation solution and the survey coordinates is the Ground Truth. The difference between a pseudo range calculated using the survey coordinates is the Corrected Pseudo Range. The actual filter mechanization used by the reference receiver is not critical to this discussion.

### *Ground Truth Method Correcting the Navigation Solution*

In this method, the Ground Truth is subtracted from the absolute location of the user's navigation solution to provide the differential GPS solution. In tests conducted at YPG, the ground truth correction was provided in Earth Centered-Earth Fixed (ECEF) coordinates.

When only four satellites are visible, only one four-satellite combination can be seleted. With five satellites, five combinations are possible. With six satellites, there are 15 combinations; seven satellites give 35 combinations; and eight satellites, the maximum expected to be visible, yield 70 combinations. Each combination requires at least three correction factors. In addition, it is possible that a user will have only three satellites in use. The Ground Truth method requires an unwieldy amount of data. We need another technique.

### *Corrected Pseudo Range Method*
### *Correcting the Pseudo Range Measurements*

If eight satellites are visible, the reference receiver need only provide corrections to the eight pseudo ranges. When the user corrects the measured pseudo ranges by the corrected pseudo ranges provided from the reference receiver, the resulting navigation solution is virtually identical to the navigation solution provided by the Ground Truth method.

## HELICOPTER DYNAMIC DIFFERENTIAL TESTS—TEST DESCRIPTION

### *User Equipment*

The user equipment consisted of a Texas Instruments High Dynamic User Equipment (HDUE). The HDUE is a five-channel set. In the normal mode, the fifth channel is responsible for taking L2 measurements so that ionospheric delay calculations can be made. In the differential mode, the fifth channel is the data link with the reference. No ionospheric measurements are possible in the differential mode. The original tests using this equipment[3] consisted of providing the ECEF Ground Truth corrections and the satellite ionospheric delay calculations via the data link. The satellite ionospheric delay terms were necessary because the reference receiver included the ionospheric terms in calculating the ground truth. A satellite constellation was commanded.

For the Corrected Pseudo Range method tests described here, the HDUE software was not modified. Instead, the data link contained zeros in the Ground Truth corrections. The sums of the ionospheric delays (if calculated by the reference) and the Corrected Pseudo Ranges were transmitted in the ionospheric terms. P-code tests and C/A-code tests were conducted. See Figure 5.

### *Reference Receiver*

The reference receiver is the Phase I Magnavox four-channel receiver located in the IRCC. A given satellite constellation was commanded. When the reference receiver was operated in P-code, ionospheric delay measurements were made. Some differential GPS C/A code tests also were conducted. Ionospheric delay was set to zero when C/A code was used.

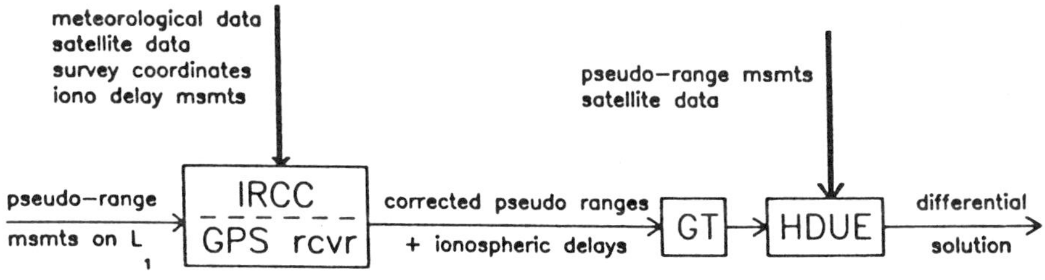


*Fig. 5—IRCC/HDUE differential block diagram*

### *Data Link*

The IRCC computer transmitted the Ground Truth correction data (zero for these tests) and the ionospheric delay data (ionospheric delay plus corrected pseudo ranges) to the GT for subsequent transmission via the L1 signal. All data was formatted into subframe 2 of the GT's navigation message. Corrections were calculated every 18 seconds but the data on the data link was updated every 30 seconds. Thus the data used by the HDUE could be as much as 48 seconds old when used by the HDUE.

### *Range Support*

The tests described herein were primarily conducted to evaluate the accuracy of the laser tracking system at YPG. In particular, the mission was to evaluate the accuracy of two lasers that had recently been installed in the North part of the range. The new lasers were added to increase the airspace of high accuracy trajectory determination from 1200 sq km to 2000 sq km. Our test demonstrated that the North Cibola lasers were not yet fully calibrated.

Unfortunately for the purpose of differential GPS evaluation, the new lasers contributed up to five meter errors. For completeness, we include a plot of a later differential test conducted when trajectory determination errors were reduced to the desired one to three meters.

## TEST RESULTS
## 10 January 1984

**Old data versus new data.** During the entire period shown in Figure 6 (13:43 to 14:00), both HDUE and IRCC and P-code tracking and corrections were made in measurement space. SV 4, which was using a quartz standard, was uploaded with new data at 13:44. The HDUE recognizes the new data immediately. IRCC software, however, does not immediately implement the new data. Built-in IRCC delays plus the 30 second update of the differential message result in the 1 ½ minutes of 10 meter horizontal and 13 meter vertical position error from 13:44 to 13:45:30. As soon as the HDUE receives the new corrected pseudo range for SV 4, horizontal errors drop to about two meters and vertical errors drop to three meters.

**C/A versus P.** In Figure 7, from 14:00 to 14:10, both IRCC and HDUE operated using P-code. From 14:16:20 to 14:29:30, both IRCC and HDUE were using C/A code. Of some concern was the apparent change in the bias of the Y-error. Analysis showed that this error was due to inaccurate laser calibrations for a new laser. Jumps in the errors such as that at 14:11 and at 14:11:30 are directly correlated to laser switches. Also, the test was conducted about dawn when the relative humidity was about 65%. This unstable atmospheric condition contributed additional trajectory errors. There is no reason to believe that the solution changed because of a change from P-code to C/A code. Note, however, that the C/A HDUE measurements do result in a noisier solution.

## TEST RESULTS
## 26 January 1984

On 26 January 1984, atmospheric conditions at YPG were more stable and additional laser calibrations had been completed. The UH-1 was flown with the

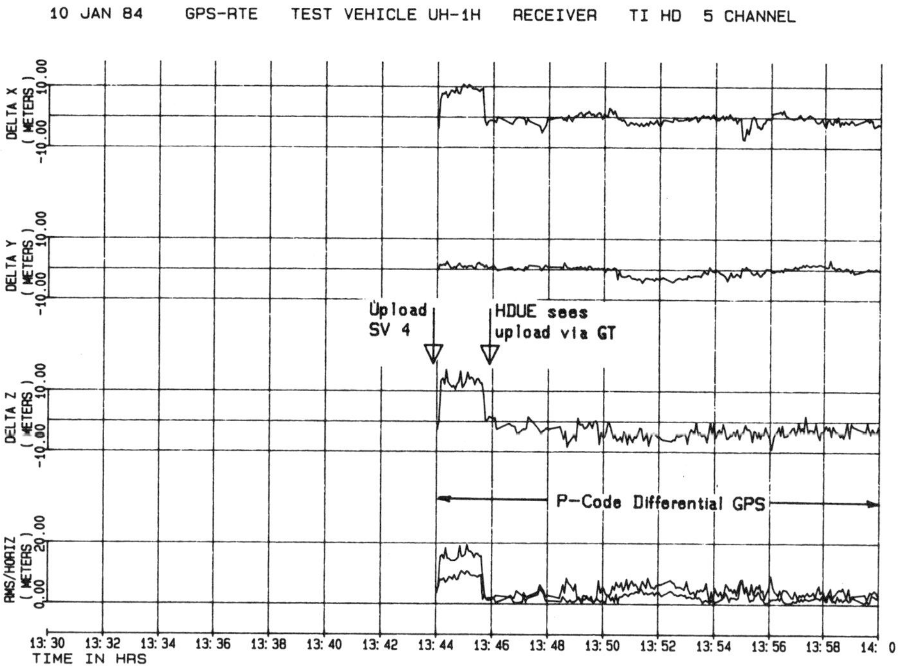


*Fig. 6—10 January 1984 differential GPS errors*

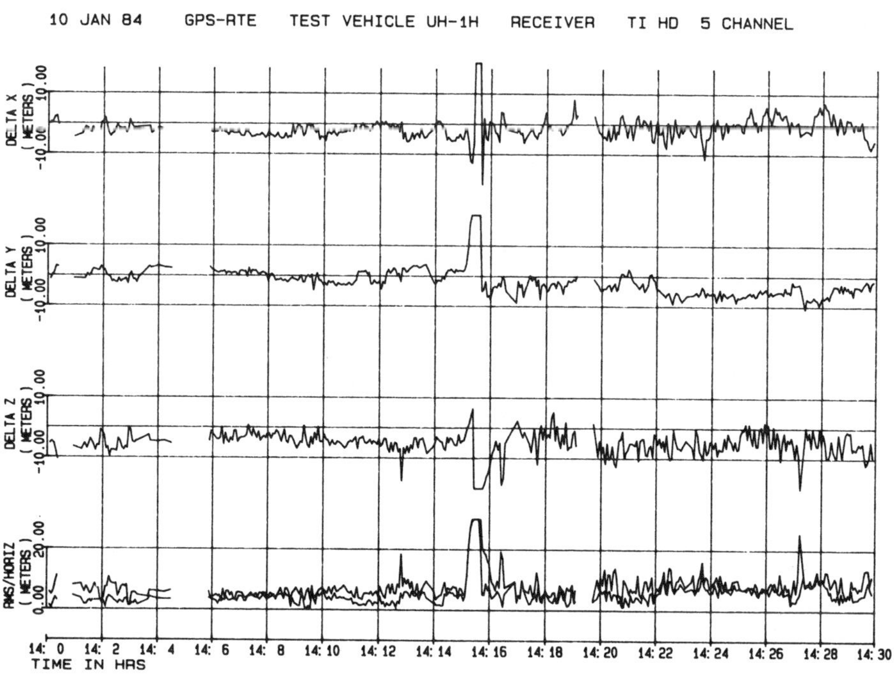


*Fig. 7—10 January 1984 differential GPS errors*

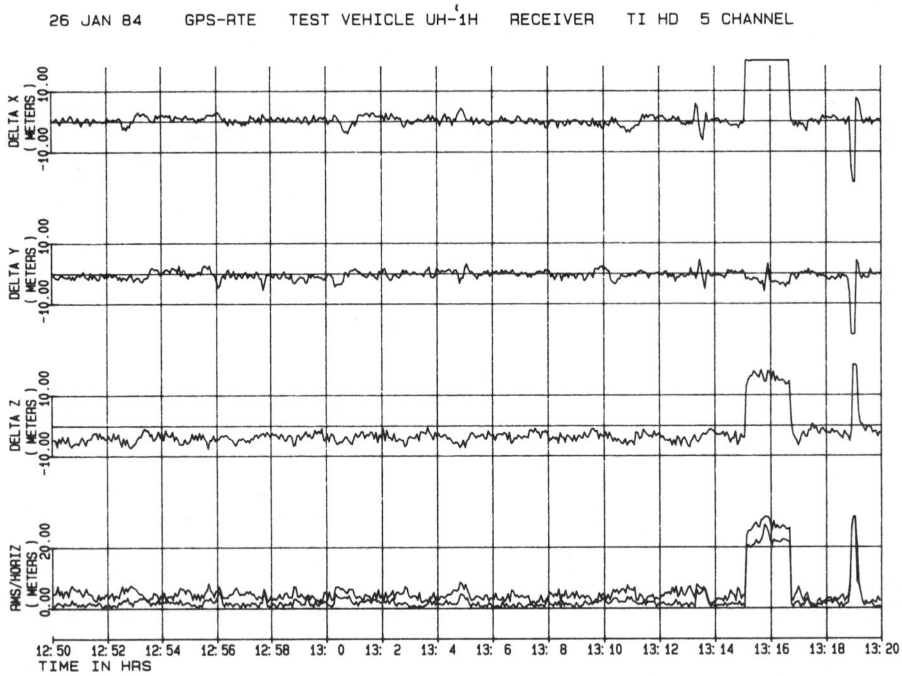


*Fig. 8—26 January 1984 differential GPS errors*

HDUE operating in differential mode using P-code. Figure 8 depicts a half hour of the flight. The large jump from 13:15 to 13:16:30 is caused by the lag in IRCC use of uploaded data. The only other aberrations are those highlighted in Figure 8. Each of these perturbations is directly correlated with helicopter dynamics as shown by the acceleration plot in Figure 9.

## STATIC DIFFERENTIAL TEST RESULTS

In a Fall 1984 test, absolute navigation errors seen by a Phase I Magnavox X-set located near the IRCC reached 200 meters. On that day, another X-set was operated at Edwards AFB, California, a distance of about 400 km. The X-set at Edwards experienced similar errors. Figure 10 presents the difference between the errors at Edward AFB and the errors from the X-set at YPG. In both cases, the X-sets operated in C/A code. No ionospheric corrections were applied. Noise in the solution was caused by an oscillator problem in the X-set located at YPG.

## SUMMARY

GPS tests have been conducted at YPG because, to evaluate a system whose specification calls for 15 meters accuracy, we needed a range with the capability to cost-effectively provide a trajectory to at least two to three meters accuracy. Validation of the range data became an issue. The development of differential GPS techniques introduced GPS accuracies equivalent to laser tracking accuracies. This has created a "who is checking whom" issue.

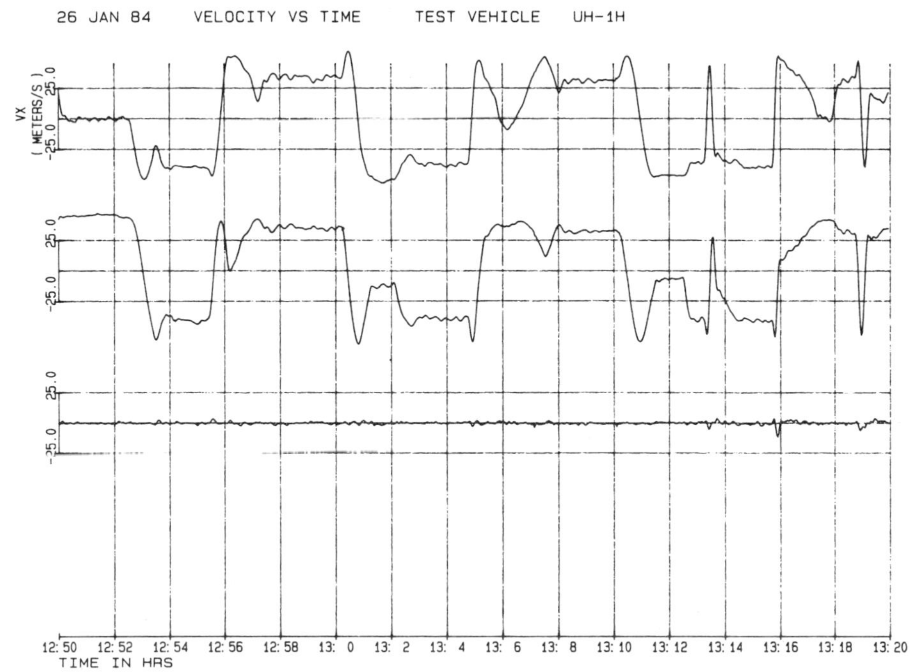


*Fig. 9—26 January 1984 UH-1 dynamics*

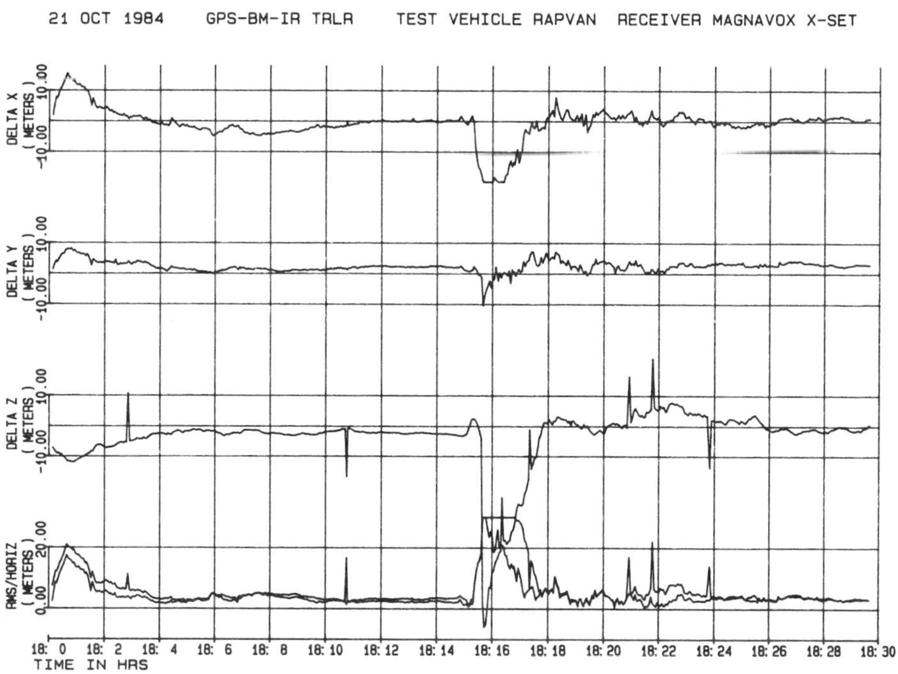


*Fig. 10—Static differential errors*

The combination of laser and GPS accuracies has opened up new horizons in weapon system evaluation. We have already contributed to isolating errors in bombing systems never before attainable. Cinetheodolites, which are expensive and limited in coverage, and radar, which does not have the accuracy required, cannot cost effectively support modern weapon system accuracy issues. Tests with the HDUE integrated with an INS aboard a Convair 880 were used but they were expensive tests. The UH-1 differential tests, however, were inexpensive and used assets that belonged to GPS and YPG. Differential GPS proved to be an excellent method to maintain quality assurance of trajectory accuracy.

The tests showed that differential GPS using C/A code is as accurate as and only slightly noiser than P-code; that corrections need be provided only on the order of every minute or so; that making corrections in measurement space was a viable technique; and that ionospheric delay calculations were not critical to differential GPS. In general, differential GPS yielded accuracies on the order of two to three meters.

## REFERENCES

1. Beser, J. and Parkinson, B. W. *The Application of NAVSTAR Differential GPS in the Civilian Community,* Global Positioning System, Papers published in Navigation, Volume II, The Institute of Navigation, 1984.
2. Lachapelle, G. et al, *Differential GPS Marine Navigation,* Proceedings of the IEEE 1984 Position Location and Navigation Symposium, November 26–29 1984, pp 245–255.
3. Teasley, S. P., Hoover, W. P. and Johnson, C. R. *Differential GPS Navigation,* Texas Instruments Inc. Technical Report, 1981.
4. Tavis, M., *Tropospheric Delay Corrections in Differential GPS Applications,* unpublished Aerospace Corp. report.

# Civil Helicopter Flight Operations Using Differential GPS

FREDERICK G. EDWARDS
NASA Ames Research Center, Moffett Field, CA

PETER V. W. LOOMIS
TAU Corporation, Los Gatos, CA
*Received June 1985*

## ABSTRACT

NASA Ames Research Center conducted a nonreal-time, Differential Global Positioning System (DGPS) flight-test experiment using two coarse-acquisition (C/A) code GPS receivers. The data were recorded concurrently at a fixed site and on board the NASA SH-3G test helicopter. The aircraft conducted several simulated mission operations, including terminal approach, while being tracked by radar and laser systems. After the flights, data recorded by the two GPS receivers and the trackers were analyzed to determine whether differential corrections could improve the navigation performance of the airborne GPS receiver. An airborne navigation error history was obtained by subtracting the reference "true" trajectory (derived from the tracking data) from the airborne GPS navigation solution. At the same time, differential GPS corrections were obtained by subtracting the ground-station GPS navigation solution from the true (surveyed) location of the ground receiver. Filtering was used to separate receiver-unique errors in the GPS data from the locally common (spatially correlated) GPS errors. The results show high correlation between airborne navigation errors and the differential corrections. The airborne navigation solution is shown to be much improved after the addition of the differential corrections. Efforts are continuing to develop a real-time differential data link between the ground station and the aircraft.

## INTRODUCTION

The NAVSTAR Global Positioning System (GPS) holds the promise of a navigation precision for suitably equipped land, sea, and air vehicles that far exceeds the capabilities of presently available positioning techniques. Because many civil helicopter operations are particularly dependent on exact position information—for example, offshore exploration, low-altitude flights into remote and mountainous terrain, and approaches into uninstrumented landing sites—the NAVSTAR system will undoubtedly find early and wide use in those applications.

Use of the GPS to support aircraft when en route and when in terminal areas has been the subject of considerable study. The military has been primarily interested in the precision code (P-code) from the NAVSTAR satellites; the P-code is highly accurate and has the potential of providing landing guidance to relatively low minimum-decision heights.[1,2] Civil users may only have access to the coarse-acquisition code (C/A code) signal, which although less accurate than the P-code, can still support en route navigation and non-precision approaches.[3] A need remains, however, for the adaptation or development of a navigation capability to support precision approaches for civil aircraft into remote areas.

Ames Research Center initiated a program to investigate the use of the C/A code in a differential mode to provide a precision-approach capability for civil operations. The program has been ongoing for several years.

As a first step, several alternative mechanizations of differential GPS (DGPS) were studied to determine the preferred configuration for supporting precision approaches into remote areas.[4,5,6] The study phase was followed by static laboratory tests using a single-channel, sequential GPS receiver (Magnavox Z-set). This receiver was installed in the NASA SH-3G helicopter and then operated during a series of flight tests designed to assess its performance in several types of helicopter operations; the results of these tests were reported in Reference 7.

Recently, NASA conducted a flight investigation to evaluate a DGPS configuration using two GPS receivers, one in a helicopter and one on the ground. During this series of tests, the data uplink between the ground system and the aircraft was not implemented and the data were acquired, time-tagged, and recorded concurrently on board the aircraft and at the ground-based GPS reference site. A differential solution was computed during post-flight data analysis. Preliminary results from these tests were reported in References 8 and 9. Future tests are planned to investigate the concept in real-time, using DGPS to support terminal approaches with the SH-3G.

The purpose of this paper is to present the results of the most recent flight tests and of the post-test analysis. The basic DGPS concept that is being investigated is discussed, the test procedures and analytical procedures are described, and the test results are presented.

## DIFFERENTIAL GPS CONCEPT

NAVSTAR GPS is subject to a number of error sources,[10] many of which are common to users over large areas. These locally common errors can be estimated by a reference station with a GPS receiver at a well-surveyed site, by comparing the receiver position estimate to the known, surveyed position. The reference station can then broadcast an appropriate correction to be incorporated in the users' navigation processors. The system has the potential of removing errors caused by wide-area ionospheric disturbances, as well as satellite orbit parameter errors. The latter are of particular interest because of the proposal by the Department of Defense to intentionally degrade the transmitted satellite ephemeris parameters as a means of preventing unauthorized use of the most accurate GPS data.[11]

The differential correction may be applied to either the position estimate (navigation domain) or the measurements (measurement domain). In navigation domain methods, ground-station measurements are processed by a navigation filter to provide an estimate of ground-station position. This position estimate is subsequently differenced from the surveyed position of the ground station, thus creating position corrections that can be broadcast to local users. In measurement-domain methods, the pseudoranges measured by the ground station are differenced from the reference slant-ranges computed, using the surveyed site position and the computed position of the satellites in order to produce a set of pseudorange errors or corrections for broadcast. These corrections are applied to the user's measured pseudoranges before the pseudoranges enter his navigation filter. In both methods, the corrections are usually processed through a low-pass filter to remove the high-frequency error component which tends to be dominated by receiver-unique errors in the ground station, mainly receiver noise. The filtered corrections reflect the larger, more slowly drifting biases associated with ionospheric delays and satellite-orbit parameter errors which are common to users over a wide area.

Three DGPS mechanizations were studied in the work reported in References 4 and 5. Based on these studies, the so called "external data link concept" appears to be the most appropriate for supporting precision approaches by helicopters. This is because of its accuracy and because of the simplicity of the airborne and ground components it requires. As a result, NASA is developing this concept for further evaluation.

The external data link scheme consists of a ground-based GPS receiver (fixed GPS monitor) operating at a known, fixed location and transmitting by RF data link to an airborne GPS receiver operating in the local area. System-bias errors common to the GPS are identified in the ground receiver navigation solution, transmitted to the airborne receiver, and subtracted from the airborne solution to achieve a highly accurate airborne navigation fix.

## TEST DESCRIPTION

### *Equipment Description*

The operational layout of facilities for the tests is shown in Figure 1, which illustrates the various components involved. These include the SH-3G helicopter with the GPS Z-set operating on board, a fixed GPS monitor installed in a van parked at a surveyed position near the runway, and the radar and laser tracker for determining the true position and velocity of an aircraft.

### *Aircraft and On-Board Research System*

The test aircraft was the NASA SH-3G helicopter (Fig. 2). The SH-3G is a twin-engine, turbine-powered single-main-rotor helicopter with a maximum gross weight of 8,860 kg. It has a maximum airspeed of 120 knots, and a cruise airspeed of 100 knots. During the flight tests, two pilots and an aircraft crew chief operated the aircraft and its major systems. The GPS navigation information was not displayed in the cockpit; the pilot used other navigation aids to fly the various flight profiles. Figure 2 also shows the relative locations of the GPS antenna, the radar transponder antenna, and the laser retroreflectors.

A single-channel, sequential GPS receiver (Z-set) was installed and flown in the SH-3G. This receiver operates only on the less precise C/A code signal from the NAVSTAR satellites. The Z-set is one component of the SH-3G airborne system being developed to support the NASA DGPS experiments. The airborne system is rack-mounted in the aircraft cabin. The Z-set is interfaced with an

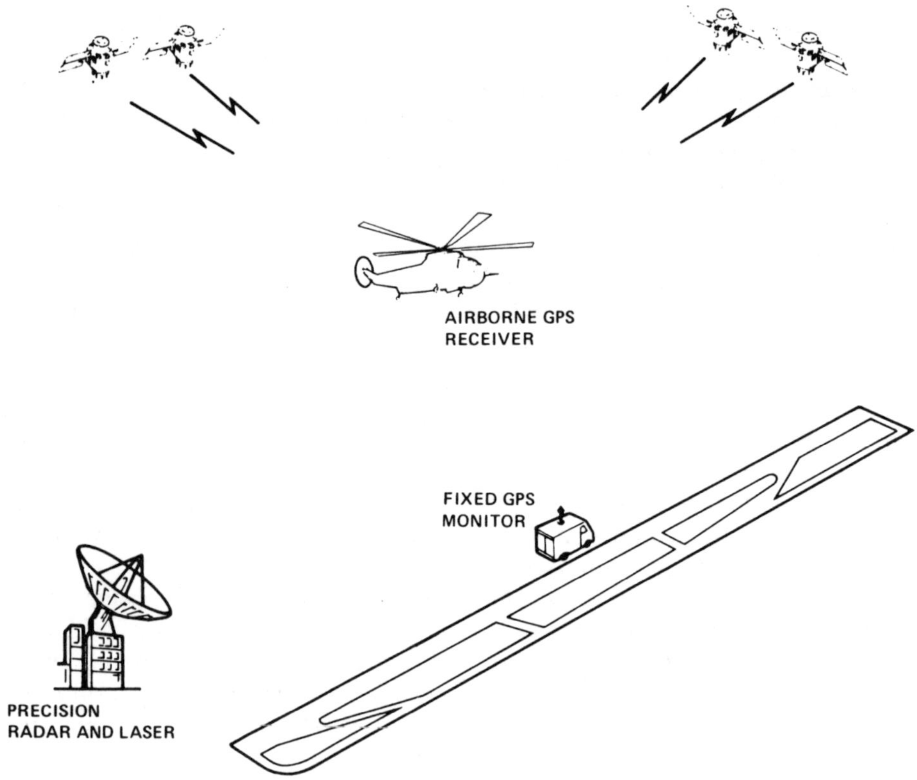


*Fig. 1—DGPS Flight Experiment*

*Fig. 2—SH-3G Helicopter*

on-board PDP-11/34M research computer through a specially built electrical interface unit which provides two-way communications with the Z-set. Through an operator terminal, an experimenter can either write information into the Z-set memory or select an array of data parameters to be acquired from the Z-set for transmittal to the computer. This data transfer does not interfere with normal operation of the Z-set. For the subject flight tests, the research computer was used for data acquisition, and reformatting, and to provide input to a telemetry system for transmittal to a ground receiving station. Normally two experimenters were on board to operate the test equipment and to coordinate operations between the ground and the helicopter. During these flights, there was no uplink telemetry capability, and there was no means of processing differential information on board the aircraft. These two capabilities are currently under development, however. On future flights, an airborne differential navigation algorithm will be installed in the research computer. Pseudorange measurements will be obtained directly from the Z-set receiver and processed in the computer. The navigation algorithm internal to the Z-set processor will be bypassed and all differential navigation processing will be done in the research computer.

### *GPS Ground Reference System*

A single-channel, sequential GPS locator supplied by the Trimble Navigation Corporation (Mountain View, CA) was used as the ground-reference receiver during the test. This receiver, a forerunner of the current Model 4000A four-channel receiver, provided high-accuracy measurements of range and range-rate, the satellite navigation message, and GPS time. The receiver operates on the C/A code transmission from the NAVSTAR satellites. It could be preset to acquire designated satellites by specifying the signal code and anticipated Doppler. The receiver sequenced through four satellites in approximately 8 sec, dwelling on each satellite for about 0.5 sec to obtain an average value of pseudorange for the tracking interval. A major portion of the sequencing interval was allocated for system management tasks, including data acquisition. The navigation receiver was interfaced with an HP 9836 desk-top computer, which was used for operator control of the receiver and to acquire and store test data on a microfloppy mass storage system. The system was installed in a van, with the system antenna mounted on the van roof. Figure 3 shows the van parked at the DGPS reference site.

### *Range Tracking System*

The flight tests were conducted at Ames Research Center's facility at Crows Landing, California (about 50 miles east of Ames). The facility has two ground-based X-band radars for position and velocity tracking of aircraft, and a Sylvania Optrac laser tracker that provides very precise position information. The SH-3G helicopter was flown from Moffett Field to Crows Landing for each day's testing activities, and then returned to Moffett. Radar tracking commenced when the aircraft arrived within visual range of the test area (about 15 miles) and continued uninterrupted until the aircraft departed the area at completion of the day's test. The aircraft was equipped with an X-band transponder to assist in initial radar acquisition and to help maintain radar tracking during aircraft maneuvering. Two laser retroreflectors were mounted on the aircraft, one on each side. The laser was used to obtain the precise position of the aircraft from the beginning of the terminal portion of an approach until touchdown.

The measured 2-$\sigma$ accuracy of the tracking radars was 4.3 m in range and 0.6 mrad in azimuth and elevation. In the vicinity of the runway coordinate

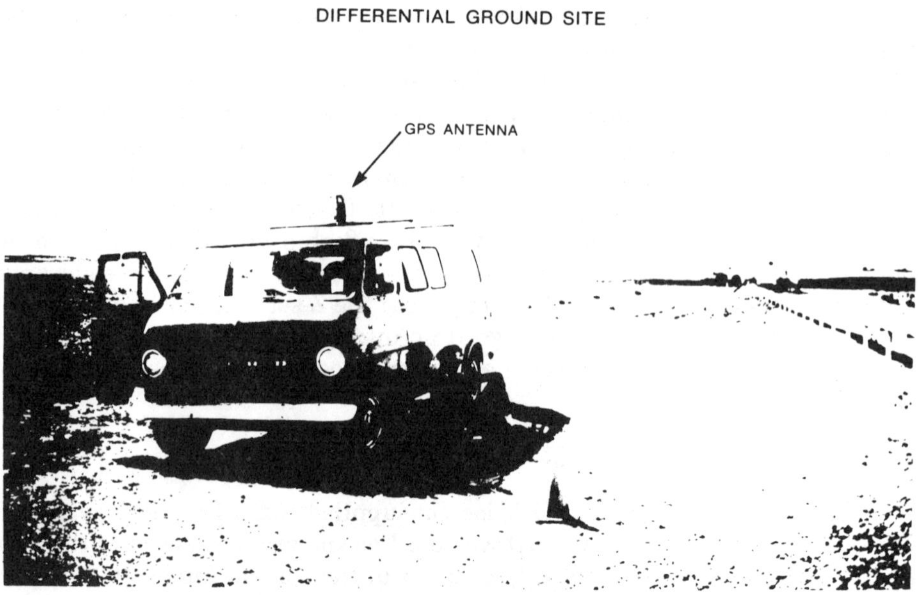


*Fig. 3—DGPS Reference Site*

system origin, this translates into about 3 m horizontally and 1 m vertically. At the extremes of flight operation, the aircraft was roughly 10 km from the origin, and the position accuracy was about 12 m along the radar line of sight (usually parallel to the runway) and 7 m perpendicular to the line of sight. Laser accuracy was about 0.3 m in range and 0.2 mrad in azimuth and elevation. During the terminal part of the approach this corresponds to position errors less than 0.5 m in all three axes.

## TEST PROCEDURES

### Flight Operations

At the beginning of each day's testing, the aircraft was parked on the Crows Landing runway at the coordinate system origin for a short period of time while the radar was tracking and the GPS Z-set navigating. This condition provided about 5–6 min of static data during each test day. These data were used for calibrating the tracking systems and for identifying biases in the GPS navigation solutions between the airborne and ground-reference system. Also during this period, a navigation message was acquired from each of the four satellites being tracked at both the ground reference site and on board the aircraft.

The flight operations included two closed racetrack paths consisting of an ascending departure, a downwind constant-altitude portion, and a descending approach to an altitude of about 50 m. Other operations included constant-altitude flyovers and high-speed, constant-altitude 360° turning maneuvers. The approaches were used to determine the accuracy of the DGPS navigation during typical helicopter approach operations. A visual approach slope indicator (VASI) light system was set up adjacent to the runway approach point to assist the pilot in flying the approach path. The GPS navigation data were not available to the pilot during these operations. The duration of each day's flight activities was about 1.5 hr.

### *Satellite Operational Status*

The satellite constellation consisted of the NAVSTAR satellites designated SV-6, SV-8, SV-9, and SV-11. The geometric arrangement of these four satellites relative to the test site is a major factor in determining the resulting position error in the GPS navigation solution. The effect of geometry on the position error is expressed by geometric dilution of precision (GDOP), a simple measure of the position-error sensitivity to range (pseudorange) error. (For a detailed discussion of this parameter and its significance, see Ref. 10.) The geometric relationship of the four satellites relative to the user location is shown in Figure 4, along with the corresponding GDOP history during the flight test.

Figure 4a presents the elevation and azimuth traces of the satellite vehicles over the duration of the 1.5-hr test; the resulting GDOP is shown in Figure 4b. The position of the satellites during the tests is indicated by the heavy arrows in Figure 4a. Time ticks are shown at 20-min intervals along the test portions of the satellite paths. The time span of the test is represented in Figure 4b by the "test operation" dimension.

The GDOP for the test satellite constellation was below 8 at the beginning of the test and remained at approximately that level throughout most of the test; it increased to above 10 only at the end of the test (Fig. 4b). These GDOP values indicate marginal satellite geometry conditions; favorable GDOP values are generally considered to be less than 6. A factor influencing accuracy that is not reflected in the GDOP is that one satellite, SV-9, is very low on the horizon during much of the test. Although a low-elevation satellite often contributes to good GDOP, it has other accuracy related draw-backs—principally, large atmospheric delays that cannot be modeled.

Each satellite is normally uploaded with current navigation data, such as the satellite ephemeris, once each day as it rises above the horizon as seen from Vandenberg AFB (the Master Control Station). Figure 4 shows when these uploads actually occurred for the specific flight described in this paper. Satellites SV-6 and SV-8 were uploaded at about 20 min into the test, and satellites SV-9 and SV-11 uploaded about midway through the test. The navigation message was collected before and after the uploads for use in the postmission processing. The plan was to use the appropriate ephemeris data during those time periods in which they were applicable.

## DATA ACQUISITION

The ground-reference station outputs a time-tagged sequence of pseudoranges and pseudorange rates for each satellite. These measurements are corrected in the receiver for satellite clock and frequency corrections before output and recording, but are otherwise raw measurements from the receiver. These data were recorded at the site and processed after the test.

The airborne Z-set data were acquired on board the aircraft and telemetered

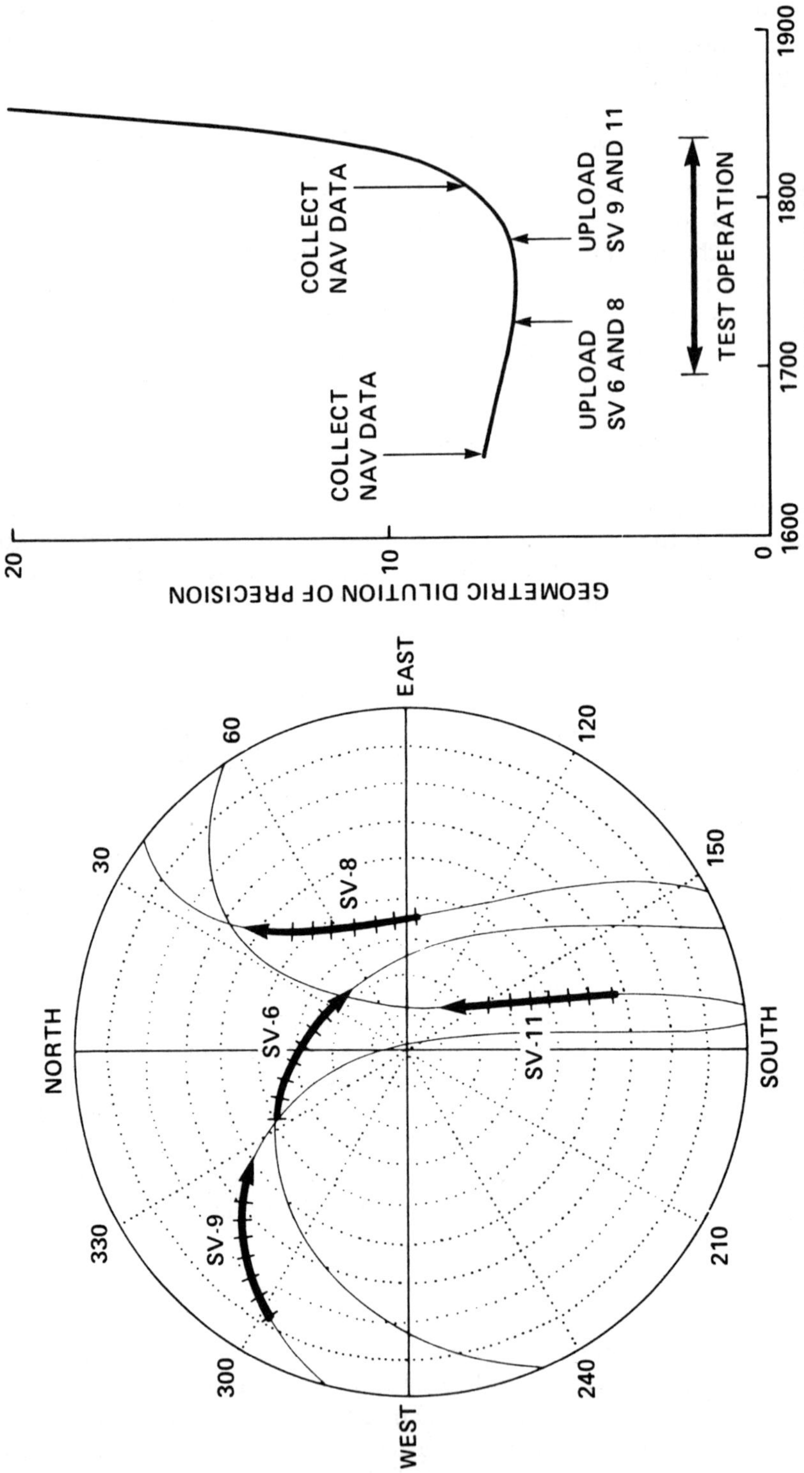


*Fig. 4—NAVSTAR GPS Satellite Constellation During Experiment (a) Elevation and Azimuth Traces of Satellites (b) GDOP of Four-satellite Constellation*

to a ground-receiving station where they were merged in real-time with the range-tracking data. The GPS data are available from the Z-set in a series of data events. A data event comprises a preselected set of data parameters. The events were available at different rates, depending on which parameters were included and the frequency of update of those parameters within the receiver or processor. A special user-defined event was prepared from a combination of parameters that appear in several other events. These data included outputs from the Z-set at the basic satellite sequencing rate of 1.2 sec per sample. This event included raw-measurement data from the Z-set receiver (e.g., signal time of transmission, time of reception, GPS time) combined with navigation-solution data from the Z-set processor (e.g., position and velocity estimates). Unfortunately, the interface between the Z-set and computer did not operate properly during the test, and the sequential satellite data for many time frames were missed. The resulting set of satellite-measurement data was in an apparent random order because of the missing time frames. The original intent of these tests was to collect data for analysis of both measurement-domain and navigation-domain DGPS corrections, but this was not possible because of the interface problem. Only a navigation-domain solution is provided. Accuracy performance for navigation-domain and measurement-domain corrections is about the same in most cases,[2,3] so the navigation-domain results presented here are representative of the performance of both methods. The problem with the interface has since been resolved and efforts are continuing to acquire data for a measurement-domain DGPS solution.

The ground-based tracking radar provided position data in a runway coordinate system (RCS), with the origin located at a point on the centerline of the Crows Landing STOL runway. Data were available for x, y, and z distance from the origin. The RCS frame is a right-hand Cartesian coordinate system, with the x-axis aligned with the centerline of the runway toward the north, the y-axis toward the east, and the z-axis downward. The information from the GPS receivers is available in both Earth-centered/Earth-fixed (ECEF) and WGS-72 coordinates of latitude, longitude, and altitude above the ellipsoid. These GPS data from both the ground-reference receiver and the airborne receiver were transformed into the RCS for the analysis.

## DATA ANALYSIS TECHNIQUES

The navigation-domain DGPS processing is illustrated in Figure 5. The figure shows that the navigation-domain correction was created through use of a Kalman filter that processed the raw pseudorange measurements and computed satellite positions from the ephemeris to produce a time-history of the ground-station position. The filter-state model comprised eight states: three components of antenna position, a clock bias, three components of antenna velocity, and a clock-bias drift. The filter was designed to follow slow drifts in position caused by the slow changes in the ionospheric delays and satellite orbit errors. The navigation position solution was expected to drift at a rate of no more than 5 m/min, so the limits on the Kalman filter drift parameters were set at twice that amount per measurement interval, or at an approximate 20-cm standard deviation. This level of variability is high enough to retain the low-frequency components of the pseudorange error, yet low enough to filter out most of the

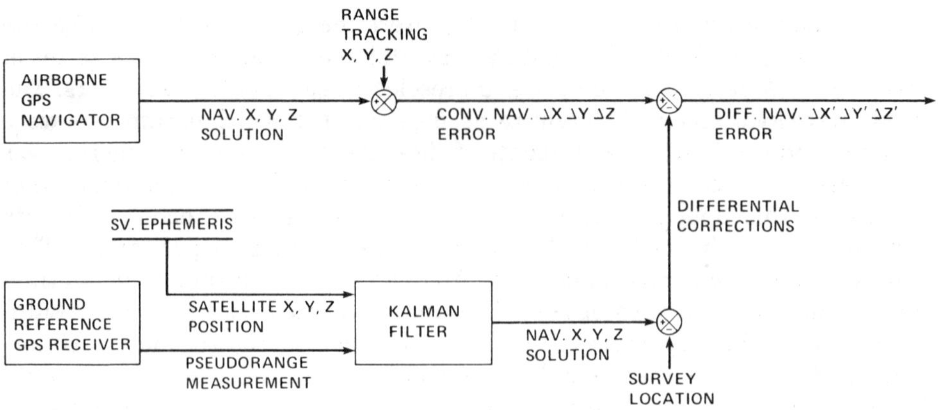


*Fig. 5—DGPS Postflight Processing*

high-frequency receiver-unique measurement noise, which has a standard deviation 20 times higher.

To provide accurate corrections for the user receiver, the filter that processes the ground data must be matched to the filter in the airborne receiver. First, the satellite ephemeris used in the ground-station solution needs to be the same as that used by the airborne receiver. This was of some concern since the satellites were uploaded at different times during the flight test (Fig. 4), and the two receivers may not have absorbed the new ephemeris data simultaneously. To alleviate this problem, and as a computational convenience, a position history for each of the satellites was reconstructed from the Z-set data recordings to ensure that the orbit data were consistent for both receivers throughout the data-processing period. This computational convenience would not be available in a real-time differential test.

Second, because the Z-set processor performs a tropospheric delay correction in arriving at the position estimates, the ground station performed a consistent tropospheric delay correction using the Z-set correction formulas. (The Z-set had no ionospheric-delay correction model, so none was included in the ground-station filter.)

No velocity-domain corrections were attempted in this test. The Z-set measures velocity by backward-differencing successive pseudoranges, measured at the beginning and at the end of the tracking interval. The velocity measure is thus correlated with the position measurement and, moreover, is of poor accuracy. The standard deviation of Z-set delta-pseudorange measurement noise is approximately 1200 cm/sec, which is quite poor considering that the velocities are changing at rates up to 5 m/sec$^2$ (0.5 g), or 6 m/sec per measurement update. In contrast, the velocity corrections coming from the ground station are of the order of 10 cm/sec and would have little positive effect on DGPS navigation accuracy in this flight test. The problem is further compounded because the velocity measurement from the Trimble receiver is a direct measurement of carrier Doppler and is not a backward difference of pseudoranges. These two types of velocity measurements have slightly different behavior under the influence of ionospheric delay variations,[12] creating a potential for systematic errors when the velocity corrections are incorporated.

The process of editing, smoothing, and time-synchronization required to merge ground radar or laser data with the airborne data is described in Reference 7 and will not be discussed here. For determining the navigation errors, the radar or laser positions were considered as "true" positions, and differences between the airborne position estimates and the positions measured by the tracking stations were attributed to errors in the airborne GPS system.

Within the precision desired for these tests, it was significant that the antenna location for the airborne GPS receiver was different from the point on the aircraft which was being tracked by the ground systems. The GPS antenna was installed at the aft end of the aircraft, atop the horizontal stabilizer and adjacent to the tail rotor (Fig. 2). The point tracked by the ground-based radar was a radar transponder antenna located on the underside of the aircraft, 9.3 m forward and 3.7 m below the GPS antenna. The aircraft was also equipped with two laser reflectors, one on each side, mounted on the sponsons. During terminal approach, the laser tracker provided very precise position information relative to these reflectors. These tracking data were corrected for the offsets so that for analytical purposes all data are referenced to the GPS antenna location.

Figure 5 shows the mechanization for determining conventional GPS and DGPS navigation errors. As shown in the figure, the conventional GPS errors are computed by differencing the Z-set navigation solution in the RCS from the radar or laser determined "true" position in the same coordinate frame. The DGPS navigation errors are determined by adding the differential corrections to the conventional errors. The differential corrections were determined as shown in the lower part of Figure 5 and as described above.

## TEST RESULTS

### *Ground-Reference Differential Corrections*

Whether using navigation- or measurement-domain differential correction methods, the corrections will be functions of the pseudorange errors measured at the ground-station receiver. In examining the data, it is instructive to begin with the unfiltered ground-station pseudorange data.

Figure 6 shows ground-station pseudorange-error histories for the four satellites. The pseudorange error is determined by differencing (1) the computed slant range between the ground site and the computed position of the satellite, from (2) the receiver-measured pseudorange as determined by scaling the signal transit time by the speed of light. These values are corrected for the receiver clock error by subtracting a filtered estimate of the receiver clock bias. The residual from the computation includes the effects of the locally common errors, such as satellite clock errors, orbit parameter errors, and atmospheric delays, and also of the receiver-unique errors, such as multipath and tracking noise. At each point in time, the set of four pseudorange errors contains the essential information for computing the differential corrections.

The plots in Figure 6 show that the behavior of the pseudorange error is very linear with respect to time, with slopes of about 10 m/hr or less. There are a few short-term excursions, but they are always less than 5 or 6 m in amplitude. It is not known whether these short-term excursions are due to errors common to local users or are receiver-unique phenomena. They tend to be overshadowed

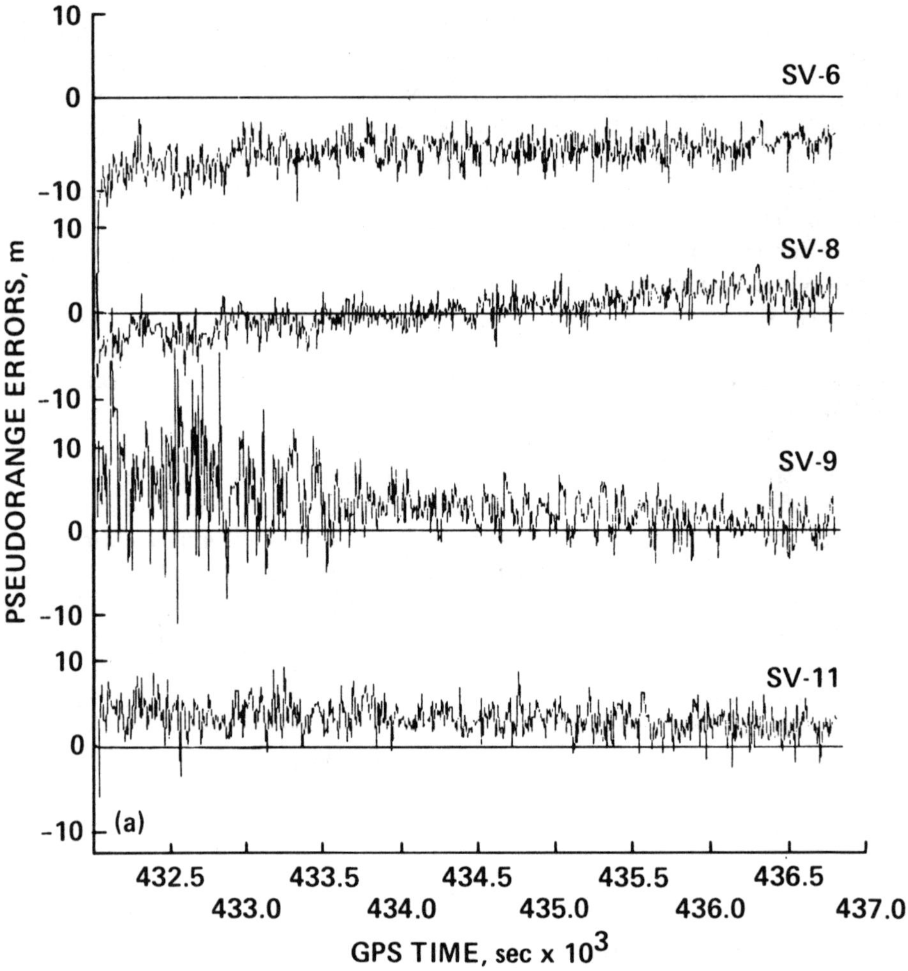


*Fig. 6—Reference Receiver Pseudorange Measurements During Test*

by the random measurement noise variation, which has an amplitude of about ±4 m for most of the satellites. The one exception is SV-9, which shows markedly different behavior—about double the random variation of the other satellites—during the first 1,000 sec. This is attributed to the low elevation of SV-9, which during the period increased from 10° to 20°. The low elevation resulted in higher sensitivity to atmospheric variation and multipath and a generally higher noise level. The high level of noise was effectively filtered out by the ground-station processing and does not seriously affect the differential corrections.

The variation of pseudorange errors plotted in Figure 6 provides some insight into the level of filtering necessary to remove the random variation in the pseudorange measurements without losing too much of the underlying dynamics. The higher-frequency random variations appearing in the measurement traces were assumed to be receiver-unique noise and to be uncorrelated with pseudorange error in the other local receivers. Actually, this was not pure speculation, for results from a separate correlation analysis (not shown here) showed that essentially zero cross-correlation exists between the pseudorange measurement errors computed from the Z-set and the Trimble receiver, for frequencies greater than 0.1 rad/sec. The effect of not removing the higher-frequency component from the differential corrections is one of doubling measurement noise variance for the user. If the differential correction filtering is too heavy (narrow-band low-pass) it removes not only the measurement noise but also medium-frequency dynamics that may contain important movements in the locally common error. Several higher bandwidth filters were used on the flight-test data to determine whether less filtering would improve differential accuracy. The effect was a small decrease in the resulting differential navigation accuracy, so the filter bandwidth was left unchanged.

Tuning of the differential correction filter will be studied in more detail before the next series of flight tests in an effort to achieve a better match of filter structure and parameters to the pseudorange error dynamics. Unfortunately, the applicability of these studies depends to a large extent on the amplitude and frequency of the pseudorange error dynamics which may be introduced by Selective Availability (SA). If SA adds large-amplitude, medium-frequency error dynamics to the signal, the differential correction processor will have to be adjusted to ensure that these dynamics are tracked and included in the corrections.

In interpreting the plots in Figure 6, it should be noted that because of the characteristics of the filter that produced the clock-bias estimate, pseudorange errors are shown relative to a benchmark value. This value (zero on the vertical scale) is a geometry-weighted average of the four pseudorange errors at the time and has essentially no effect on the navigation solution.

The differential navigation correction time-histories for the flight test are shown in Figure 7 in the RCS frame. The $\pm 4$ m pseudorange measurement noise that appears in Figure 6 has been effectively removed by the filter in the differential correction processing. The vertical axis correction is uniformly larger than the horizontal throughout the interval, indicating that GPS conventional navigation is worst in that direction. In contrast to the behavior of the pseudorange errors, the differential corrections are not very time-linear, especially in the vertical coordinate. This is due to the changes in satellite geometry, which affect not only GDOP but also the functional relationship between pseudorange errors and the navigation-domain corrections. Note (Fig. 7) that as the GDOP becomes worse toward the end of the 5,000-sec test period, the differential corrections become larger, especially in the vertical coordinate.

The differential correction histories shown in Figure 7 are based on the assumption that continuous and instantaneous corrections are computed by the ground station and transferred to the user with none of the delays expected in a real-time computer/data-link system. Realistically, this process takes several seconds, during which time the differential-correction accuracy degrades as satellite geometry and pseudorange errors change. As shown in the figure, the rate of change of the differential corrections is very slow for this set of flight-test data, and delay in uplinking the corrections results in a navigation error growth rate of only a few centimeters per second. At this rate, uplinking once per minute would hold this source of error in the differential corrections to the order of 2 m. Unfortunately, SA may increase the error dynamics to the

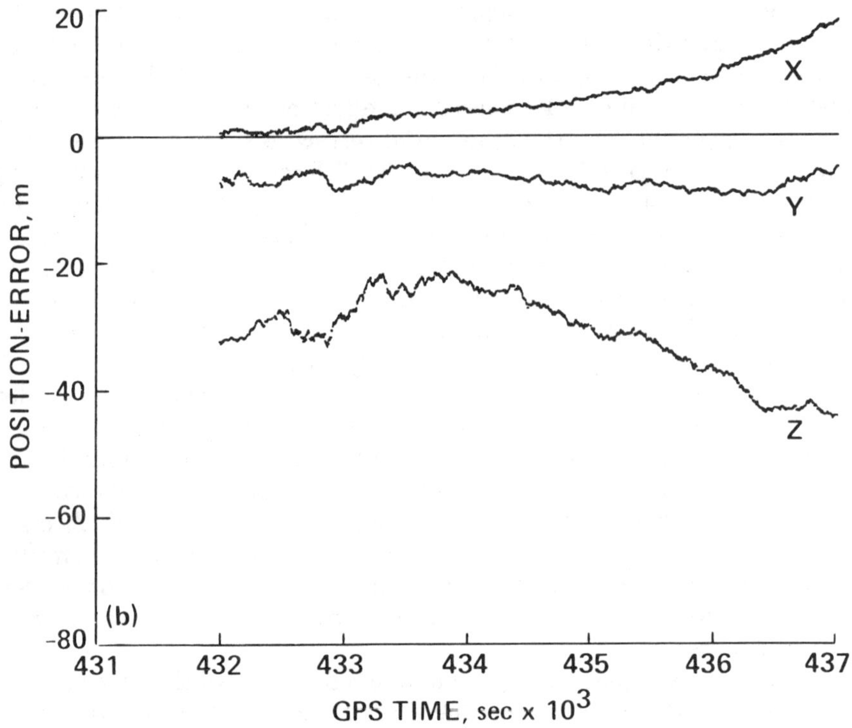


*Fig. 7—Differential GPS Corrections*

point where more frequent uplinks are required. A study by Kalafus[11] indicates that SA as proposed may have differential position-correction error growth rates as high as 1 m/sec for three-dimensional navigation and, therefore, require much higher uplink rates to meet accuracy requirements. As a consequence, uplink rates will remain open to change until more is known about SA.

### *Airborne Navigation GPS Solution*

Figures 8 and 9 show the GPS navigation errors in the lateral (x,y) directions and the vertical (z), respectively, for both conventional GPS and DGPS solutions. The time-histories of the x, y, and z navigation errors are shown for the entire 1.5-hr flight duration of the test. Also included on the left side of each figure is the time-history of the GPS ground-reference corrections. These curves are repeated from Figure 7 for direct comparison with the airborne navigation errors. It is interesting that the differential corrections are mirror reflections of the low-frequency variation of the airborne navigation errors. This of course should not be too surprising and is the necessary result if the differential corrections are to compensate for the bias errors.

The high-frequency oscillations in the airborne Z-set solution are primarily the result of the unmodeled aircraft dynamics in the Z-set Kalman filter. These variations correlated directly with the high dynamic turning maneuvers which occurred at intervals during the test. The eight-state Z-set filter does not model accelerations and as a result tends to "wander off" during the turning maneu-

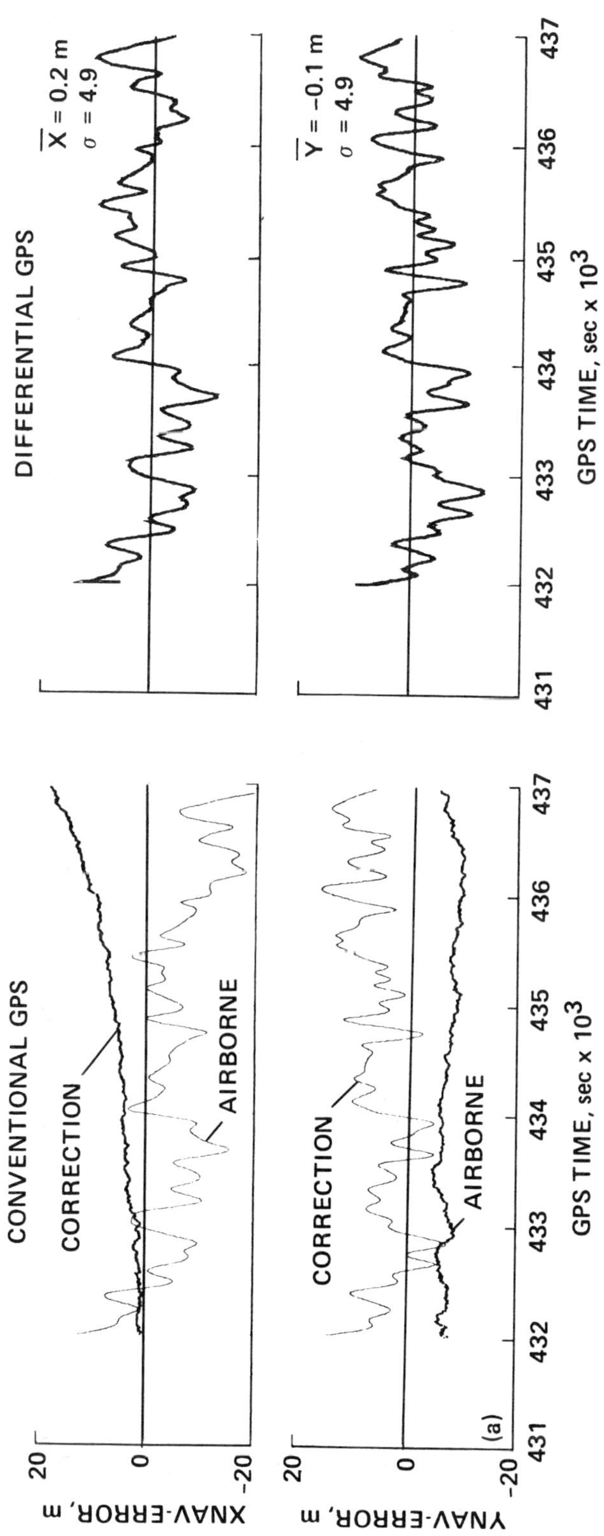


*Fig. 8—GPS Lateral Navigation Errors*

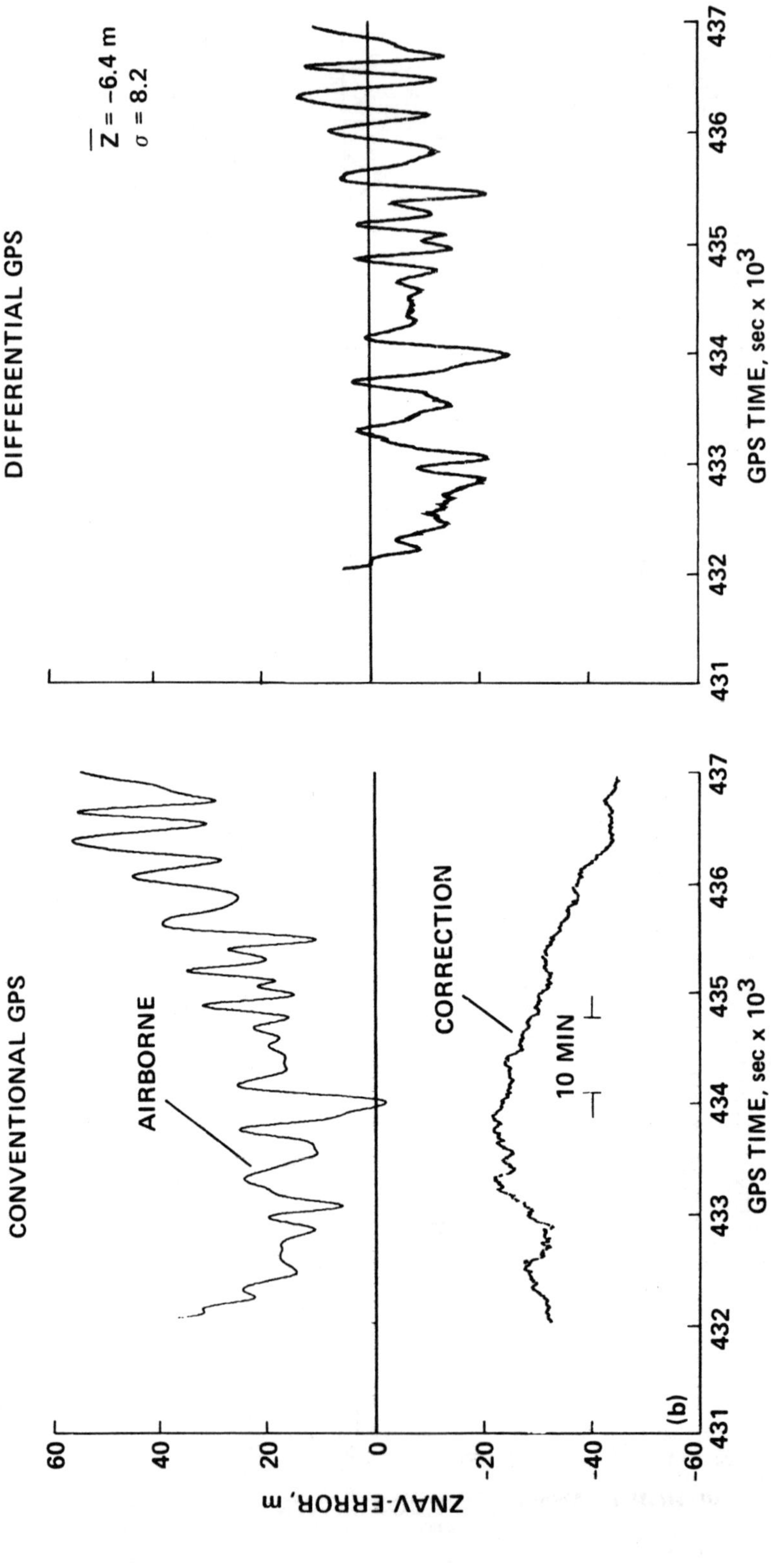


*Fig. 9—GPS Vertical Navigation Errors*

vers. It is also possible that a portion of the high-frequency variations results from the Z-set internal receiver noise which was found to be higher than that of the Trimble receiver (Fig. 7). This is due in part to the wider bandwidth tracking filter required for the Z-set receiver to operate successfully in a dynamic environment.

When the differential corrections are applied to the conventional solution, the resulting differential solution provides excellent performance, as shown in the plots at the right of Figures 8 and 9. For each axis, there is a significant reduction in the resulting navigation error. The improvement is quite noticeable on the vertical axis, where the conventional navigation error of 40 m is reduced to −6.4 m. Bias in both horizontal axes is reduced to less than 1 m. Since there is very little high-frequency correlation between the conventional solution and the differential correction signal, the standard deviation of the errors is essentially unchanged after correction.

Shown in Figure 10 is the total navigation error for both conventional and differential GPS solutions. This measure is the magnitude of the three-dimensional position error shown in the previous figure. It is apparent from the plot that the performance of the DGPS is significantly better than that of the conventional GPS. It is equally significant that the error in the differential solution is essentially independent of satellite geometry (GDOP) during this period, whereas the conventional error follows the variation in GDOP closely. The total conventional error increases gradually from about 20 m to over 40 m during the test, as GDOP (see Fig. 4) increases from about 7 to well over 10.

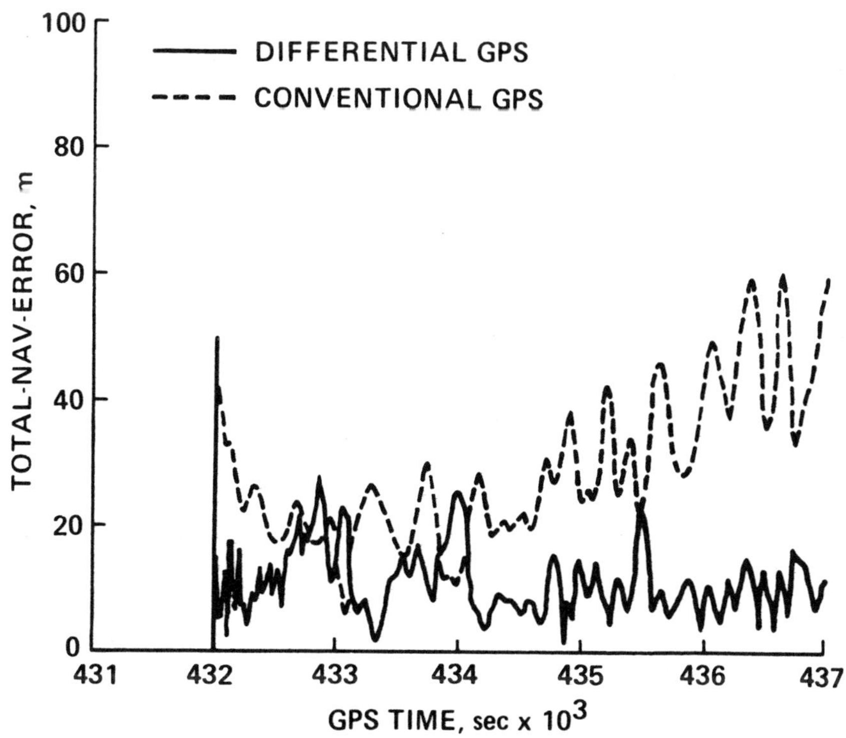


*Fig. 10—Total Navigation Error Comparison*

The total differential error has a mean value of about 10 m during the entire mission and is unaffected by the increasing GDOP. Even with poor satellite geometry, DGPS provides exceptional performance.

This illustrates quite graphically that DGPS is so effective that the locally common navigation errors are diminished to the point where the receiver-unique errors dominate. Since these receiver-unique errors (noise, multipath, and filter mismodeling in the case of airborne receivers) tend to be independent of GDOP, the effect of GDOP on navigation error is effectively masked. This is in contrast to conventional GPS navigation, in which the dominant source of navigation error is pseudorange-measurement error, which is highly GDOP-sensitive. Except when GDOP is extremely high (30 or above) or when receiver-unique errors are extremely low (as may happen with a static user receiver), GDOP-sensitive terms will not dominate differential navigation accuracy.

### *Terminal Approach DGPS Navigation*

One of the primary purposes of the tests was to assess the use of DGPS to support terminal-approach operations. Thus, a series of approaches was included in each day's test program as part of the closed "racetrack" operations. Each racetrack operation included a departure climb-out, a downwind leg, and a terminal approach to an altitude of about 50 m above the runway. Vertical and horizontal profiles of a typical terminal approach are presented in Figure 11.

Figure 11 shows the aircraft lateral ground track and the altitude profile

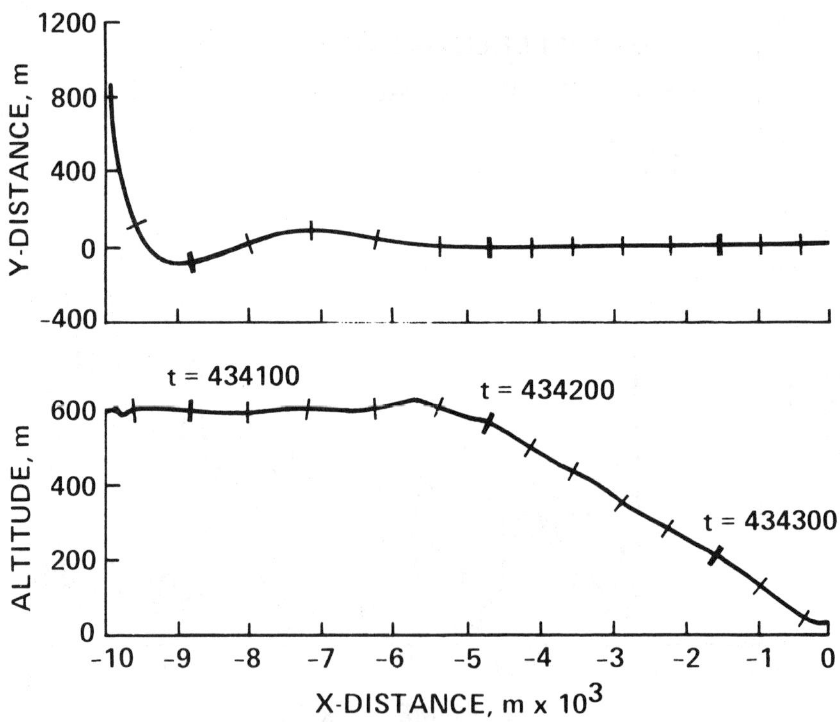


*Fig. 11—Helicopter 6-degree Approach Trajectory*

from final turn until the aircraft passed over the approach point (RCS origin). The aircraft descended on a 6° approach at an airspeed of about 60 knots to runway 17 at the Crows Landing test area. A 4-min segment of data surrounding the terminal approach was excerpted from the 1.5-hr total mission and is shown in Figure 11, with time ticks at 20-sec intervals. In addition to range radar tracking, very precise laser tracker data were acquired for each terminal-approach operation and used in computing the navigation errors.

Figure 12 shows the navigation errors in the RCS for the approach. The three traces on each plot correspond to the conventional GPS navigation from the airborne solution, the differential corrections from the ground-reference solution, and the differential GPS navigation solution as determined from combining the two during postflight analysis. The aircraft position time-history derived from the laser tracker data was used as a reference (true position) for computing the conventional navigation errors. Adding the corrections to the conventional

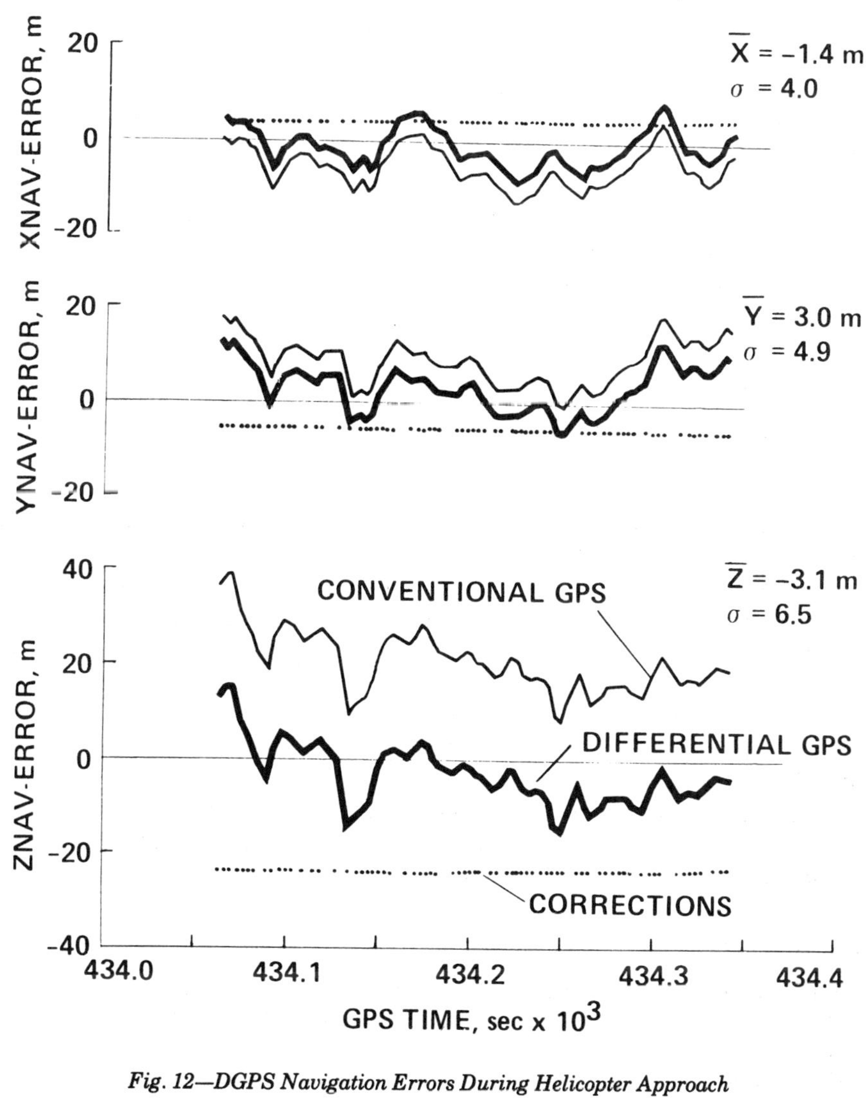


*Fig. 12—DGPS Navigation Errors During Helicopter Approach*

navigation error produces the DGPS error time-history. The mean error for the DGPS solution in each axis is smaller than that for the conventional solution, but the standard deviation is essentially unchanged. The mean and standard deviations of the error in the z-axis are larger than those in the x- and y-axes, both before and after correction. The resulting bias errors for the x-, y-, and z-axes are −1.4, 3.0, and 6.5 m, respectively.

The nearly flat differential corrections shown in Figure 12 illustrate the heavy low-pass filtering occurring in the ground-station filter. One result of this is that the standard deviation of the navigation error was unchanged by the correction. One might suspect that some of the higher-frequency behavior of the differential corrections currently being filtered out might actually reflect locally common errors. Adjusting the ground-station filter parameters to pass higher frequencies showed that this was not correct, however, for the resulting solution actually tended to increase the standard deviation of the differential solution error. This behavior indicated that little or no higher-frequency correlation exists between data sets from the ground and airborne receivers. It is assumed that the Z-set errors are internal receiver errors or that they are due to idiosyncrasies of the adaptive navigation filter.

Aircraft dynamics during the approach were moderate, but they did correlate with the higher-frequency oscillations shown in the three conventional navigation error traces (Fig. 12). Once again it is indicated that these navigation errors result from unmodeled vehicle dynamics in the airborne filter. This suggests that even though the differential performance is quite remarkable, it may be improved still further through more precise vehicle modeling in the airborne filter algorithm.

## CONCLUSIONS

The postflight processing of data from a nonreal-time differential GPS (DGPS) flight experiment conducted with two C/A-code, sequential receivers revealed the following:

1. Application of the differential corrections provides significant improvement over conventional GPS navigation solutions in all axes.
2. The differential correction signal contains only a very small high-frequency component (above 0.1 rad/sec) that can be attributed to common system errors and that would correlate with other GPS receivers operating in the local area.
3. The slow variation in the differential corrections implies that slow update rates (about once per minute) are acceptable for DGPS in the absence of selective availability.
4. DGPS reduces the effect of errors that are sensitive to satellite geometry.
5. Precise filter modeling of the vehicle dynamics may improve DGPS terminal-approach performance for helicopters.

## REFERENCES

1. Macmillian, A. J., *Navstar/Global Positioning System Flight Test Program Overview,* AIAA/SETP et al., First Flight Testing Conference, Las Vegas, Nev., Nov. 1981.
2. Kruczynski, L. R. et al., *Global Positioning System Differential Navigation Tests*

*at the Yuma Proving Ground,* Proceedings of the National Technical Meeting of the Institute of Navigation, Jan. 1981.

3. Connor, J. T., Sawtelle, E. M., and Till, R. D., *Fixed-Wing and Rotary-Wing Flight Testing of Navstar GPS as a Civilian Navigation System,* IEEE National Telecommunication Conference, New Orleans, La., Nov. 1981.
4. Cardall, J. D., Cnossen, R. S., *Civil Application of Differential GPS Using a Single-channel Sequential Receiver,* NASA CR-166168, 1981.
5. Beser, J., Parkinson, B. W., *The Application of Navstar Differential GPS to Civil Helicopter Operations,* Navigation, Vol. 29, No. 2, Summer 1982.
6. Denaro, R. P., Cabak, A. R., *Simulation and Analysis of Differential Global Positioning System for Civil Helicopter Operations,* NASA CR-166534, 1983.
7. Edwards, F. G., Hamlin, J. R., *Operation of a Single-Channel Sequential Navstar GPS Receiver in a Helicopter Mission Environment,* The 39th Annual National Forum of the American Helicopter Society, May 1983.
8. Denaro, R. P., *Application of Differential GPS to Civil Helicopter Terminal Guidance,* Sixth Digital Avionics Conference, Dec. 1984, Baltimore, Md.
9. Denaro, R. P., Beser, J., *Helicopter Flight Test Demonstration of Differential GPS,* Proceedings of the National Technical Meeting of the Institute of Navigation, Jan. 1985.
10. Milliken, R. J., Zoller, C. J., *Principle of Operation of Navstar and System Characteristics,* Navigation, Vol. 25, No. 2, Summer 1978.
11. Kalafus, R. M., *Differential Operation of Navstar GPS,* Navigation, Vol. 30, No. 3, Fall 1983.
12. Jorgensen, P. S., *Ionospheric Measurements from Navstar Satellites,* SAMSO-TR-79-29, 1979.

# Performing Centimeter-Level Surveys in Seconds with GPS Carrier Phase: Initial Results

BENJAMIN W. REMONDI

National Ocean Service, NOAA, Rockville, MD
*Received September 1985*

## ABSTRACT

Two kinematic surveys were performed, using TI-4100s, which substantiate the author's earlier claim that centimeter-accuracy relative surveys can be accomplished within seconds. The first test was carried out in Arkansas using wide-bandwidth tracking. Although noisy, these data permitted the concepts and the data processing software to be verified. Based on results of this survey, a definitive kinematic survey experiment was designed and executed at the National Bureau of Standards, Gaithersburg, MD. The results of this experiment show that centimeter-level relative surveys can be performed in seconds. A priori geodetic coordinates of the initial location of the "roving" antenna are not required since they, too, can be determined within seconds.

## INTRODUCTION

This paper is a continuation of a developmental activity documented in a number of earlier papers.[1,2,3,4] The research is rooted in preliminary discussions at the National Geodetic Survey between Bruce Douglas, James Lucas, and the author, in 1982, with regard to decimeter-level aircraft trajectory determination using the Global Positioning System (GPS) carrier (beat) phase. It was later realized that other investigators had achieved preliminary results in this area.[5] By 1983, when centimeter-accuracy static surveys became a routine activity,[2,6,7] the problem presented itself differently. That is, it seemed productive and natural to spend 1 to 3 hours performing static relative surveys for base lines 10 to 100 km long using L-band carrier phase measurements. However, it seemed unnatural to use that same technology to measure a 50-meter base line vector. This provided the motivation to develop technology for performing kinematic centimeter-level relative surveys in seconds.

Although the theory and programs are similar to a moving platform trajectory application, such as aircraft trajectory determination, there are differences. The following are examples of these differences: (1) Loss of lock is a more serious problem on a moving platform application than on a kinematic survey, so wider bandwidth tracking must be sustained; in kinematic surveying the bandwidth can be collapsed when the target survey mark is reached and subsequently reopened before departure, in order to maintain lock. (2) In kinematic surveying the initial coordinates of the roving antenna can be precisely determined by the simple artifice of moving what is normally the stationary antenna to the initial location of the roving antenna while, at the same time, moving the roving antenna from its initial location to the location of the stationary antenna. This procedure will, hereinafter, be referred to as "exchanging" or "swapping" antennas. Such an exchange requires continuous phase tracking of four or more GPS satellites during its execution and would take a minute, or less, to perform; the antennas might be swapped back to their initial positions for a redundant measurement. This procedure could be performed at the beginning and end of an aircraft trajectory application but it would be more cumbersome. (3) Although the trajectory is determined as a byproduct of kinematic surveying, it is of secondary interest. Consequently, incomplete modeling of receiver processes which result in systematic errors while traveling between marks, but do not result in systematic errors when static mode is achieved (at the survey mark), can be tolerated. When the trajectory, itself, is the object of interest, more attention to the receiver processes is required.

Throughout this paper a somewhat restricted definition of kinematic survey will be employed. It will be assumed that the carrier phase tracking of four or more GPS satellites will be maintained at every instant of the experiment and, specifically, without loss of lock. Kinematic surveying, here, will be restricted to *relative* kinematic surveying, where, except for antenna "exchanges," one antenna will remain stationary and one antenna will be a rover. It should be clear that many of these constraints are not required but are being imposed to facilitate the discussion.

In the following discussion, two kinematic survey experiments are described and the results of the data processing are reported. The objective, here, is to demonstrate that it is possible to perform centimeter-accuracy relative surveys within seconds based on the GPS carrier phase observable. The geodetic instruments employed in these experiments were Texas Instruments, Inc. TI-4100 GPS receivers.

## THE MODEL

The modeling equations upon which the results reported here are based have been documented.[3,4] For this reason only a brief theoretical presentation will be included.

Suppose GPS orbital data and carrier frequencies are errorless. If a GPS receiver is scheduled to take carrier phase measurements at time $t_i$, but these are actually taken at time $t_i + \delta t_i$, then the transmission time of this phase event can be computed as:

$$t^j_{T_i} = t_i + \delta t_i - \tau^j(i)$$

where $\tau^j(i)$ is the signal transit time from the GPS transmitter to the GPS receiver (corrected for Earth rotation during transit). Let us symbolize a "one-way" carrier phase measurement between receiver R and GPS satellite S at true time $t_i + \delta t_i$ by $m^S_R(i)$. Then let us generate a triple difference observation based on satellites j and k, receivers 1 and 2, and receipt times i and i+1 as follows:

$$\begin{aligned} OT_{1,2}(j,k,i,i+1) = {} & [(m^k_2(i+1) - m^k_2(i)) - (m^k_1(i+1) - m^k_1(i))] \\ & - [(m^j_2(i+1) - m^j_2(i)) - (m^j_1(i+1) - m^j_1(i))] \end{aligned}$$

It has been shown in the cited references that a simplified model of this generated observation is (f represents frequency):

$$CT_{1,2}(j,k,i,i+1) = f\cdot [(\tau_1^k(i+1) - \tau_2^k(i+1)) - (\tau_1^k(i) - \tau_2^k(i))] - f\cdot [(\tau_1^j(i+1) - \tau_2^j(i+1)) - (\tau_1^j(i) - \tau_2^j(i))]$$

OT and CT represent the observed and computed triple differences, respectively. This model reflects a GPS receiver which measures satellite phase minus receiver phase; the signs in the model would be reversed for a receiver that measures receiver phase minus satellite phase.

## THE ARKANSAS KINEMATIC SURVEY (JULY 4, 1985)

The National Geodetic Survey's TI-4100 survey team was instructed to establish (with compass and tape) five survey marks at an airport during their survey operation in Arkansas. They used approximately 100-foot spacings between marks and oriented the figure approximately in the cardinal directions. Figure 1 depicts the configuration of marks (A to E).

The stationary GPS receiver was to occupy Mark A for the duration of this experiment (Arkansas, User 4). The moving GPS receiver was to commence tracking at Mark B and occupy Marks B through E according to the sequence B-C-D-E-B-C-D-E-B. L-band carrier phase tracking was maintained on four GPS satellites for the duration of the test. This experiment was later repeated (Arkansas, User 3) using a reduced tracking bandwidth.

The designators "User 3" and "User 4" indicate the dynamic level the receiver software is able to accommodate. User 3 represents 8 Hz tracking, and User 4 represents 16 Hz tracking. In comparison, the bandwidth setting for static mode is typically 0.75 Hz. To translate these into an intuitive realm, 16 Hz

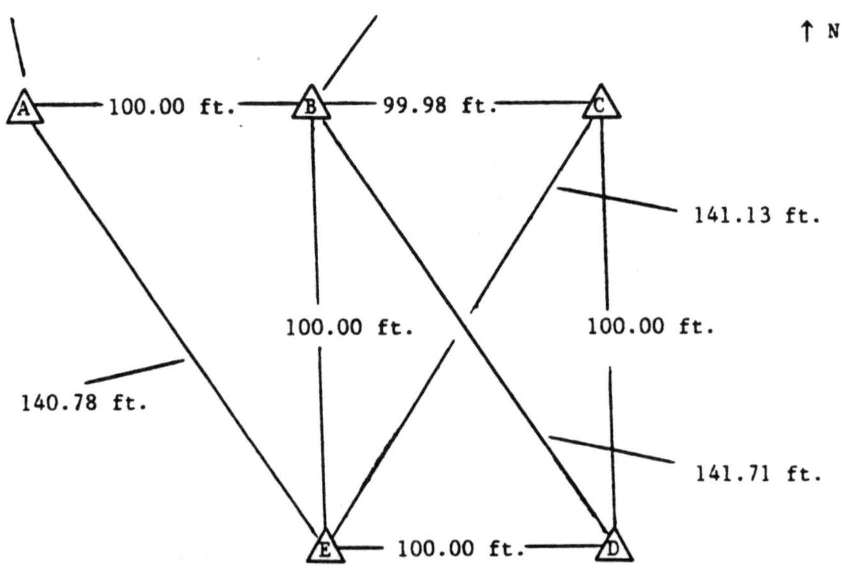


*Fig. 1—Arkansas Test Network, West Helena, AR*

tracking permits approximately 4 g of acceleration (22 Hz is the bandwidth limit before cycle ambiguities predominate).[8,9] Although 4 g of acceleration is quite accommodative for the kinematic surveying application, this level can be exceeded with sudden jerks. Tracking at 8 Hz permits a smooth increase to approximately 1 g of acceleration. In practice it is very difficult to maintain carrier tracking while in motion at 8 Hz.

Figure 2 shows the results of processing data from the Arkansas User 4 experiment. The $L_2$ plots are not included because they are very similar to the $L_1$ plots; however they are are nearly twice as noisy. Figure 2a displays the reduced horizontal trajectory. Figures 2b, 2c, and 2d show the northerly, easterly, and vertical components as a function of time. Figure 2e displays the noise penalty one must pay for wide-bandwidth tracking. Figure 3 is similar to Figure 2 but comprises the results of the Arkansas User 3 experiment. One can see that halving the bandwidth has halved the noise, as expected. In the User 4 experiment the square was circumnavigated twice, whereas in the User 3 experiment the loop was made just once. This accounts for the two north to south and east to west cycles in Figures 2b and 2c and the eight vertical cycles in Figure 2d but only half as many in Figure 3. The horizontal trajectory plots of Figures 2a and 3a reflect this as well. The odd-shaped trajectory seen in Figure 2a reflects the fact that the antenna cable got caught in the survey van.

This experiment was performed to provide data to the investigator to verify processing software and was not intended to be a definitive survey. In fact, geodetic truth was neither desired nor established. The correct geodetic coordinates of the marks could be ascertained, however, if desired. This was not of primary interest because the kinematic survey performed at the National Bureau of Standards (NBS) was designed to answer the question of accuracy. The Arkansas experiment was considered successful in that:

(1) It established 16 Hz as the appropriate bandwidth for this particular geodetic instrument based on its current design.
(2) It pointed to the need to be able to collapse the bandwidth once the antenna reached static mode over a geodetic mark, to reduce the noise.
(3) It pointed to the need to include, in the survey design, an additional step for the determination of the initial geodetic coordinates of the moving receiver.
(4) It pointed to the need to revisit the moving receiver's initial mark (or *any* precisely known mark) between visits to unknown marks. This is considered a temporary measure; it is expected to be unnecessary when the modeling is complete and the technique is mature.
(5) It allowed the data processing software to be verified, and suggested enhancements.

Two final comments with regard to the Arkansas survey. The lack of closure in the vertical component is a result of two factors: (1) coordinates for Mark B were imprecisely known—only the length was precisely known; (2) the simplified model used in the data reduction, reported here, does not account for a small linear drift in the vertical component and a much smaller linear drift in the horizontal components. This point will be discussed later.

## THE NBS KINEMATIC SURVEY (AUGUST 12/13, 1985)

On August 12/13, 1985 a kinematic survey similar to the Arkansas survey was performed at the National Bureau of Standards (NBS), Gaithersburg, MD.

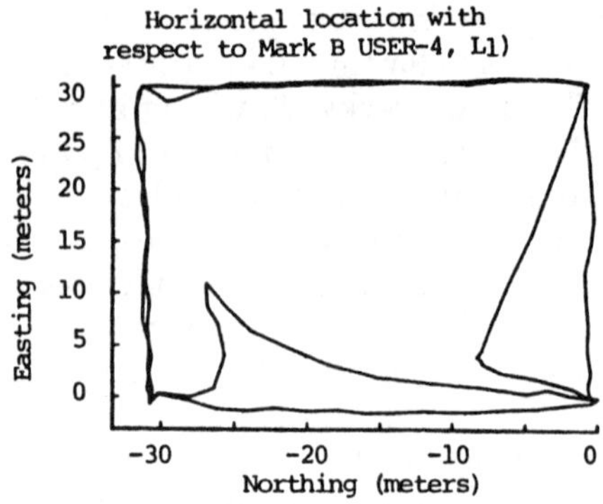


Figure 2a.

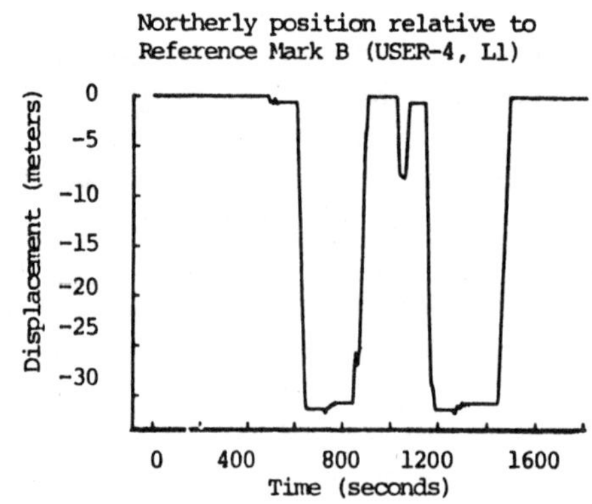


Figure 2b.

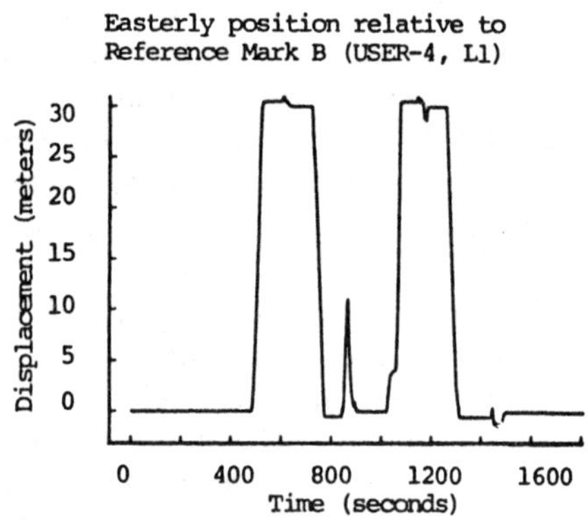


Figure 2c.

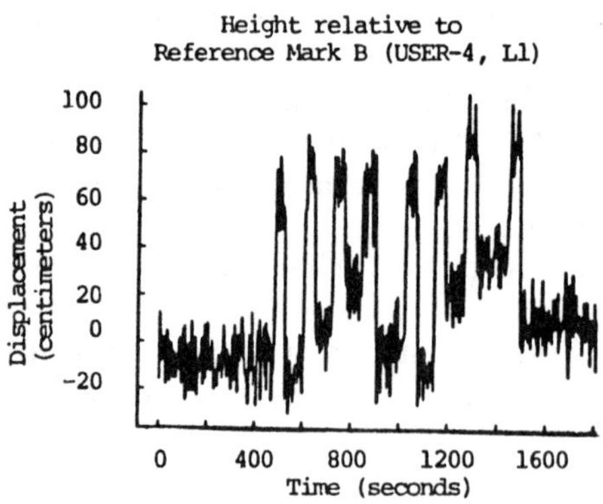


Figure 2d.

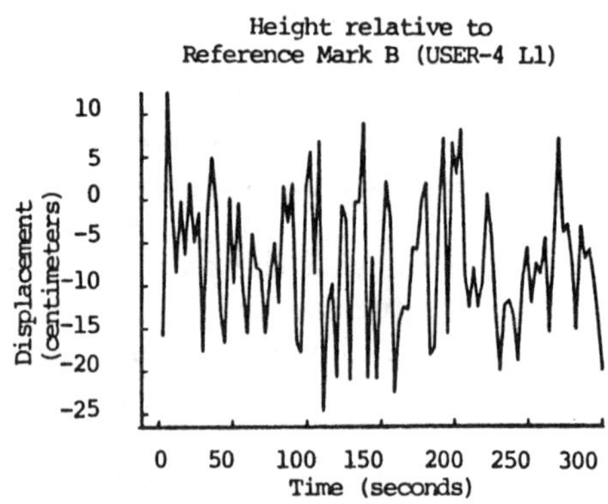


Figure 2e.

*Fig. 2—Arkansas "User 4" Results*

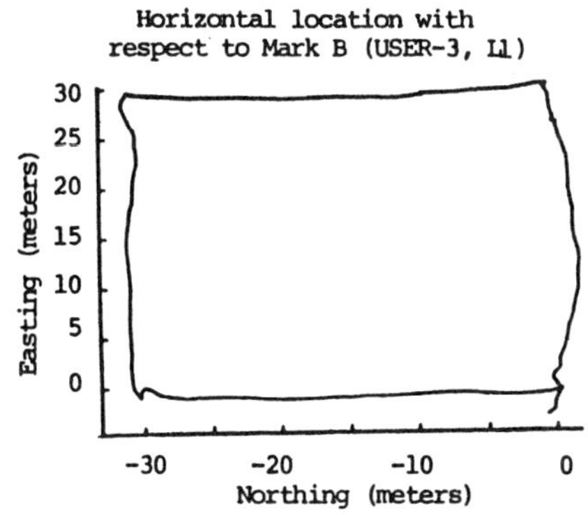


Figure 3a.

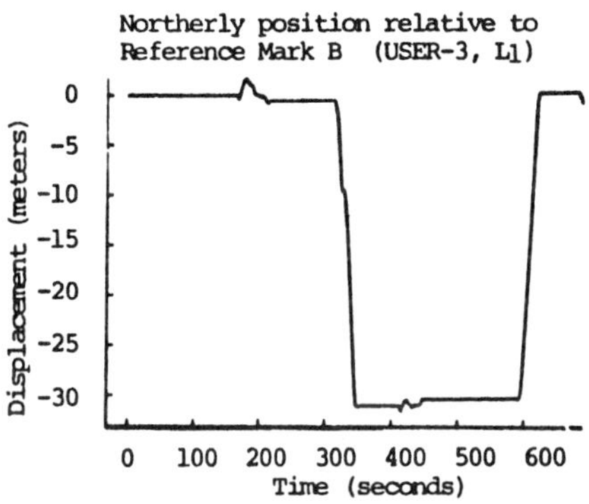


Figure 3b.

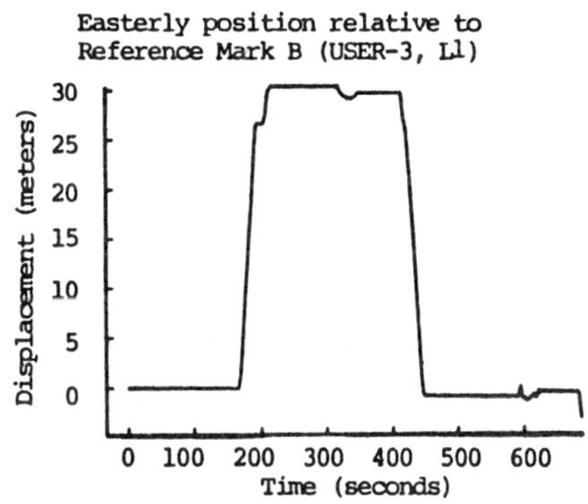


Figure 3c.

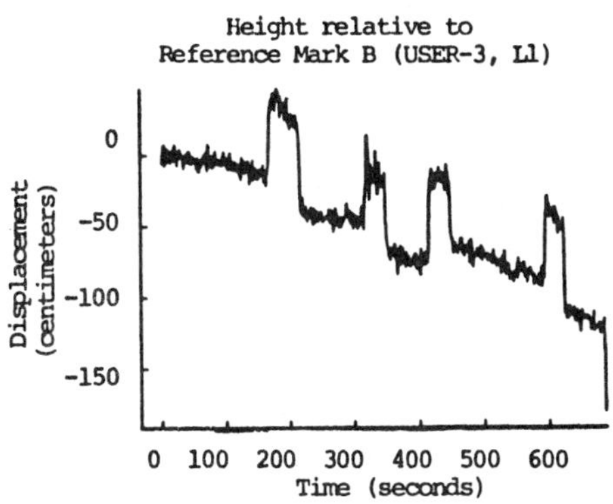


Figure 3d.

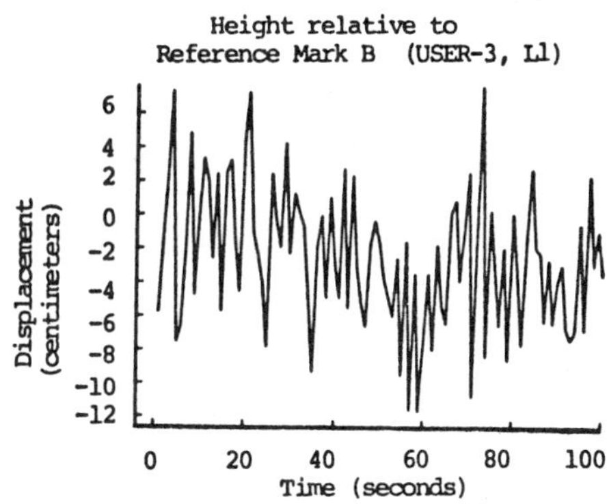


Figure 3e.

*Fig. 3—Arkansas "User 3" Results*

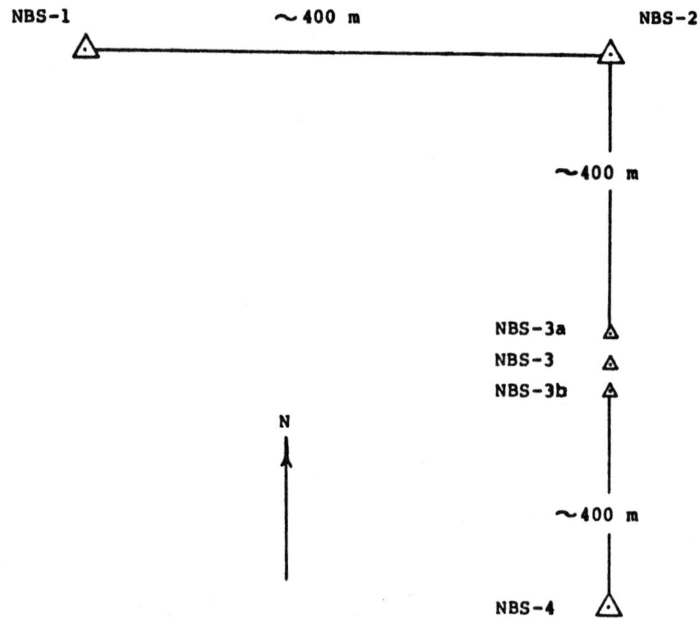


*Fig. 4—Kinematic Survey Test Network at the National Bureau of Standards, Gaithersburg, MD*

Table 1—Geodetic truth (in WGS-72) at the antenna phase center

| Survey Mark | North latitude | East longitude | Ellipsoid height |
|---|---|---|---|
| NBS-1 | 39.133689736 | 282.786495208 | 106.822 |
| NBS-2 | 39.13385466 | 282.790696732 | 105.381 |
| NBS-3 | 39.130829886 | 282.790787626 | 105.774 |
| NBS-3a | 39.130985798 | 282.790782946 | 105.774 |
| NBS-3b | 39.130674044 | 282.790760425 | 105.635 |
| NBS-4 | 39.12760489 | 282.790224959 | 95.564 |

The relationship of the survey marks is depicted in Figure 4. They are within 1 km of each other. Marks NBS-1, NBS-2, NBS-3, and NBS-4 were precisely known a priori. Marks NBS-3a and NBS-3b were established on August 12 using terrestrial methods, specifically as part of this experiment. NBS-3 is in an open field and, for this reason, was chosen as the stationary site. Marks NBS-3a and NBS-3b were established to allow the demonstration of an important aspect of this kinematic surveying method: the ability to determine the starting coordinates of the moving receiver within seconds. We would be free from this minor inconvenience if it were possible for the stationary antenna and the roving antenna to share a common starting point. This aspect of the experiment proved to be completely successful in that the reduced geodetic coordinates of NBS-3b were correctly determined at the millimeter level. The precise coordinates for the NBS survey marks, relevant to this test, are included in Table 1. In all cases the coordinates are known to within a few millimeters.

Receiver UTO9 was initially situated at NBS-3b; receiver UTO4 was initially situated approximately 17 m north at NBS-3. The applications software in the TI-4100 was a modified version of the Arkansas User 4 software with an enhancment provided by Mrs. Laura Hawkins of NSWC which allowed one to toggle between 16 Hz (high dynamic (HD) mode) tracking and 0.75 Hz (low dynamic (LD) mode) tracking. Data were recorded every 5 seconds so as to limit the number of observations. It was felt that minimizing the random component was of little concern at this stage; this improvement will be there when it is required. Table 2 shows the log of times and events.

A few comments are in order. Tracking began in LD mode. For the remainder of the experiment, every departure was preceded by a toggle to HD and every arrival was followed by a toggle back to LD. The only events that are unusual, and deserve special mention, were the four occasions where the antennas of UTO9 and UTO4 were exchanged. Only one of these was required, but repeat measurements seemed appropriate. These antenna swaps provide a quick and accurate way of determining the initial coordinates of the roving antenna. Such a swap yields exactly twice the vector from one mark to the other. This swapping event would normally be the only motion that the otherwise stationary antenna would undergo for the remainder of the survey. The speed of travel between the survey marks did not exceed 10 to 15 miles per hour since the terrain was rough. The procedure was to remove the antenna from the tribrach and hand it to an individual on top of the survey vehicle. Care was exercised so as not to block the GPS signals. It should be mentioned that the receiver display was monitored at all times for the possibility of loss of lock. This was tedious and a loss-of-lock buzzer would be required in an operational mode. No effort was made to orient the antennas in the same direction. This fact, alone, could lead to several millimeters of error.

Figure 5a shows the computed trajectory including the antenna swaps referred to above. As before, only the $L_1$ plots are displayed since the $L_2$ plots are similar. These swapping events yield an exaggerated trajectory of UTO9. The distortion ends as soon as the stationary receiver returns to its site (NBS-3). Figure 5b shows the northerly component of the trajectory including the distortions during the swapping events. Figure 5c provides an expanded scale of the northerly component. Notice, for example, that the 17 m swapping event appears as a 34 m event (first peak). The subsequent event was not a swap; it reflects the transition of UTO9 to NBS-3a while UTO4 was fixed. Just by coincidence NBS-3a is approximately 34 m north of NBS-3b whereas NBS-3 is approximately 17 m north of NBS-3b. The result is that the 17 m swap appears nearly identical to the 34 m transition from NBS-3b to NBS-3a. Figure 5d provides yet another scale change. The transitions between 16 Hz and 0.75 Hz tracking are apparent.

Figure 6 displays information similar to Figure 5 for the easterly and vertical components. The apparent spikes in Figure 6b reflect two facts: (1) NBS-3 is almost directly north of NBS-3b; (2) during the swap UTO9 traveled wide to the east (from NBS-3b to NBS-3) whereas UTO4 traveled wide to the west (from NBS-3 to NBS-3b). The motions, again, appear to be distorted because in the computations UTO4 was assumed to be stationary—even though it was not.

## RESULTS OF THE NBS KINEMATIC SURVEY

Only the geodetic coordinates of NBS-3 were assumed in the data reduction. Even this was not required since the point position solution, based on pseudor-

Table 2—Log of events for kinematic survey at National Bureau of Standards

| | Event | Date/Time |
|---|---|---|
| (1) | UTO4 and UTO9 begin tracking in LD mode. | 1985/08/12/23/09/55 |
| (2) | UTO4 and UTO9 go into HD mode. | 23/17/00 |
| (3) | UTO4 and UTO9 go into LD mode. | 23/18/50 |
| (4) | UTO4 and UTO9 go into HD mode. | 23/20/05 |
| (5) | UTO4 departs from NBS-3 for NBS-3b. | 23/21/00 |
| (6) | UTO9 departs from NBS-3b for NBS-3. | 23/21/00 |
| (7) | UTO4 arrives at NBS-3b. | 23/21/40 |
| (8) | UTO9 arrives at NBS-3. | 23/21/40 |
| (9) | UTO4 and UTO9 go into LD mode. | 23/21/45 |
| (10) | UTO4 and UTO9 go into HD mode. | 23/22/35 |
| (11) | UTO4 departs from NBS-3b for NBS-3. | 23/22/55 |
| (12) | UTO9 departs from NBS-3 for NBS-3b. | 23/22/55 |
| (13) | UTO4 arrives at NBS-3. | 23/23/30 |
| (14) | UTO9 arrives at NBS-3b. | 23/23/30 |
| (15) | UTO4 and UTO9 go into LD mode. | 23/23/45 |
| (16) | UTO9 goes into HD mode. | 23/24/35 |
| (17) | UTO9 departs from NBS-3b for NBS-3a. | 23/25/00 |
| (18) | UTO9 arrives at NBS-3a. | 23/25/45 |
| (19) | UTO9 goes into LD mode. | 23/25/55 |
| (20) | UTO9 goes into HD mode. | 23/27/30 |
| (21) | UTO9 departs from NBS-3a for NBS-4. | 23/28/00 |
| (22) | UTO9 arrives at NBS-4. | 23/31/55 |
| (23) | UTO9 goes into LD mode. | 23/32/10 |
| (24) | UTO9 goes into HD mode. | 23/33/45 |
| (25) | UTO9 departs from NBS-4 for NBS-3b. | 23/34/00 |
| (26) | UTO9 arrives at NBS-3b. | 23/36/45 |
| (27) | UTO9 goes into LD mode. | 23/36/55 |
| (28) | UTO9 goes into HD mode. | 23/38/40 |
| (29) | UTO9 departs from NBS-3b for NBS-3a. | 23/39/00 |
| (30) | UTO9 arrives at NBS-3a. | 23/40/00 |
| (31) | UTO9 goes into LD mode. | 23/40/05 |
| (32) | UTO9 goes into HD mode. | 23/41/00 |
| (33) | UTO9 departs from NBS-3a for NBS-3b. | 23/41/10 |
| (34) | UTO9 arrives at NBS-3b. | 23/42/15 |
| (35) | UTO9 goes into LD mode. | 23/42/20 |
| (36) | UTO4 and UTO9 go into HD mode. | 23/43/40 |
| (37) | UTO4 departs from NBS-3 for NBS-3b. | 23/44/00 |
| (38) | UTO9 departs from NBS-3b for NBS-3. | 23/44/00 |
| (39) | UTO4 arrives at NBS-3b. | 23/44/40 |
| (40) | UTO9 arrives at NBS-3. | 23/44/40 |
| (41) | UTO4 and UTO9 go into LD mode. | 23/44/45 |
| (42) | UTO4 and UTO9 go into HD mode. | 23/45/30 |
| (43) | UTO4 departs from NBS-3b for NBS-3. | 23/46/00 |
| (44) | UTO9 departs from NBS-3 for NBS-3b. | 23/46/00 |
| (45) | UTO4 arrives at NBS-3. | 23/46/30 |
| (46) | UTO9 arrives at NBS-3b. | 23/46/30 |
| (47) | UTO4 and UTO9 go into LD mode. | 23/46/45 |
| (48) | UTO9 goes into HD mode. | 23/48/20 |
| (49) | UTO9 departs from NBS-3b for NBS-2. | 23/49/00 |
| (50) | UTO9 arrives at NBS-2. | 23/51/55 |
| (51) | UTO9 goes into LD mode. | 23/52/05 |

Table 2—Log of events for kinematic survey at National Bureau of Standards (Cont'd).

| Event | Date/Time |
|---|---|
| (52) UTO9 goes into HD mode. | 23/53/35 |
| (53) UTO9 departs from NBS-2 for NBS-1. | 23/54/00 |
| (54) UTO9 arrives at NBS-1. | 23/56/40 |
| (55) UTO9 goes into LD mode. | 23/56/50 |
| (56) UTO9 goes into HD mode. | 23/57/45 |
| (57) UTO9 departs from NBS-1 for NBS-3a. | 23/58/00 |
| (58) UTO9 arrives at NBS-3a. | 1985/08/13/00/01/30 |
| (59) UTO9 goes into LD mode. | 00/01/45 |
| (60) UTO9 goes into HD mode. | 00/06/45 |

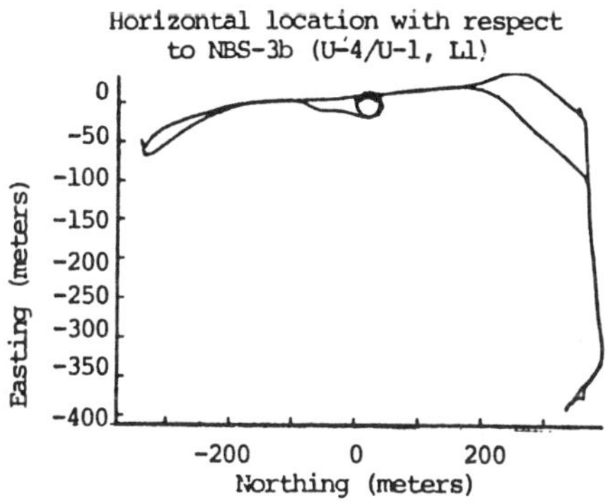


Figure 5a.

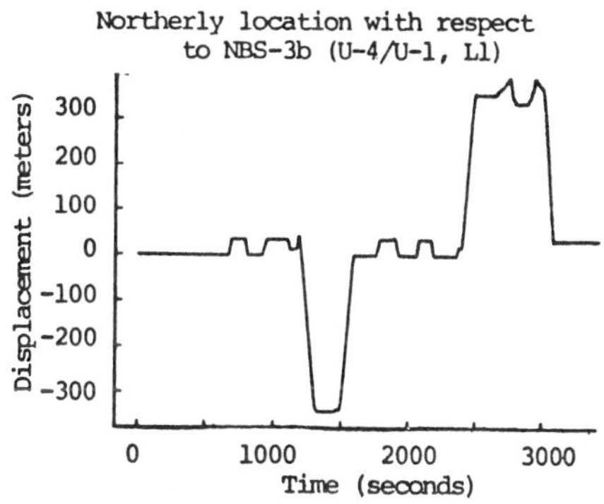


Figure 5b.

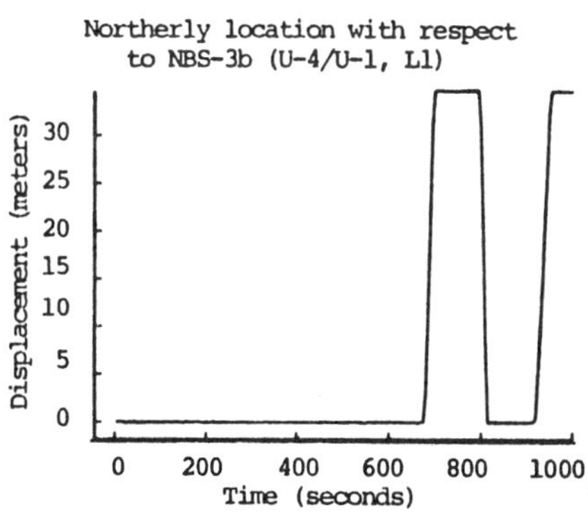


Figure 5c.

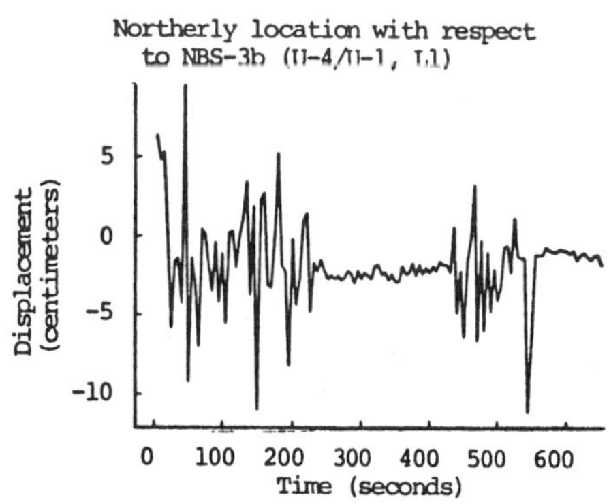


Figure 5d.

*Fig. 5—High Dynamic/Low Dynamic Results from the National Bureau of Standards Test*

anges, would have been satisfactory. Precise NBS-3 coordinates were used for ease in verifying the reduced geodetic positions of the other geodetic marks.

Starting with NBS-3 coordinates, the first data reduction task was to determine the precise coordinates of NBS-3b based on the swapping events where,

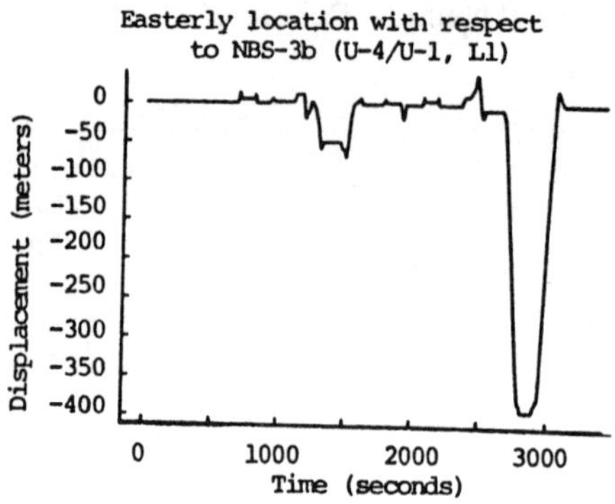


Figure 6a.

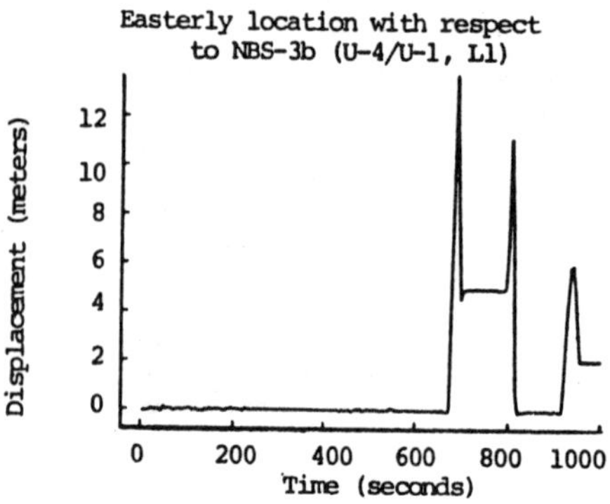


Figure 6b.

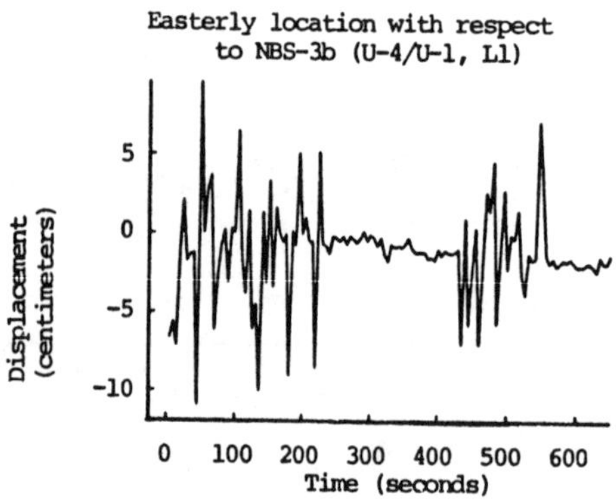


Figure 6c.

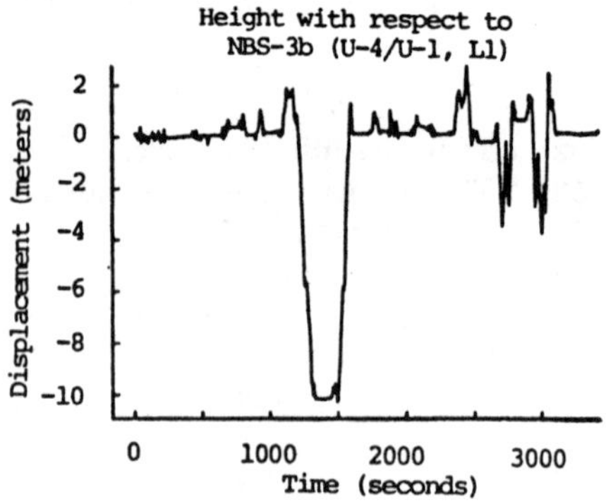


Figure 6d.

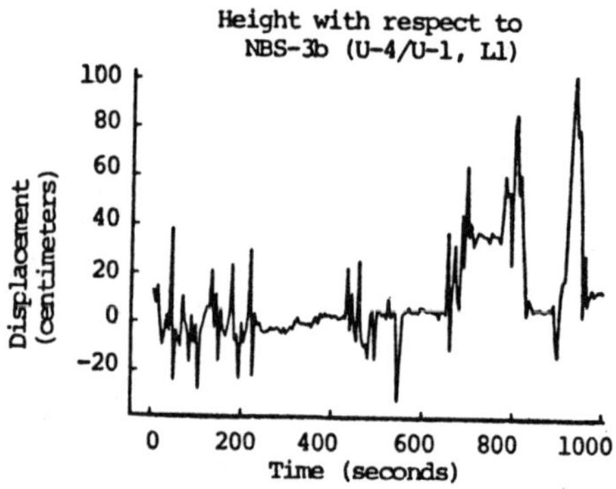


Figure 6e.

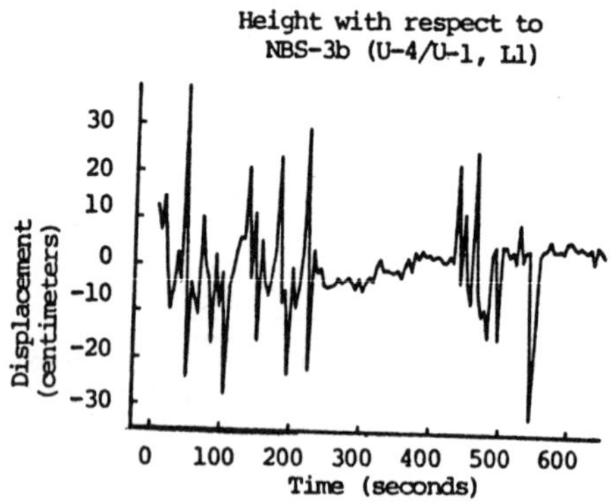


Figure 6f.

*Fig. 6—Additional High Dynamic/Low Dynamic Results from National Bureau of Standards Test*

recall, NBS-3b was the starting location of the kinematic survey. Geodetic coordinates of NBS-3b are subsequently required for the determination of the geodetic locations of the rest of the survey marks visited.

Table 3 includes the four independent determinations of NBS-3b based on these antenna exchanges. Included is the average of these determinations and a comparison with geodetic truth. The data reduction agreed with the geodetic truth at the 2-millimeter level. This is considered fortuitous since the variation was at the 1-centimeter level. It should be pointed out that the ability to determine the initial vector of the roving antenna within seconds is a new and significant result. The theoretical implication is that the integer biases[2,6,7] can be determined within seconds and applied throughout the survey. A more mundane implication is that it is possible to add an azimuth mark within seconds (depending on distance).

The next phase of the data processing requires approximate point position coordinates of the stationary site (NBS-3) and precise relative coordinates of the initial location of the roving antenna. Using the model and the least-squares algorithm given in the cited references, the geodetic locations of the roving antenna were determined at each measurement epoch. These solutions are tabulated, along with their times, only for those times when the receiver achieved narrow-bandwidth tracking over the target geodetic marks. About 10 to 15 seconds are required for the tracking bandwidth to collapse from 16 Hz to 0.75 Hz, as the TI-4100 is currently designed. For this reason, one may depart from a survey mark from 15 to 30 seconds after arrival. Table 4 lists the "true minus computed" values.

An examination of these results reveals a linear drift in the coordinates of NBS-3b. This drift is likely the product of an over-simplified model, and unmodeled clock drift is the suspected error source. Since the coordinates of NBS-3b are now precisely known, one is justified in removing the nearly linear drift in the coordinates of NBS-3b and subsequently applying this correction to the remainder of the survey as well. Although such a correction is expected to yield improved results for similar surveys, the primary purpose, here, is to demonstrate the accuracy one can expect with an improved model and the corresponding processing software. Naturally this will be the next refinement. Table 5 presents the results of Table 4 after correcting for this linear drift. The resulting root-mean-square error is 0.6 cm in latitude, 1.0 cm in longitude, and 2.6 cm in height.

The presented results are based on $L_1$ carrier phase measurements. The

Table 3—Solutions of the initial location of the roving antenna

| Event | N. latitude | E. longitude | Ellipsoid height |
|---|---|---|---|
| swap 1. | 39.130674105 | 282.790760403 | 105.636 |
| swap 2. | 39.130674411 | 282.790760348 | 105.644 |
| swap 3. | 39.130673975 | 282.790760475 | 105.630 |
| swap 4. | 39.130674044 | 282.790760445 | 105.626 |
| average | 39.130674058 | 282.790760415 | 105.634 |
| truth | 39.130674044 | 282.790760425 | 105.635 |
| error | 1.4 mm | 1.0 mm | 2.0 mm |

Table 4—True minus computed geodetic coordinates for NBS test

| Date and time | Survey mark | ΔN lat (cm) | ΔE long (cm) | Δht (cm) |
|---|---|---|---|---|
| 8/12/23/09/55 | NBS-3b * † | 0.0 | 0.0 | 0.0 |
| 23/19/35 | NBS-3b * | 0.3 | 1.9 | −5.3 |
| 23/24/20 | NBS-3b * | 1.8 | 1.2 | −6.2 |
| 23/26/55 | NBS-3a | 1.9 | 1.9 | −7.9 |
| 23/33/10 | NBS-4 | 0.9 | 3.3 | −5.9 |
| 23/38/00 | NBS-3b * | 2.7 | 2.5 | −10.0 |
| 23/40/45 | NBS-3a | 3.6 | 2.7 | −14.8 |
| 23/43/15 | NBS-3b * | 3.4 | 1.9 | −12.4 |
| 23/47/45 | NBS-3b * | 3.7 | 1.1 | −10.0 |
| 23/53/00 | NBS-2 | 6.0 | 0.2 | −16.1 |
| 23/57/25 | NBS-1 | 4.4 | 0.1 | −15.4 |
| 8/13/00/04/00 | NBS-3a | 5.0 | 1.0 | −9.7 |

* Initial location of roving antenna.

† Imposed initial coordinates based on results from the antenna exchanges (This point was not included in the linear fit reported in Table 5.)

Table 5—True minus computed geodetic coordinates for NBS test corrected for the linear drift observed at NBS-3b

| Date and time | Survey mark | ΔN lat (cm) | ΔE long (cm) | Δht (cm) |
|---|---|---|---|---|
| 8/12/23/19/35 | NBS-3b | −1.1 | 0.1 | 0.1 |
| 23/24/20 | NBS-3b | 0.0 | 0.6 | 0.3 |
| 23/26/55 | NBS-3a | 0.2 | 0.2 | 0.8 |
| 23/33/10 | NBS-4 | −1.9 | 1.6 | 2.5 |
| 23/38/00 | NBS-3b | −0.7 | 0.7 | −0.7 |
| 23/40/45 | NBS-3a | −0.1 | 1.0 | −4.7 |
| 23/43/15 | NBS-3b | −0.5 | 0.2 | −1.7 |
| 23/47/45 | NBS-3b | −0.8 | −0.6 | 1.6 |
| 23/53/00 | NBS-2 | 1.0 | −1.5 | −3.2 |
| 23/57/25 | NBS-1 | −1.1 | −1.6 | −1.5 |
| 8/13/00/04/00 | NBS-3a | −0.9 | −0.8 | 4.6 |

results from processing $L_2$ are similar but the contribution due to noise is about twice as large. The $L_2$ linear drift in the NBS-3b coordinates is approximately 1.6 times as large as that of $L_1$ and in the opposite direction. Notice that $f_1/f_2 = 1.283$ and the ratio of the bias frequency plan ($-6000$ Hz for $L_1$; $+7600$ Hz for $L_2$) is 1.27. The product of these quantities is 1.625. For this reason a timing error is suspected. This modeling deficiency will doubtless be corrected but does not obscure the central fact that centimeter-level accuracy is achievable in seconds.

## CONCLUSIONS

Centimeter-level relative positioning in seconds has been achieved. Furthermore, centimeter-accuracy determination of the initial relative vector of the roving antenna, within seconds, has been demonstrated. These results are new and significant. Considering the improvements which would obviously result from tracking 6 to 9 SVs, while using more than one fixed geodetic receiver, and when data are recorded at a high rate, one must conclude that the future is bright for this technique. When the model is upgraded, the artifice of revisiting the initial location of the roving antenna and removing the small linear drift will be unnecessary. Since the trajectory of the roving antenna, in this application, is of no particular interest, the main results reported here do not require the *recording* of data during the transition between marks. In fact, it makes sense to record measurements at a high rate at the survey marks once narrow-bandwidth tracking begins. Since the number of measurements can be kept small, they can be transferred on a real time basis. Not only does this permit real-time centimeter-level surveying in seconds, but also it would allow the user to navigate to within centimeters of a desired location in real time!

## ACKNOWLEDGMENTS

The author acknowledges the technical support provided by Stanley Meyerhoff, Laura Hawkins, and Michael Sims of the Naval Surface Weapons Center, the technical advice provided by Dennis Henson of Texas Instruments, Inc., and the technical guidance provided by Harold Beard, Dennis Hoar, and Michael Mathwig of NGS in the planning and execution of the kinematic surveys.

## DISCLAIMER

Mention of a commercial company or product does not constitute an endorsement by the National Oceanic and Atmospheric Administration. Use for publicity or advertisement purposes of information from this publication concerning proprietary products or the test of such products is not authorized.

## REFERENCES

1. Hofmann-Wellenhof, B. and Remondi, B. W., *Determination of the Trajectory of a Moving Platform Using GPS Carrier Phase.* Federation Internationale des Géomètres, joint meeting Study Groups 5B and 5C on Inertial, Doppler, and GPS Measurements for National and Engineering Surveys. Munich, Federal Republic of Germany, July 1–3, 1985.
2. Remondi, B. W., *Using the Global Positioning System (GPS) Phase Observable for Relative Geodesy: Modeling, Processing, and Results,* Ph.D. Dissertation (Center for Space Research, The University of Texas at Austin, Austin, Texas), 1984.
3. Remondi, B. W., *Performing Centimeter Accuracy Relative Surveys in Seconds using GPS Carrier Phase.* Proceedings of the First International Symposium on Precise Positioning with the Global Positioning System, (NOAA), Rockville, Maryland, April 15–19, pp. 789–797, 1985.
4. Remondi, B. W., *Global Positioning System Carrier Phase: Description and Use,* NOAA, Technical Memorandum NOS NGS-42, 1985.
5. Evans, A. G.; Hermann, B. R., and Fell, P. J.,: *Global Positioning System Sensitivity Experiment,* Navigation, Vol. 28, No. 2, pp. 74–84, Summer, 1981.
6. Bock, Y.; Abbot, R. I.; Counselman, C. C.; Gourevitch, S. A.; King, R. W., and Paradis, A. R.; *Geodetic Accuracy of the Macrometer™ Model V-1000.* Bulletin Geodesique Vol. 58, No. 2, pp. 211–221, 1984.

7. Goad, C. C. and Remondi, B. W., *Initial Relative Positioning Results Using the Global Positioning System.* Bulletin Geodesique, Vol. 58, No. 2, pp. 193–210, 1984.
8. Henson, D., Texas Instruments, Inc., Lewisville, TX., Private communication, 1985.
9. Sims, M., Naval Surface Weapons Center, Dahlgren, Va. Private communication, 1985.

# Operational Benefits and Design Approaches for Combining JTIDS and GPS Navigation

WALTER R. FRIED

Hughes Aircraft Company, Fullerton, CA

*Received January 1984*

## ABSTRACT

This paper describes the Joint Tactical Information Distribution System (JTIDS) Navigation Function and the Global Positioning System (GPS), emphasizing their technical similarities and differences. Based on these, the paper discusses the operational benefits, performance, and design approaches for combining the navigational and time data from these two systems. Both JTIDS and GPS are time synchronous, spread spectrum systems which provide high accuracy navigation data through the use of time-of-arrival measurements and pseudo-ranging techniques. Both are subject to Geometric Dilution of Precision (GDOP) effects, though under different conditions; and both are highly resistent to interference, though to different degrees. JTIDS is more a tactical or local system, while GPS is global in nature. Exchanging data between the two systems, under certain constraints, therefore has the potential of great performance and operational benefits with respect to system availability, under a variety of operational scenarios and interference conditions. These benefits and proposed methods for combining the data from the two systems are described in the paper.

## *INTRODUCTION*

Both the Joint Tactical Information Distribution System (JTIDS) and the Global Positioning System (GPS) have been under development by the U.S. Department of Defense for a number of years, and both are currently in the full-scale development (FSD) phase, preparatory to production and operational deployment. JTIDS is a tactical, integrated communication-navigation-identification (CNI) system which includes a high accuracy navigation function. The function has been called Relative Navigation (RELNAV); however, it operates simultaneously in both an absolute, geodetic (latitude-longitude-altitude) mode and in a relative grid (U-V-W) mode. It uses a spread spectrum waveform, time-of-arrival (TOA) measurements and pseudo-ranging techniques to provide continuous position, velocity and time information. The principles of operation and system details of JTIDS/RELNAV have been described in papers by Fried[1,2,3,20], Stow,[4] Bivin and Soloway,[5] Rome and Stambaugh,[6] Westbrook and Snodgrass,[7] and others. The architecture of JTIDS as a whole has been described in papers by Ellingson,[8] Dell-Imagine,[9] Newman,[10] and Rubin.[11] GPS is a satellite-based, global, absolute navigation system, which also uses a spread spectrum waveform, TOA measurements and pseudo-ranging techniques to provide continuous position, velocity and time information. The principles of operation and system design details of GPS are described in a series of papers which have been published as a compendium by the Institute of Navigation.[12]

Since these two systems provide similar navigational data, they have frequently been considered competitive systems. One of the purposes of this paper is to show that they are, in fact, highly complementary systems, and that combining the data from the two systems has both technical and operational benefits. On the other hand, since both systems operate in the L-band and use similar spread spectrum and ranging techniques, it has also been suggested that the two systems could be integrated in a hardware sense; i.e., circuit functions could be time-shared and one common navigation filter could be used[15,16]. Although that approach may have some advantages, it also exhibits a number of important disadvantages.

The overall objective of this paper is to describe the operational and technical benefits for combining the navigation data from GPS and JTIDS, and approaches for accomplishing this. Brief descriptions of JTIDS RELNAV and GPS are provided first; their similarities are highlighted and some of their limitations are discussed. The complementary nature of their operational characteristics is then described, and different methods of combining the data from the two systems are discussed. Both, advantages and disadvantages of actual hardware integration of the two systems are considered. Finally, the operational and performance benefits of combining the data from the two systems are discussed, showing the synergistic nature of this approach.

### *The JTIDS RELNAV Function*

JTIDS is a synchronous, jam-resistant, spread spectrum, secure communication, navigation and identification system. There are two basic implementations of JTIDS: Time Division Multiple Access (TDMA) and Distributed Time Division Multiple Access (DTDMA). In both implementations, a group (net) of users operate on a common, precise time base through use of stable clocks which are synchronized to a time master, called Net Time References (NTR), by means of either an active round trip timing (RTT) technique or a fully passive technique. The latter is actually a by-product of the RELNAV function.

In TDMA, time is divided into epochs (12.8 minutes long), cycles (12 seconds long) and time slots (7.8125 ms long). Each of these time slots are designed to carry a complete message (see Figure 1). The time slot contains a fixed, large number of 6.4 microsecond pulses which are frequency-hopped, and each pulse is divided into 32 chips (representing 5 bits) which are phase coded using MSK (minimum shift keying). The frequency-hopping and phase shift keying uses pseudo random (PR) codes. Thus, TDMA uses two spread spectrum dimensions, i.e., frequency and code (direct sequence spreading). A more detailed description of the JTIDS TDMA waveform and equipment implementation is given in Reference 9.

DTDMA is very similar to TDMA except that it makes use of one more dimension, i.e., time-hopping, in addition to the PR frequency hopping and PR code. This is accomplished by taking the pulses (and hence bits) which represent one complete message and *distributing* them pseudo-randomly over a longer

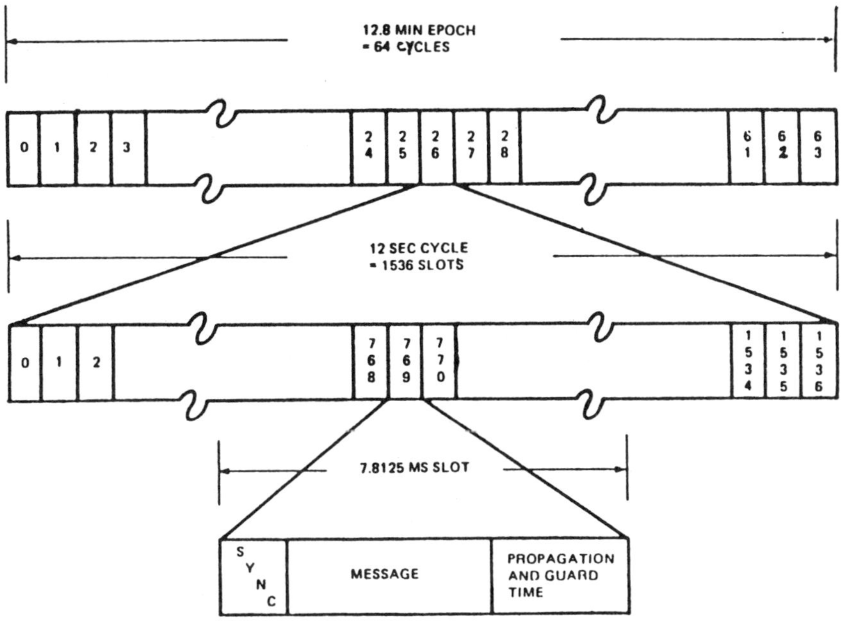


*Fig. 1—JTIDS message timing (TDMA).*

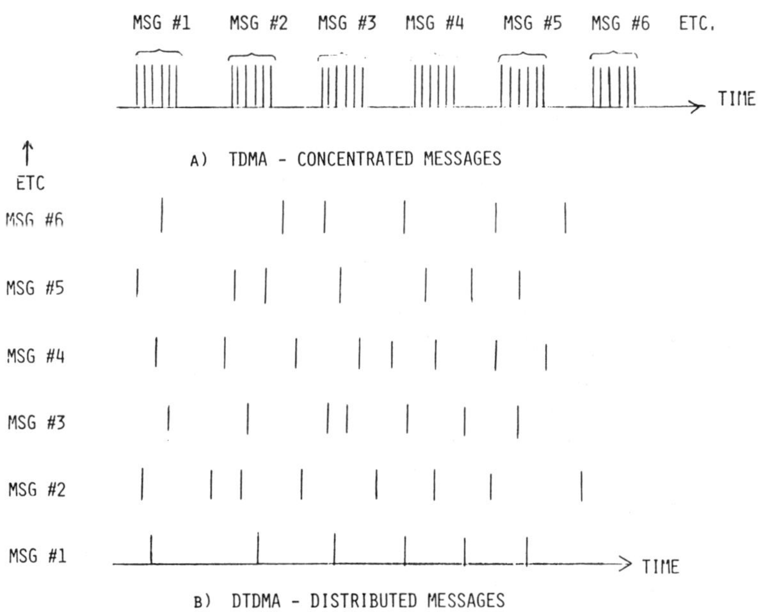


*Fig. 2—TDMA and DTDMA operation.*

time interval, interleaved with pulses conveying other messages, without guard time (see Figure 2). A more detailed description of the DTDMA waveform and equipment implementation is given in Reference 11.

The JTIDS RELNAV function, which will be briefly described next, is essentially identical for the TDMA and DTDMA implementations. (In DTDMA the TOA must be referenced and extrapolated to a particular time, which is done *before* it is provided to the RELNAV software function.)

Two message formats have been used for JTIDS: one is the Interim JTIDS Message Specification (IJMS) and the other is the TADIL-J Message Standard. Again, the operation of RELNAV is the same in both, except that, in the former, the so-called P-messages (position and status messages) and in the latter, the so-called PPLI (Precise Participant Location and Identification) messages are used for the RELNAV function.

The primary purpose of the JTIDS RELNAV function is to determine own unit position and velocity in both absolute and relative coordinates from data received from other terminals. This, in turn, results in placing participating units accurately within a common grid, thereby improving the ability for track correlation of multiple tracks for both airborne and shipborne units. The RELNAV function also provides own terminal time synchronization.

In operation, all active units in a JTIDS net periodically transmit the position messages mentioned earlier, which contain own unit position and velocity (in both coordinates) and the position and time qualities of the transmitting unit. A navigating user receives these messages on a sequential basis and *passively* ranges to the transmitting terminals (sources) by means of the measurement of the time-of-arrival (TOA) of these messages. A sequence of such passive range measurements on messages from sources having higher position and time qualities permits a determination of the user's position, velocity and time offset (see Figure 3). Since range and time errors are not separable in a single passive ranging event, at least four such pseudo-range measurements are required to determine the user's three-dimensional position.

The ranges are called pseudo ranges, since the range measurements are not *true* ranges, but are made with respect to the user's own clock time, rather than on an absolute basis. The basic RELNAV observation equation is given by (Ref. 1):

$$R_o = c(\mathrm{TOA}) = [(X_s - X)^2 + (Y_s - Y)^2 + (Z_s - Z)^2]^{1/2} - b \cdot c \qquad (1)$$

where $R_o$ is the observed (pseudo) range; $X_s$, $Y_s$, $Z_s$ are the source's transmitted position coordinates; $X$, $Y$, $Z$ are the user's own position coordinates; $b$ is the user's time bias and $c$ is the speed of light. With the source positions known from the data in the received messages, several measured pseudoranges lead

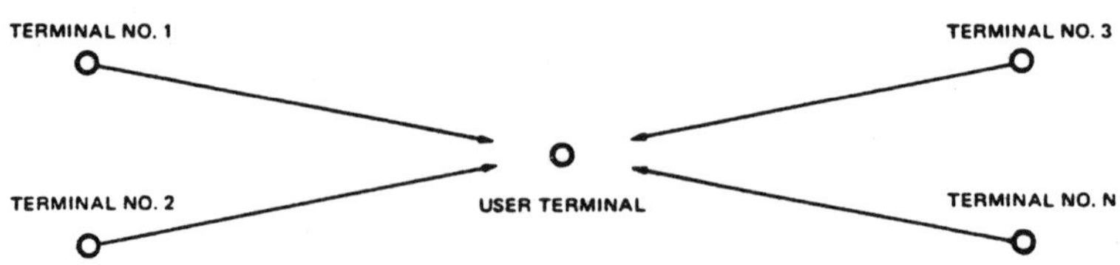


- SEQUENTIAL PASSIVE TIME-OF-ARRIVAL (TOA) MEASUREMENTS (PSEUDO-RANGES) FROM 3 OR MORE SOURCES BY THE USER TERMINAL LEAD TO A DETERMINATION OF OWN POSITION AND TIME
- PSEUDO-RANGES ARE MEASURED WITH RESPECT TO OWN CLOCK TIME
- ROUND TRIP TIMING (RTT) IS USED BY SOME UNITS TO MAINTAIN THE HIGHEST TIME QUALITY
- THE RELNAV FILTER COMBINES THE TOA DATA WITH DEAD RECKONING (E.G. INERTIAL) DATA FOR EXTRAPOLATION, PREDICTION AND CALIBRATION OF THE DEAD RECKONER ERRORS

*Fig. 3—Basic JTIDS RELNAV principles.*

to a determination of the user's own position coordinates and the time bias. Since the date is obtained sequentially, recursive (e.g., Kalman) filtering is used in typical RELNAV implementations and dead reckoner (e.g., inertial) data is mixed with the TOA data for smoothing and prediction purposes (Figure 4). The RELNAV Kalman filter can then inherently estimate the dead reckoner errors using the accurate TOA data as reference data, thereby continuously "calibrating out" these dead reckoner errors.

It is also possible to operate RELNAV using TOA-derived data only, i.e., without interconnection to any dead reckoner (see Reference 1). This form of RELNAV operation is more applicable to relatively low dynamic vehicles, such as ships, ground vehicles, or low dynamic aircraft. In high dynamic vehicles, this mode of operation would lead to excessive lags in position determination during maneuvers.

The major RELNAV processing functions are shown in Figure 5. All received P-messages are screened in the source selection function, which first rejects those sources whose position and time qualities are inferior to those of the user. From the P-messages which have survived these tests, those which will improve the user's position based on the user's own estimated error ellipse are selected. The associated message TOAs and source position data are then passed to the filter function for processing. Similarly, P-messages are selected on the basis of potential improvement in the user's knowledge of time, altitude and azimuth.

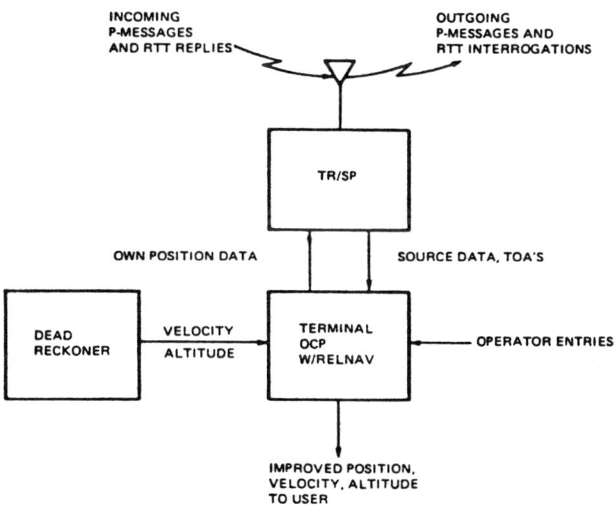


*Fig. 4—JTIDS RELNAV operation.*

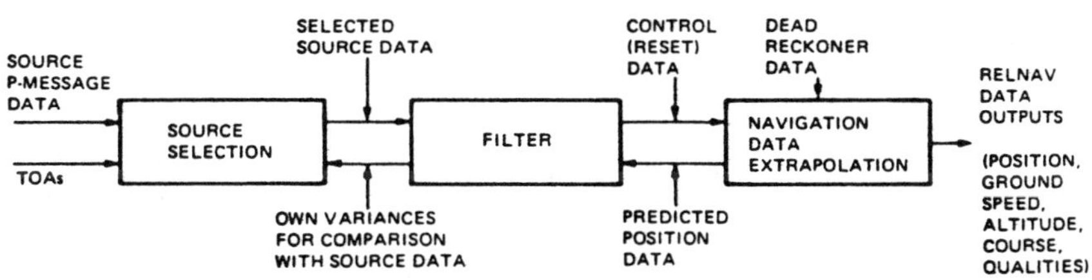


*Fig. 5—Major RELNAV processing functions.*

(See References 2, 3 and 20). In the source selection function, the filter-estimated position and time variances are used for comparison with the equivalent reported source variances (Figure 5). The RELNAV filter then calculates predicted TOAs from extrapolated (predicted) position obtained from the navigation data extrapolation (dead reckoning) function and from the received source position data. It compares the measured and predicted TOAs to generate residuals (observables). Using each of these observations, the filter updates the various states of the filter, such as position, velocity and time errors, as well as the covariance matrix. After several such observations have been used, the filter applies corrections (controls) to the navigation data extrapolation function, which continues to extrapolate the navigation data with these corrections, for use in transmitted P-messages, vehicle avionics and for subsequent filter prediction purposes. Those control data from the filter which represent clock corrections, i.e., time bias and frequency errors, are provided to the terminal clock via the synch maintenance function of the terminal's computer program. The reported time and position qualities of the sources are converted to equivalent variances and are used to modify the appropriate terms in the filter gain calculation for each observation. The user's own position and time variances which are calculated by the filter are, in turn, converted to position and time qualities, to be transmitted on the outgoing position messages, along with RELNAV estimated position, velocity and altitude.

In addition to the basic passive ranging operation just described, the RELNAV function also uses an active or round trip timing (RTT) technique available in JTIDS terminals, provided the unit is an active (non-radio silent) terminal. This technique independently generates an excellent measure of the user's own time bias. In the RTT Technique, the user transmits a message to a selected unit of higher time quality (called the donor); the donor measures the time-of-arrival (TOA) of that message and reports it back to the interrogating user unit in the return message. The latter then measures the TOA of the donor's reply message. The difference between the donor-reported TOA and the user-measured TOA (divided by 2 and corrected for a known fixed transponding delay) is a very accurate measure of the time offset of the user with respect to the clock time of the donor. If the donor is the NTR, the measured time offset is the desired time bias with respect to system time. Thus, to optimize RELNAV system operation, it is intended to permit some units to RTT relatively frequently (i.e., whenever their filter calls for it), in order to maintain the highest time quality, so that they, in turn, can be used as "navigating sources" by other units, which may have to be radio-silent for tactical reasons.

Details of the RELNAV architecture and operation, e.g., the definition of the geodetic and grid coordinate system used, the hierarchy of position and time qualities, and the error characteristics of the system, are described in Reference 2 and will not be repeated in this paper. Suffice it to say, however, that the geodetic coordinate system is based on the WGS-72 ellipsoid, which is the same as that used in GPS, and the transmitting terminals' qualities are based on the related filter covariances. Figure 6 indicates how position and time knowledge can be propagated within a network of participants beyond the horizon (beyond line-of-sight), even though the Lx frequency band is used in JTIDS. This is accomplished by virtue of the fact that users who improve their position and time knowledge through passive ranging to other units (sources) within line-

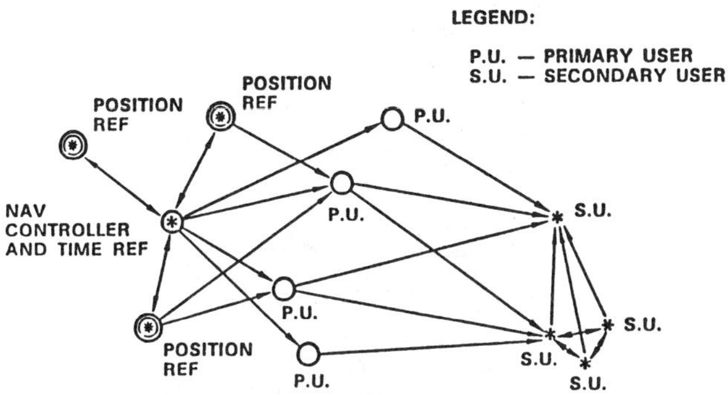


*Fig. 6—JTIDS RELNAV architecture.*

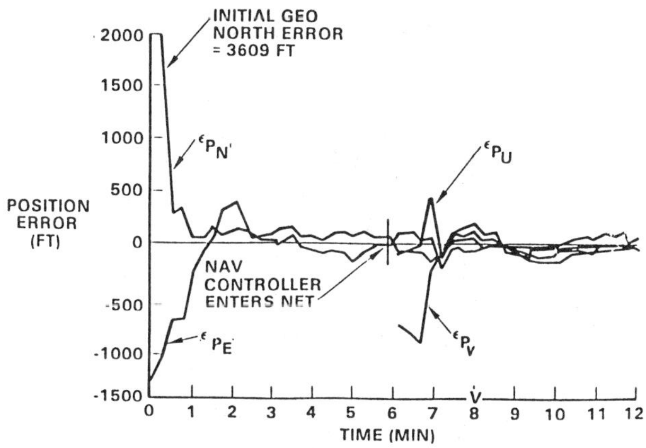


*Fig. 7—Changing pure geo-to-dual grid scenario simulation results.*

of-sight to them, in turn become navigating sources for still other units which are in line-of-sight to them, but not to the original sources. (In Figure 6, Primary Units are active units and Secondary Units are basically passive units.)

Details of a system implementation and the Kalman filter of a RELNAV-Inertial mechanization using the HIT (Hughes Improved Terminal) JTIDS equipment and the AN/ASN-109 (LN-31) Inertial System are described in Reference 3. Some early RELNAV simulation results are presented in that paper. Figures 7 through 10 show some later computer simulation results for scenarios where the user transitions from pure geodetic to dual grid operation and from pure relative to dual grid operation, as well as dual grid and pure relative scenarios. For example, from Figure 7 it is seen that the geodetic errors converge to very small values after approximately 2.5 minutes of operation. Similarly, the relative errors converge shortly after the navigation controller has entered the net. In Figure 8, the reverse was tested, starting out and converging in pure grid operation, with all terminals moving, and with a Geodetic Position Reference (GPR) entering the net after 6.3 minutes. Again, stable performance is exhibited throughout the simulated mission showing high position accuracies (near 100 feet). These simulation scenarios actually were not particularly optimum for geodetic navigation, since only one GPR was present in the net; however, the motion of the users served to bring down the

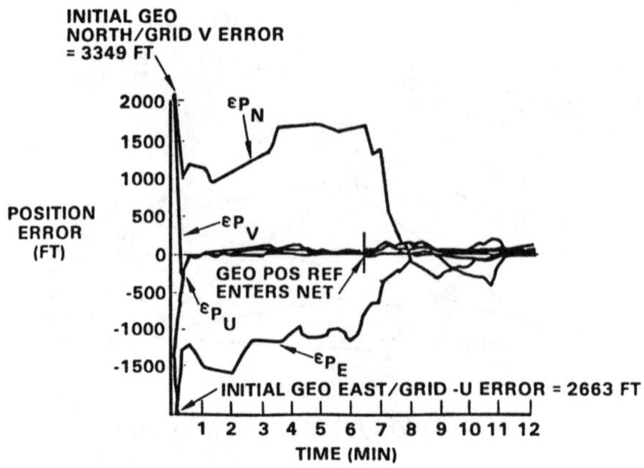


*Fig. 8—Changing pure relative-to-dual grid scenario simulation results.*

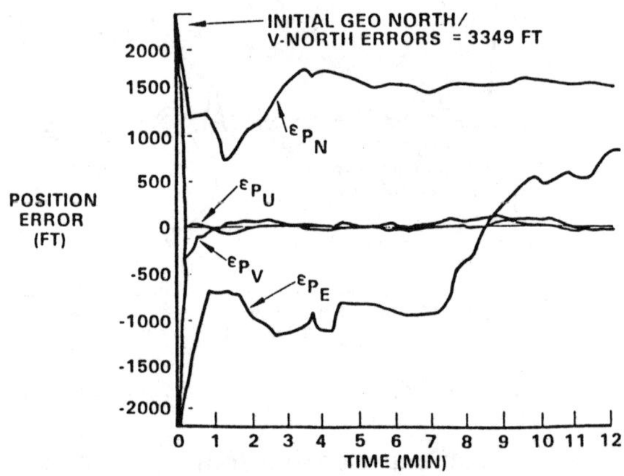


*Fig. 9—Pure relative grid operation simulation results.*

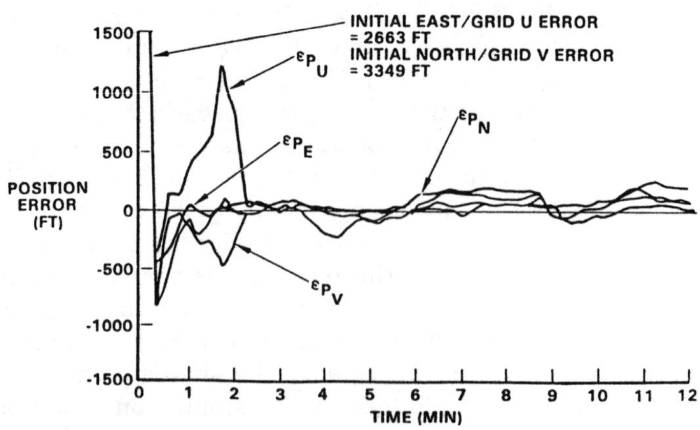


*Fig. 10—Dual-grid RELNAV operation simulation results.*

errors. Figures 9 and 10 show scenarios for pure relative grid operation and dual grid operation, using different user-to-source geometries. This system was also tested with real time software and actual JTIDS and inertial equipment hardware, showing good performance results.

## *GPS Navigation*

The NAVSTAR Global Positioning System (GPS) is a satellite based, global, geodetic navigation system which also uses passive pseudoranging techniques, as represented by the observation equation (Equ. 1) of the previous section. In the current design, 18 satellites will be deployed in a suitable orbit configuration. Each of these satellites continuously transmits unique, pseudo-noise-coded signals, which are precisely time synchronized to system time by means of atomic clocks located on the satellites. The satellites also transmit data on their ephemerides (positions), clock errors and time-of-day. A user receives the signals from four satellites, measures the time-of-arrival (TOA) and hence the pseudo-ranges to the satellites. From these, his computer calculates his three-dimensional position and his time bias with respect to system time. Through measurement of pseudo-range rate or Doppler (using the so-called delta range technique), the user's velocity components and clock frequency offset can also be calculated. The system operates in the L-band and uses the direct sequence spread spectrum technique in order to provide resistance to jamming. The general architecture of the system is depicted in Figure 11.

The overall GPS system consists of three segments, i.e. the satellite segment, the user segment, and the control segment. The control segment tracks the satellites to determine their ephemerides and atomic clock errors and uploads data to each satellite for re-transmission to the user. The tracking is accomplished by means of four monitor stations which continuously make pseudo-range measurements on signals from the satellites, whenever they are visable, in a similar manner as a high quality receiver. The data are sent from the monitor stations to a master control station (MCS). The MCS computer calculates satellite ephemeris and both satellite and monitor station clock parameters. The satellite ephemeris and clock parameters are sent to an upload station which transmits these data to the satellite when it is overhead. In order to increase the reliability and availability of the system, two Control Stations (rather than one) are being planned for installation within the United States. (One has been in existence for some time.) The system operates at two L-band frequencies (1227 and 1575 MHz) in order to permit automatic ionospheric

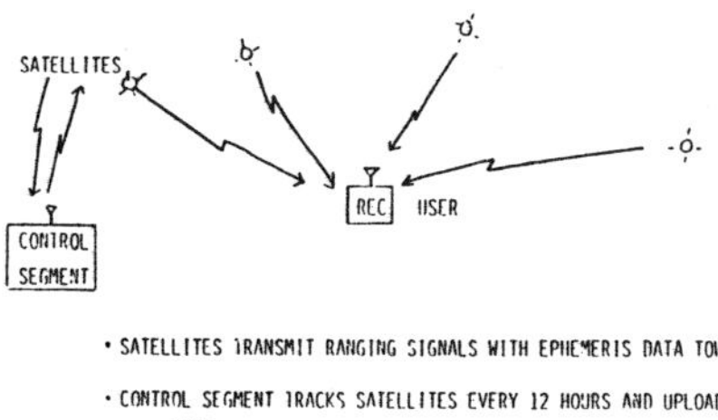


- SATELLITES TRANSMIT RANGING SIGNALS WITH EPHEMERIS DATA TOWARD EARTH
- CONTROL SEGMENT TRACKS SATELLITES EVERY 12 HOURS AND UPLOADS REQUIRED DATA
- USER RECEIVER SELECTS 4 SATELLITES IN VIEW, MEASURES TOA's AND CALCULATES POSITION

*Fig. 11—GPS operation.*

propagation delay compensation. The system uses code division multiplexed, continuous wave, transmission. It employs two coded waveforms; the P-code for maximum accuracy and antijam performance and the C/A code for fast acquisition. The user requires passive (receiving) equipment only. Two basic types of user sets have been designed: simultaneous tracking and sequential tracking. The former has four receivers which simultaneously track the carrier signals from four satellites; the latter has only one receiver which is time-shared (commutated) between four satellites for carrier tracking.

Figure 12 shows a simplified block diagram of a typical user receiver. The papers in Reference 12 describe the operation, system architecture and error characteristics of GPS in great detail, and these will not be repeated in this paper. Reference 13 describes predicted accuracy results for an earlier, 24-satellite, configuration, showing that horizontal position errors of less than 25 feet (90% probability) can be expected. Reference 14 shows *actual* test results with units operating at Yuma Proving Ground with a limited number of satellites in orbit. These results showed that, in the P-code mode, the system is capable of achieving horizontal position accuracies of the order of 30 ft. A considerable amount of more recent analysis, simulation and testing has demonstrated good performance results in various configurations.

## SIMILARITIES AND DIFFERENCES OF JTIDS RELNAV AND GPS

The similarities of the JTIDS RELVAV and GPS concepts are striking. Both systems use a common precise time base for all participants, and the participants maintain time synchronization at all times. Both systems use multiple pseudo-ranging (time-of-arrival measurements) from sources whose position is known, in order to determine own-unit position, velocity and time offset, as represented mathematically by Equ. 1. Both systems use modern estimation techniques, notably Kalman filtering, in order to generate best estimates of the desired quantities, as well as the related covariances and, therefore, the qualities of these estimates. Both systems are subject to the well-known geometric dilution of precision (GDOP) effect, which makes the accuracy performance dependent on the relative geometry between user and sources. Both systems are basically positioning systems and hence will benefit from interconnection to a dead reckoning system, such as an inertial system, notably on high dynamic platforms, in order to avoid undesirable lag effects. Both systems use forms of spread spectrum modulation techniques which result in high resistance to jamming interference and high time-of-arrival measurements accuracy. Both

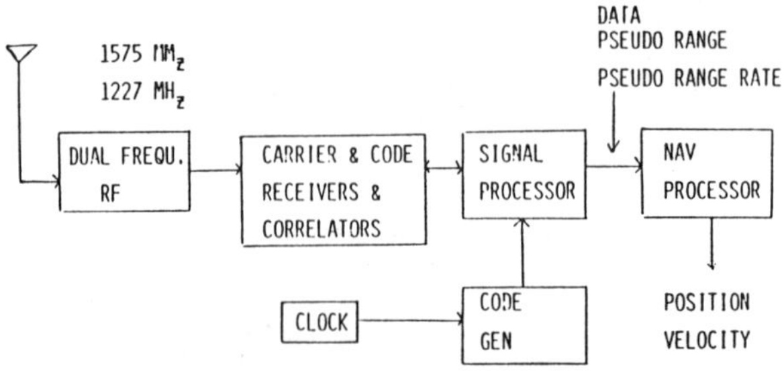


*Fig. 12—GPS user equipment.*

systems operate in the L-band, although in somewhat different portions of the band and therefore basically depend on line-of-sight radio propagation.

The differences of the characteristics of the two systems are significant, however; and it is these which ultimately lead to the conclusion that the systems are complementary and not competitive. By the nature of the current design, GPS is a worldwide (global) system, while JTIDS is a more local or tactical system. In GPS the sources are satellites in fixed orbit configurations, while in JTIDS RELNAV, the sources can be fixed ground sites, aircraft, ships or other vehicles. The GPS navigation message signals are transmitted simultaneously from all the satellites on a continuous wave basis, whereas JTIDS terminals transmit their position messages generally on a periodic, sequential basis, using a pulse waveform. In GPS, all signal transmissions are referenced to one common system time; while in JTIDS, each RELNAV net can use a different net time reference. Since the GPS satellites are generally in fixed orbits, while the JTIDS RELNAV source locations are quite dynamic, the geometric source-user relationships in GPS, and hence GDOP, are more stable, i.e. tend to vary over a smaller range than those in a typical JTIDS scenario. However, this difference becomes less pronounced as the number of units in a JTIDS net increases. This greater stability in source-to-user geometric relationship of GPS should give rise to higher average accuracy performance than JTIDS whenever the GPS satellite signals are available to the GPS receiver. The GPS waveform uses only direct sequence spreading, and the levels of the signals received from the satellites, which are located 12,000 miles away, are extremely low (near-166 dBW). In contrast, the JTIDS waveform uses frequency-hopping as well as direct sequence spreading, the transmitted powers are considerably higher; and the propagation ranges are much shorter (several hundred miles.) Thus, the anti-jam performance of JTIDS should exceed that of GPS in a typical tactical environment. The loss of one or two GPS satellites for any reason could degrade GDOP somewhat and hence system accuracy performance. However, the use of in-orbit spare satellites is being considered in order to reduce the effects on system performance of the loss of any satellites. The large number of dynamically deployed JTIDS RELNAV sources in a typical JTIDS net gives rise to inherent redundancy which protects against the loss of one, or even a few, participating units acting as navigating sources. Both GPS and JTIDS benefit greatly from interconnection with an Inertial Navigation System (INS), for several reasons, including increased anti-jam peformance and continuity of accurate data during high dynamic conditions.

## LIMITATIONS OF JTIDS RELNAV AND GPS

From the characteristics discussed in the previous section, the limitations of JTIDS RELNAV and of GPS can be summarized. The performance of JTIDS RELNAV is highly dependent on the availability of active transmitting sources, having high position and time quality and located in favorable geometry. If the average transmission rate of the position (PPLI) messages from the sources is too low for some reason, the resulting accuracy of the RELNAV solution will degrade in high dynamic platforms, unless a high performance dead reckoner (e.g., inertial system) is interfaced with the RELNAV filter. If the navigating unit leaves the RELNAV net in which it has been operating, its knowledge of position will begin to degrade as a function of the performance and calibration condition of the interconnected inertial or other dead reckoning system. One interesting and highly desirable property of JTIDS RELNAV, however, is the fact that loss of any particular unit, even the current net time reference (NTR), will not catastrophically affect system performance, since JTIDS is a distributed or basically nodeless system. (Ultimately, however, a new NTR must be designated in order to maintain system performance.) As discussed earlier, if a GPS user receiver should lose the required number of satellite signals, the accuracy peformance would degrade as time progresses.

## BENEFITS OF COMBINING GPS AND JTIDS RELNAV DATA

Virtually all of the limitations of the two systems discussed in the previous section can be eliminated or greatly reduced by combining the data from GPS and JTIDS. This is achieved, on one hand, as a result of the technique redundancy afforded by the two separate systems operating over different propagation paths, in different frequency bands and using different sources; and, on the other hand, as a result of the similarity of the two systems, i.e., dependency on time synchronuous operation, use of pseudo-ranging, Kalman filtering and interface to a dead reckoning systems.

If a sufficient number of GPS satellites are in view of the user and his GPS receiver's peformance is not degraded by jamming, the GPS position and time accuracy performance should be excellent. On the average, it may be superior to the geodetic position accuracy of JTIDS RELNAV. Under this condition, position and time data can be fed to the JTIDS RELNAV Kalman filter as observation inputs of high quality (low variance), and the RELNAV output data will coverge to the GPS data.

If the required number of satellites were to become unavailable for any reason, the JTIDS RELNAV function will very likely be able to take over, in view of the dependence of JTIDS RELNAV on entirely different signal sources. The RELNAV filter would thus start out with high position accuracy and, depending on the availability of good geodetic sources, the quality of the dead reckoner and the user dynamics, would maintain this accuracy for some time. If the user were to enter an area in which GPS signals are again available, i.e., improved jamming environment and/or visibility of satellites, the high accuracy position and time knowledge available from RELNAV should markedly reduce the GPS time-to-first-fix. Also, it may be desirable to obtain the required satellite ephemeris and clock data via the jam-resistant JTIDS communication link.

Conversely, the initialization of the JTIDS RELNAV filter would be greatly enhanced if the initial position and time are known with GPS accuracy. Convergence time of the RELNAV filter to the desired accuracy should be markedly decreased. Simulation results have shown that the accuracy and stability of even the RELNAV *relative* grid coordinates benefit from good geodetic position and velocity knowledge of certain users, notably the navigation controller; particularly shortly after grid origin acquisition.

From the above consideration, it is seen that combining JTIDS and GPS data is a highly synergistic process, i.e., the benefits obtained represent more than the "sum" of the contributions of each. Specifically, with the combination of

GPS and JTIDS RELNAV data, the probability should be very high of having *continuous availability* of high accuracy position and velocity knowledge, which is so important for weapon delivery, over-ocean grid lock, station keeping, etc.

## METHODS OF COMBINING JTIDS AND GPS

There are at least three alternative approaches for combining JTIDS and GPS:

(1) Hardware and Software Integration
(2) Software Integration (common filter)
(3) Exchange of Data

Complete integration of the JTIDS and GPS hardware and software was investigated by Draper Laboratory,[15] Hughes Aircraft Company, and others. Since the two systems operate in adjacent frequency bands in the L-band and both use modern spread spectrum technology, hardware integration is technically feasible. However, since the detailed technical requirements (such as sensitivity, IF bandwidth, correlator performance, etc.) are significantly different for the two systems, the overall advantages of using some common hardware are somewhat limited. (This may change sometime in the future, as new technologies such as VHSIC, Charge Coupled Devices, etc., become more mature.) In addition, having available on the user platform two *independent* pieces of hardware (including power supplies, for example) using different "techniques" (propagation paths, etc.) leads to a large increase in reliability/availability.

The use of a common navigation filter (common software) for the two systems has also been investigated.[16,17] Again, basic technical feasibility exists. However, inherently the size of the Kalman filter state vector will increase markedly, since both JTIDS and GPS related variables would have to be estimated. This, in turn, would lead to large covariance matrices, and thus require a large computer processing power and memory. It would also tend to reduce software maintainability, in view of the changes typically experienced in systems as complex as GPS and JTIDS. Finally, having independent and smaller software processes provides inherent reliability advantages.

The third approach, i.e., exchanging the normal output data between GPS and JTIDS has very low risk associated with it, as well as high potential reliability and availability. Figure 13 shows a functional block diagram of a possible system configuration for interconnecting JTIDS and GPS and exchanging data between them. On many modern platforms, digital data are transmitted between different equipments via a multiplex data bus, such as MIL-STD-1553. As indicated on the figure, it is assumed that a central bus controller would control the data bus traffic between the equipments. For example, GPS position, altitude, and associated variances could be received by the JTIDS equipment over the data bus. Based on the relative variances of JTIDS RELNAV and GPS data, the RELNAV source selection function would make the decision as to whether to use the GPS data. If the GPS position data is selected, the source selection function would feed the data to the RELNAV Kalman filter as an external geodetic position update. The Kalman filter would process the difference between the GPS position and the estimated (and extrapolated) RELNAV position as the observation (measurement residual), using the associated GPS position variances in the filter gain computation.

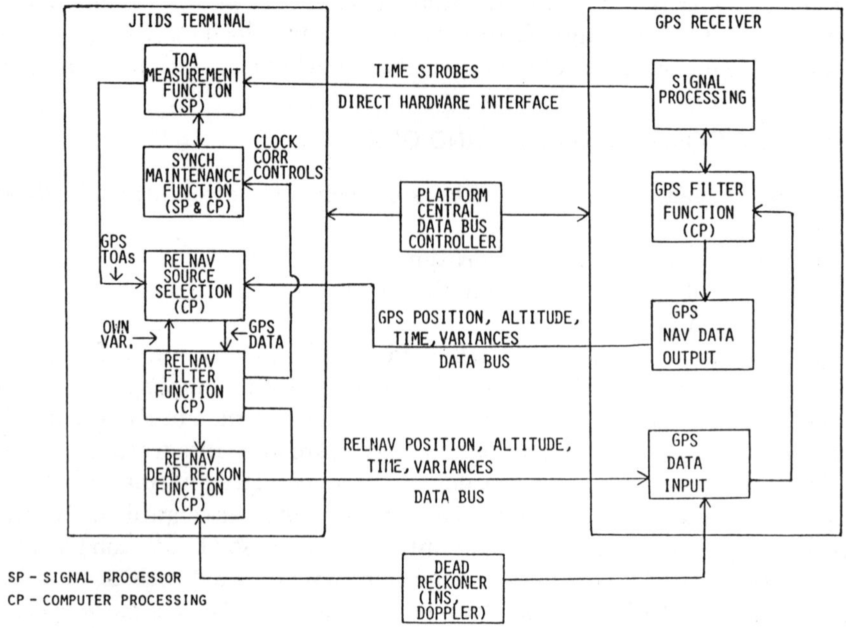


*Fig. 13—JTIDS—system interconnection.*

If only GPS position is used to update JTIDS, this is the only processing required in the JTIDS RELNAV software. This approach is currently implemented in the Stage 1 U.S. Navy JTIDS terminal program, primarily in order to optimize schedule and risk. GPS position variances could be converted to the RELNAV quality values (0 to 15) using the standard conversion table, before sending it to the RELNAV Function. Alternatively, a fixed, quality level value for GPS can be used.

For the inverse process of feeding JTIDS geodetic position data to the GPS receiver via the data bus, similar processing within the GPS navigation software could be implemented. Even if only horizontal position is exchanged between JTIDS and GPS, large performance benefits can be obtained, as discussed earlier. This can be achieved even if GPS and JTIDS operate on an entirely different time base.

Exchanging the third dimension of position, i.e. altitude, can also be very beneficial on many platforms, and has some interesting features associated with it. Although both the GPS and JTIDS RELNAV Kalman filters typically are inherently designed to estimate altitude, the actual observability of altitude is quite different in the two systems. Because of the normally favorable geometry of four visible satellites, altitude typically has good observability from TOA measurements in GPS. In contrast, although the RELNAV filter *can* estimate altitude from TOA measurements, the observability is typically poor, in view of the existing source-to-user geometries. This is so because the ranges between the sources and the users (aircraft, ships, ground stations, etc.) are normally much larger (e.g., several hundred miles) than their differences in altitude (less than 10 miles). Therefore, in RELNAV implementations on aircraft, altitude (e.g., baro-inertial or barometric altitude) is provided as an input to the JTIDS equipment. If the geometry turns out to be favorable (as automatically sensed by the filter), the RELNAV filter will improve on altitude knowledge as a result of the TOA measurements. Thus, if GPS were to provide good platform altitude data to the JTIDS RELNAV function, this would inherently improve RELNAV computations. As indicated in Figure 13, altitude would typically be provided via the platform data bus, along with GPS horizontal position. Again, the data could be accepted on the basis of relative variance information, and it could also be provided by JTIDS to GPS when the requirement exists, such as re-acquisition of GPS satellite signals after temporary signal loss by the receiver.

Finally, it would be beneficial to exchange time data between GPS and JTIDS, as discussed earlier. In order to implement this, it would be required to operate GPS and the JTIDS RELNAV circuits on the same time base. (In DTDMA, other metachannels or circuits used strictly for communication could operate on different time bases, if this is desired.) To achieve the common time base operation of GPS and JTIDS, it would be necessary to synchronize JTIDS Net Time References (NTRs) to GPS system time. In the future, this may be Universal Time (UT). There should be no problem in doing this, since the major platforms which are likely to be assigned as NTRs (e.g., large surveillance aircraft, attack aircraft, naval aircraft carriers, etc.) will have GPS receivers installed on them.

A mechanization for accomplishing the exchange of time data between GPS and JTIDS is also shown in Figure 13. Part of this is accomplished via a special hardware interface. Time-of-day and time variance (quality), are available in both GPS and JTIDS software. However, in order to exchange time-of-day data to the required accuracy, a data bus having very high resolution would be required. Currently-used airborne multiplex bus configurations have not been designed for that purpose. A time strobe can be provided over a direct hardware interface from GPS to JTIDS. (A port is typically available in JTIDS terminals for synchronization from an external clock.) If this GPS time strobe is received by the JTIDS terminal, the terminal TOA measurement function could be used to perform a TOA measurement on the time strobe pulse and send it to the RELNAV function. The RELNAV source selection function would make the decision on whether or not to use the GPS time data, based on relative time variance information. If the source selection function decided to accept the GPS time data, it would correct the measured GPS TOA for the known equipment delays (using constants from memory) and then provide the corrected TOA, and the GPS and JTIDS time-of-day values (if available) to the RELNAV filter function. The filter would then difference the respective time-of-day values, subtracting the corrected TOA from this value and use the resultant as the observation residual for filter update of the clock states by means of an appropriate observation vector. (The process is somewhat similar to the RTT observation used in current JTIDS mechanizations.)

In summary, Figure 13 depicts a mechanization for exchanging all three variables, i.e., horizontal position, altitude and time between GPS and JTIDS. It is obviously possible to implement only some (and not all) of these functions initially, or on particular platforms, and yet obtain significant benefits from doing so. However, the technology does exist for exchanging all three variables, by means of relatively small modifications to the software in the two systems and thereby obtaining the maximum benefit from exchanging data between GPS and JTIDS.

## CONCLUSION

Combining of the JTIDS RELNAV and GPS data has been shown to have a large number of operational and technical benefits. JTIDS is a secure, highly jam-resistant, tactical communication, navigation and identification system, which is scheduled to be installed on a large number of Air Force, Navy, Army and Marine Corps platforms. Its navigation function, which is both geodetic and relative, is highly accurate, but it has a number of well-known limitations, primarily relating to scenario geometry. GPS is the latest, global, satellite based navigation system which is scheduled to be employed on most platforms of the U.S. military services. Far from being competitive, exchanging data between the two systems on the same platform provides the user with greatly increased availabiltiy and reliability of navigational and positional information. Many types of navigation systems are currently in use by the military and civil community, most of which represent technologies which were developed some time ago. The performance characteristics and operational benefits of these systems have been studied and compared in the past (for example, see Ref. 18 and 19). GPS and JTIDS RELNAV represent implementations of the most modern technology available to system designers. The synergism of JTIDS and GPS should provide to the military services, and some day to civil users, both accuracy performance and reliability of positioning never before available.

## REFERENCES


1. Fried, W. R., "Principles and Simulation of JTIDS Relative Navigation," IEEE Transactions, Vol. AES-14, No. 1, January 1978.
2. Fried, W. R., "JTIDS Relative Navigation—Architecture, Error Characteristics and Operational Benefits," IEEE NAECON Proceedings, May 1979.
3. Fried, W. R. and Loeliger, R., "System Configuration and Algorithm Design of the Inertially Aided JTIDS Relative Navigation Function," 1979 IEEE NAECON Proceedings.
4. Stow, R. C., "Tactical Hybrid Navigation Scheme for Mixed Community Coordination," Proceedings of IEEE EASCON '73, September 1973.
5. Bivin, J. and Soloway, G., "Testing of an ITNS/DAHRS System," Proceedings of the Institute of Navigation National Aerospace Meeting, Alamogordo, N.M., 12–13 May 1975.
6. Rome, H. J. and Stambaugh, J. S., "Evaluation of the Accuracy of Community Relative Navigation Organization Concepts." Proceedings of the Institute of Navigation National Aerospace Conference, Naval Air Development Center, Warminster, PA, April 1976.
7. Westbrook, E. A. and Snodgrass, R. C., "Relative Navigation by Passive Ranging in a Synchronous Time Division Multiple Access Data Net," the MITRE Corporation, MTR-2996, prepared for Electronic Systems Division, U.S. Air Force, March 1975.
8. Ellingson, C. E. and Kulpinski, R. J., "Dissemination of System Time," IEEE Transactions on Communications, Vol. COM-21, NO. 5, May 1973, pp. 605–624.
9. Dell-Imagine, R. A., "JTIDS—An Overview of the System Design and Implementation," Proceedings of the IEEE Position Location and Navigation

Symposium, San Diego, CA. IEEE Publication 76-CH1138-7 AES, pp. 212–218, November 1976.
10. Neuman, D. E., "Seek Bus, A Time Division Multiple Access System," AIAA Digital Avionics Conference, April 1975.
11. Rubin, J., "The Full Scale Development of U.S. Navy DTDMA JTIDS Integrated Communication-Navigation-Identification Terminals," Proceedings of the 1982 IEEE Position Location and Navigation Symposium, Atlantic City, N.J., December 1982.
12. "Global Positioning System," Compendium of Papers Published in *Navigation,* Journal of the Institute of Navigation, 815 Fifteenth Street, Suite 832, Washington, D.C. 20005; 1980.
13. Lassiter, E. M. and Parkinson, B., "NAVSTAR/GPS: Operational Status," Proceedings of the International Navigational Congress, 1976, Boston, Mass. August 1976.
14. Henderson, D. W. and Strada, J. A., "NAVSTAR Field Test Results," Proceedings of the ION National Aerospace Symposium, 6 March 1979; and in Reference 12.
15. Stonestreet, W. M. and Douglas, D. A., "Design of an Integrated GPS-JTIDS-INS System," IEEE-NAECON Proceedings, Dayton, Ohio, May 15–17, 1979.
16. Kriegsman, B. A. and Stonestreet, W. M., "A Navigation Filter for an Integrated GPS-JTIDS-INS System for Tactical Aircraft," Proceedings of the 1978 IEEE Position Location and Navigation Symposium, San Diego, CA, November 1978.
17. Widnall, W. S., et al, "Decentralized Relative Navigation and JTIDS/GPS/INS Integrated Navigation Systems" Avionics Laboratory, Air Force Wright Aeronautical Laboratories, WPAFB, Ohio, Report AFWAS TR-81-1264, January 1982.
18. Fried, W. R., "A Comparative Performance Analysis of Modern Ground-Based, Air-Based and Satellite-Based Radio Navigation Systems," *Navigation,* Vol. 24, NO. 1, Spring 1977.
19. Kayton, M. and Fried, W., Ed., "Avionics Navigation Systems," John Wiley and Sons, New York, 1969.
20. Fried, W. R. and Loeliger, R., "Principles, System Configuration and Algorithm Design of the Inertially Aided JTIDS Relative Navigation Function," *Navigation,* Vol. 26, NO. 3, Fall 1979.

# Satellite Selection Criteria During Altimeter Aiding of GPS

BARRY A. STEIN

The MITRE Corporation, McLean, Virginia
*Received June 1985*

## ABSTRACT

The 21 satellite constellation envisaged for GPS yields occasional geometric configurations that result in poor navigation solutions. It therefore may be necessary to use altimeter aiding in these cases in order to improve the geometry. During altimeter aiding, only three of the visible satellites will be used in conjunction with the altimeter. It is of paramount importance to select the three satellites that will yield the best possible navigation solution when combined with the altimeter data. This paper presents a closed form algorithm for satellite selection. It differs from other satellite selection algorithms in that they use four visible satellites and do not consider the altimeter. Also, the usual problem of computationally inverting a $4 \times 4$ matrix is avoided. Computer simulations show that the algorithm avoids selecting any of the satellite configurations that would yield a poor navigation solution. A comparison has been made between using the true GDOP closed form solution and the maximum volume of a tetrahedron. The latter is shown to be an integral part of the former. Results indicate that it yields sufficiently accurate results under the assumed conditions.

## INTRODUCTION

The Global Positioning System (GPS) is expected to become operational in 1988. The constellation will consist of 18 satellites and 3 spares. The spares will be operational since it has been ascertained that keeping them turned off will not significantly increase their longevity.[1] It is therefore assumed that the operational constellation will consist of 21 functioning satellites as depicted in Table 1.

It is intended that GPS be the primary radionavigation aid for aircraft in the future. It will be used as a stand-alone navigator as well as in conjunction with other navigation equipment. The instrument of interest herein is the altimeter, specifically the use of a baroaltimeter to aid GPS.

The present GPS constellation shows performance degradations for short periods of time at certain locations on the earth at specific times every day of the year.[2] These degradations are caused by the poor geometry of the satellites visible at the location of concern. Specifically, the satellites are coplanar and can not provide sufficient independent information to yield a meaningful position and time fix. This is true for cases involving not only the minimal four satellites being visible, but also when more than four satellites are visible.

The same type of degradation can occur when a satellite fails, or when a

Table 1—Operational Orbit Description Longitude Relative to Earth and Astronomical Coordinates[1]

| Number | Orbit Plane | Longitude of the Ascending Node, deg | Right Ascension of the Ascending Node, deg* |
|---|---|---|---|
| 1 | 1 | 0, 180 | 30 |
| 2 | 1 | 240, 60 | 30 |
| 3 | 1 | 300, 120 | 30 |
| 4 | 2 | 260, 80 | 90 |
| 5 | 2 | 320, 140 | 90 |
| 6 | 2 | 20, 200 | 90 |
| 7 | 3 | 340, 160 | 150 |
| 8 | 3 | 40, 220 | 150 |
| 9 | 3 | 100, 280 | 150 |
| 10 | 4 | 60, 240 | 210 |
| 11 | 4 | 120, 300 | 210 |
| 12 | 4 | 180, 0 | 210 |
| 13 | 5 | 140, 320 | 270 |
| 14 | 5 | 200, 20 | 270 |
| 15 | 5 | 80, 260 | 270 |
| 16 | 6 | 220, 40 | 330 |
| 17 | 6 | 280, 100 | 330 |
| 18 | 6 | 160, 340 | 330 |
| *Spares* | | | |
| 19 | 1 | 195, 15 | 30 |
| 20 | 5 | 215, 35 | 270 |
| 21 | 3 | 25, 205 | 150 |

*Referenced to astronomical coordinates of 1950.0 as of 1 July 1985, 0 hr 0 min GMT and regressing at −0.04009 deg/day.

satellite is shadowed by the local terrain or by the vehicle carrying the GPS antenna. In these cases there is more to the concern than the coplanar condition of four satellites. There is also the distinct possibility of only three active satellites being visible. The navigation/time solution is therefore no longer possible in three dimensions. The case of satellite failure will not be considered in this paper.

It is at this point that the use of baroaltimeter aiding becomes of interest. It is theoretically possible to use the baroaltimeter as a source of information (altitude) that when combined with GPS data would yield the desired navigation solution when at least three active satellites are visible. There are certain conditions, however, that if encountered would still yield degraded navigation solutions comparable to the aforementioned degradations due to poor constellation geometry without altimeter aiding.

It has been shown [2,3] that these degradations vary with time and location, and are associated with certain "key" satellites. It is the expressed purpose of this paper to derive a satellite selection algorithm that will always avoid selecting this "key" satellite and hence guarantee minimal performance degradation during altimeter aiding. This will be accomplished by selecting the satellite combination that will yield the best possible navigation solution when combined with altimeter data.

## APPROACH

The use of baroaltimeter aiding of GPS is the same for the poor geometry and satellite shadowing cases from a mathematical point of view. The baroaltimeter is simply to be used as a source of additional data that will make the navigation solution mathematically possible. To illustrate, if there are three or more active usable satellites, and they are coplanar, the baroaltimeter will provide additional information that will facilitate the navigation solution as in Figure 1. Specifically, the baroaltimeter will provide altitude data. There is, however, a limitation to the usefulness of this data. If the satellites are coplanar, and the local vertical is in the plane of the satellites, the altitude data will not help in providing a solution. In fact if the satellites are exactly coplanar (which the three satellite case is by definition) and the local vertical is precisely in the satellite plane, the navigation solution for three dimensions does not exist (see Figure 2). It is therefore desirable to select three satellites that are not coplanar with the local vertical.

The actual severity of the degradation or failure can be ascertained by calculating the Geometric Dilution of Precision with altimeter aiding ($GDOP_a$). The quantity $GDOP_a$ can be calculated by a variety of methods. Those considered here are:

1. Matrix Inversion by Computer
2. Closed Form Algorithm
3. Volume of a Tetrahedron

Any of these can be used as a satellite selection criteria. The drawback is the computational time required since all of the above would require a number of

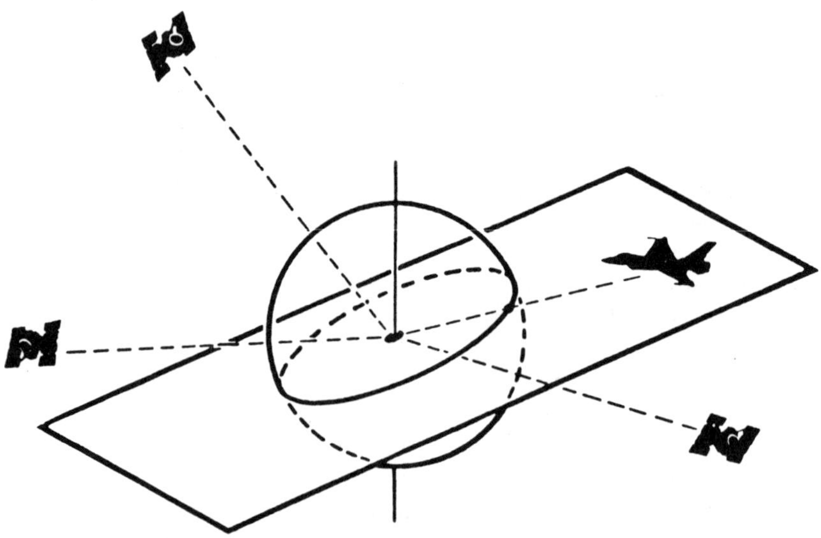

*Fig. 1—Successful altimeter aiding*

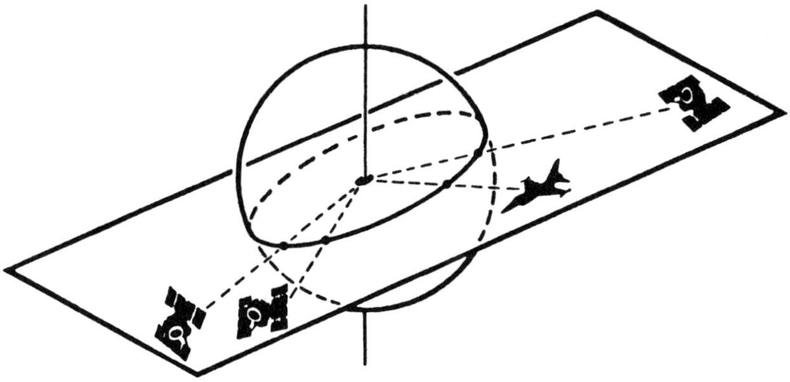

*Fig. 2—Unsuccessful altimeter aiding*

iterations to determine the best $GDOP_a$. The number of these iterations is dependent upon the number of satellites visible. Fortunately, the poor geometry cases are most frequent when only four satellites are visible. It is therefore only necessary to have four iterations to determine the appropriate three satellites needed to minimize $GDOP_a$ in most cases.

It remains to be shown that both the Closed Form Algorithm and the Volume of the Tetrahedron approaches are equivalent in their selection of the best set of satellites, and in certain ways superior to the Matrix Inversion by Computer approach. This is desirable since the latter approach is the most time consuming and requires the most computer memory. Indeed, the Volume of the Tetrahedron approach is the most desirable since it requires the least number of computer operations. It remains to be seen, however, if it is a sufficient solution at all times since it is an approximation. The purpose of this paper is to address these questions.

A computer program that simulates the true orbital motions of the entire GPS constellation was previously written. This program was modified to produce $GDOP_a$ calculations at any point on the earth. It has the ability to search the northern hemisphere for performance degradation by stepping through latitude, longitude and time in any increments desired. Mask angles can be input in order to simulate a variety of shadowing possibilities.

The three approaches for calculating $GDOP_a$ have been included and are calculated simultaneously. This will facilitate comparisons and allow the optimal satellite selection criteria to be chosen.

## TECHNICAL ANALYSIS

The effects of geometry on the performance of any radio-navigation system is expressed in terms of Geometric Dilution of Precision (GDOP). For the case of altimeter aiding of GPS, it has been shown[3] that this can be expressed as:

$$GDOP_a = \sqrt{\text{trace}(G_a{}^T R G_a)^{-1}} \qquad (1)$$

where

$$\mathbf{G}_a = \begin{bmatrix} 0 & 0 & -1 & 0 \\ l_1 & m_1 & n_1 & -1 \\ l_2 & m_2 & n_2 & -1 \\ l_3 & m_3 & n_3 & -1 \end{bmatrix} \tag{2}$$

and

$$\mathbf{R} = \begin{bmatrix} 1/u^2 & 0 & 0 & 0 \\ 0 & 1 & 0 & 0 \\ 0 & 0 & 1 & 0 \\ 0 & 0 & 0 & 1 \end{bmatrix} \tag{3}$$

Note,

$l_i$, $m_i$, $n_i$ = the direction cosines from the user to the ith satellite; subscript "a" refers to altimeter aiding; and u is the ratio of the random altimeter error to the random pseudorange error as given by:

$$u = \frac{\sigma_a}{\sigma_p} \tag{4}$$

It is now possible to expand the expression for $GDOP_a$ and obtain a closed form solution. Noting

$$(G_a{}^T R G_a)^{-1} = \frac{adj(G_a{}^T R G_a)}{det(G_a{}^T R G_a)} \tag{5}$$

$$= \frac{adj(G_a{}^T R G_a)}{det(G_a{}^T) det(R) det(G_a)} \tag{6}$$

$$= \frac{d}{D^2} \tag{7}$$

Substituting (7) into (1) yields

$$GDOP_a = \frac{\sqrt{t}}{D} \tag{8}$$

where

$$t \equiv trace\ (d).$$

Algebraic manipulation yields

$$\begin{aligned} t = {} & a_{33}[(a_{11} + 3)a_{22} - a_{42}{}^2 - a_{21}{}^2 + 3a_{11} - a_{41}{}^2] \\ & - a_{43}{}^2(a_{11} + a_{22}) + 2a_{43}(a_{31}a_{41} + a_{32}a_{42}) \\ & - a_{32}{}^2(a_{11} + 3) + 2a_{21}a_{31}a_{32} \\ & - a_{22}(a_{31}{}^2 - 3a_{11} + a_{41}{}^2) \\ & - a_{11}a_{42}{}^2 + 2a_{41}a_{21}a_{42} - 3a_{31}{}^2 - 3a_{21}{}^2 \end{aligned} \tag{9}$$

where the $a_{ij}$s are functions of the direction cosines from the user to the three satellites being used in conjunction with the altimeter. These relationships are given by the following equations:

$$\begin{aligned} a_{11} &= l_1^2 + l_2^2 + l_3^2 \\ a_{21} &= l_1m_1 + l_2m_2 + l_3m_3 \\ a_{22} &= m_1^2 + m_2^2 + m_3^3 \\ a_{31} &= l_1n_1 + l_2n_2 + l_3n_3 \\ a_{32} &= n_1m_1 + n_2m_2 + n_3m_3 \\ a_{33} &= n_1^2 + n_2^2 + n_3^2 + 1/u^2 \\ a_{41} &= -(l_1 + l_2 + l_3) \\ a_{42} &= -(m_1 + m_2 + m_3) \\ a_{43} &= -(n_1 + n_2 + n_3) \end{aligned} \tag{10}$$

Similarly, by combining (6) and (7), it can be shown that

$$D^2 = [(l_2 - l_1)(m_3 - m_2) - (l_3 - l_2)(m_2 - m_1)]^2/u^2 \tag{11}$$

The solution for $GDOP_a$ is therefore completely given in closed form by (8) through (11).

There are several interesting factors to note about the expression for $D^2$ given in (11). It is immediately obvious that it is independent of the $n_i$ direction cosines. This is of interest since the $n_i$ direction cosines refer to the local vertical direction, and hence would yield information that is already being obtained through the use of the altimeter. This is of further consequence when it is realized that the expression for D is directly related to the volume of the tetrahedron that is formed by four points. In this case, the four points are defined by the unit vectors from the user to the three satellites and the altimeter which is assumed to be a pseudosatellite located at the center of the earth.

Furthermore, it is known that the volume of the tetrahedron is a fairly reliable indicator of GDOP. This can be seen from the relationship

$$V = \frac{1}{6}(\mathbf{A} \times \mathbf{B}) \cdot \mathbf{C} \tag{12}$$

for the tetrahedron depicted in Figure 3. Note that the following definitions are used:

$V$ ≡ volume of the tetrahedron
$\mathbf{e_a}$ ≡ unit vector from user to altimeter (satellite at center of earth)
$\mathbf{e_1}$ ≡ unit vector from user to satellite 1
$\mathbf{e_2}$ ≡ unit vector from user to satellite 2
$\mathbf{e_3}$ ≡ unit vector from user to satellite 3
$\mathbf{A}$ ≡ vector from altimeter (satellite at center of earth) to satellite 1
$\mathbf{B}$ ≡ vector from satellite 1 to satellite 2
$\mathbf{C}$ ≡ vector from satellite 2 to satellite 3

It can be shown that

$$(\mathbf{A} \times \mathbf{B}) \cdot \mathbf{C} = \det(G_a) \tag{13}$$

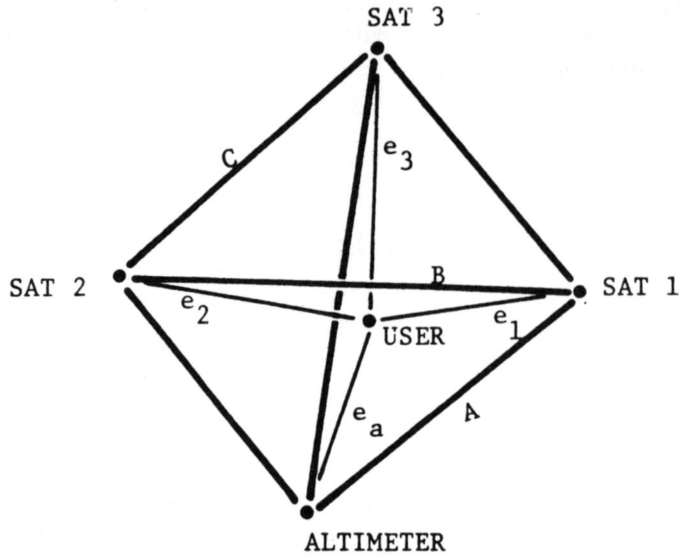


*Fig. 3—Geometry of tetrahedron*

which yields

$$(\mathbf{A} \times \mathbf{B}) \cdot \mathbf{C} = [(l_2 - l_1)(m_3 - m_2) - (l_3 - l_2)(m_2 - m_1)] \tag{14}$$

It is seen that this is similar to the expression given in (11). Therefore combining (11), (12), and (14) yields:

$$V = \frac{1}{6} Du \tag{15}$$

The relationship between the volume of the tetrahedron and $GDOP_a$ is therefore given by

$$GDOP_a = \frac{u\sqrt{t}}{6V} \tag{16}$$

If it is assumed that t is nearly constant, it would be possible to minimize $GDOP_a$ by maximizing the volume of the tetrahedron. The validity of the constant t assumption must, however, be verified by computer simulations comparing true $GDOP_a$ with the volume of the tetrahedron. In reality, it is known that t is not constant, but its variations may be small when compared to the divergences in D.

## ASSUMPTIONS FOR SIMULATION

The computer simulation program was implemented with the 21 active satellite constellation. The area covered was the entire northern hemisphere for a 12 hour time period. The properties of the constellation are such that these results would be repeated in the next 12 hour time period, but shifted approximately 180°. The result for the northern hemisphere is mirrored in the southern hemisphere, but shifted and inverted approximately 90° over the second 12 hour time period.

In order to shorten computer run time, it was decided to define a possible degradation to exist only when the minimum number of four satellites were visible. When this minimal condition occured, each of the four satellites was "removed" from the constellation in turn and the baroaltimeter substituted in its place. It was thus possible to determine whether the key satellite was indeed identified by the particular implementation of $GDOP_a$ and hence rejected. There are, of course, locations and times when five or more satellites would still have geometry that yields a poor navigation solution and in which the substitution/addition of a baroaltimeter would not be of assistance. Therefore, the results presented are basically "best case" in the sense that there are assuredly other locations and times where the baroaltimeter aiding mode will not be of use.[3]

Initial assumptions included a uniform mask angle of 5° and a ratio of random altimeter error to random pseudorange error (u) of two. A degradation situation was defined as $GDOP_a$ greater than six. The search was conducted for 10° increments of latitude and longitude over the hemisphere at 10 minute time intervals.

It should be noted that the assumptions of a 5° mask angle and a u of two in certain situations is a best case scenario. There are many cases of flying over land areas at low altitude that would result in mask angles as high as 20°. Also, the many currently produced altimeter models have a fairly wide range of error probability and are a function of altitude. Therefore, the assumed value of $u = 2$ may be conservative.

## RESULTS

The computer simulations yielded very useful results. The primary result was that all three approaches yielded the same selection of satellites. Also, the closed form solution and the matrix inversion by computer technique agreed to the 5th significant digit for $GDOP_a$ calculation. This is significant since the closed form solution takes less computer run time and uses less computer memory. It can, therefore, be used for $GDOP_a$ calculations and these calculations can then be compared in order to select the three satellites that will yield the most accurate navigation solution during altimeter aiding.

Lastly, the volume of the tetrahedron satellite selection approach agreed with the closed form solution totally. The tetrahedron approach by its very nature does not yield true values of $GDOP_a$, but could indeed be used as the satellite selection algorithm.

## CONCLUSIONS

It is clear that the Closed Form Algorithm and the Volume of the Tetrahedron approach as derived in this paper are sufficient solutions to the satellite selection criteria problem. Either of these, properly implemented, will avoid the problems of navigation solution degradation during altimeter aiding of GPS. Also, the Closed Form Algorithm will yield an accurate value for altimeter aided Geometric Dilution of Precision. Careful note must, however, be taken of the assumptions. In particular, if only three satellites are visible due to a satellite failure, the possibility still exists for performance degradation.

## REFERENCES

1. Kruh, P., et al, *A Strategy for Buildup to the Operational NAVSTAR/GPS Constellation,* Proceeding of the Institute of Navigation/National Aerospace Conference, March 1983.
2. Kalafus, R. M., *Service Outages in GPS Associated with Satellite Failures,* presented at IEEE PLANS 84 Conference, San Diego, CA, November 1984.
3. Stein, B. A., and Giordano, D. E., *Limitations in Altimeter Aiding of the NAVSTAR/GPS Constellation,* presented at ION National Technical Meeting, San Diego, CA, January 1985.

# Combining LORAN and GPS—The Best of Both Worlds

PAUL BRAISTED, RALPH ESCHENBACH, ANIL TIWARI

Trimble Navigation, Mountain View, CA

*Received January 1986*

## ABSTRACT

LORAN and GPS are both well known navigation systems, and both have advantages and disadvantages. By combining both, the navigator can get the advantages of both and minimize the disadvantages of each.

When GPS becomes fully operational, it will replace virtually all other navigation systems. Until that time (late 1988), there will be substantial lapses of coverage. It is during this time that LORAN provides the backup. This combination is the first system that can provide the transoceanic vessel with real time, accurate navigation information at both ports and periodic accurate navigation information enroute.

This paper will review the two systems and then show how the whole is greater than the sum of the parts. In particular, data will be presented to show how the combined product can yield better navigation information than either system by itself.

## SYSTEM DESCRIPTIONS

Complete system descriptions exist elsewhere in the literature [1,2,3,4], so only a pertinent overview will be given here. This overview will allow the reader to see the merits of combining the two systems.

### *LORAN*

LORAN-C was developed during the late 1950's and early 1960's by the Department of Defense. The system is comprised of about 15 chains or GRIs (Group Repetition Interval) which cover about 70% of the coastlines in the northern hemisphere. A single chain will cover between one and two thousand miles of coastline to a range of about one thousand miles from the coast. Even with this coverage, there are still large gaps in the mid-continent and mid-oceanic regions.

A single chain is made up of a master and two or more slaves. A receiver computes its position by measuring the relative time of arrival of the signal from the master and one slave. This Time Difference (TD) places the user on one Line-Of-Position (LOP). A measurement from another slave gives two TDs and the intersection determines the user's position.

This all sounds good, but there are several serious problems which make the system less than 100% reliable. These are discussed in the following sections.

***ASF***—Additional Secondary phase Factor is an error correction term which is applied to the raw TD measurement to correct for the reduced propagation velocity of the LORAN signals over land. This term varies as a function of the conductivity of the path over which the signal passes. Most receivers model this conductivity and attempt to correct for this error, but correction below 0.2 nautical miles is rarely achieved. Also this term varies seasonally and this seasonal effect is rarely modeled.

***Cycle Slip***—No receiver is immune to the difficulties involved with selecting the correct cycle from which to do the time measurement. Under most conditions there is no problem selecting the correct third cycle, but when the signal is weak this can be a very difficult task. When a cycle slip occurs, a position error will result with a magnitude of up to 5 miles. Most receivers allow the user to force the cycle choice to a different cycle. This forces the receiver to choose a particular cycle, but the user must know where he is in order to make this choice.

***Ambiguity***—An ambiguity occurs whenever the two LOPs cross in two locations. When this occurs, the receiver tries a third LOP (if it is available) to resolve the ambiguity. If this is not available, the user must intervene and specify which is the correct solution. Once again, the user must know where he is in order to get the LORAN tracking correctly.

***GRI***—Although not a serious problem, the user must know the Group Repetition Interval (GRI) for his area in order to have application, GRIs can change rapidly, and the user must continuously be aware of these changes.

In spite of all of these problems, LORAN is a very useful and effective system. It would be nice, however, if these problems could be reduced or even eliminated. As we will see, GPS can be used to minimize these problems.

## GPS

The Global Positioning System (GPS) was developed by the US Department of Defense starting around 1973. This is a satellite-based system where a user measures the pseudorange from four satellites and then computes his position and time. When the system is fully operational, in late 1988, it will give better than 50 meter absolute accuracy and RMS errors of less than 10 meters. At present, however, this level of accuracy is only available about five hours a day. This five hour period moves earlier every day by four minutes.

By assuming that the user is on the surface of the earth, only three satellites need to be used to determine position. This extends the usable time from four to six hours as seen in Figure 1. This two dimensional (2D) coverage will not be continuous until after 1987, and until then the GPS system alone will not be adequate for the practicing navigator.

The signal transmitted by the satellites is at 1.575 GHz, and is much easier to demodulate than the analog zero crossing of LORAN, but a problem still exists. This is called the integer millisecond problem. Since the code is one millisecond long, a receiver cannot determine which millisecond is being tracked with respect to the satellite. To help solve this problem more quickly, the user can supply an approximate starting position to the GPS receiver.

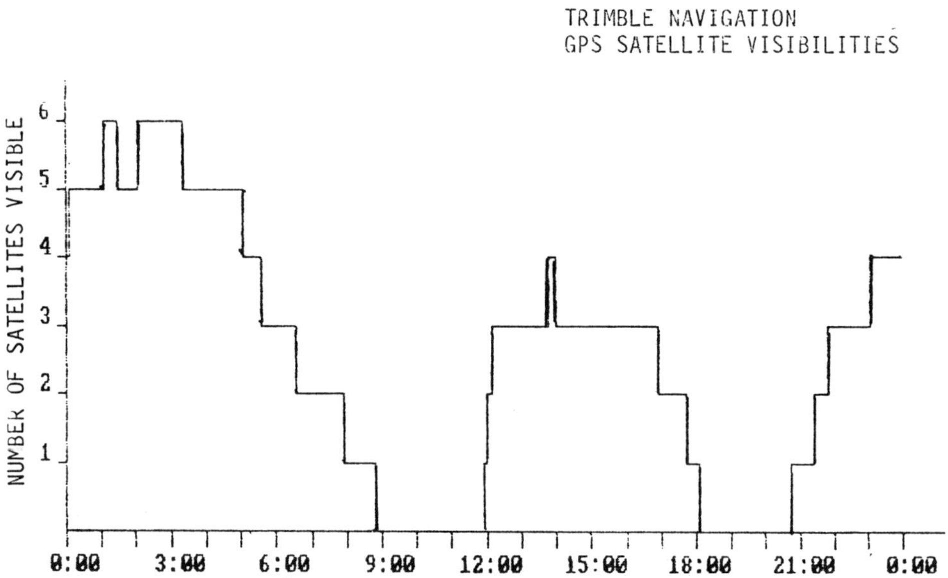


*Fig. 1—GPS Satellite Coverage at Long Beach, CA., January 21, 1986*

## SYMBIOSIS

This section looks at how the two systems can assist with the deficiencies of the other. Let's first look at how the GPS can aid the LORAN.

### *GPS aids LORAN*

***ASF Corrections***—As mentioned before, the LORAN signal propogation velocities vary as a function of surface conductivity. The models used in most receivers attempt to correct for these errors, but it is impossible to do this very accurately because of the daily and seasonal variation in the conductivity. By using the GPS as an accurate reference, the mean correction for a given day can be computed and used to correct the LORAN fixes when the GPS system is not available.

As shown in the typical data in Figure 2, the mean position error between the two systems is about 200 meters, but the RMS of each of the LORAN fixes is about 20 meters and the RMS of the GPS is about 5 meters. The LORAN data was taken every half hour for a 24 hour period.

***Cycle Slip***—Due to many causes, a LORAN Receiver may have a difficult time determining the correct cycle of the LORAN pulse. One of the primary causes is a very weak signal. This can be caused by signal blockage for extreme range to the transmitter. If GPS signals are available, the correct cycle can be determined very accurately, and the Loran can be trusted to give correct answers even when GPS subsequently disappears.

***Ambiguity***—Once again, GPS can be used to eliminate an ambiguity caused by only being able to receive two slave stations. As with cycle slip, once the ambiguity has been resolved, the LORAN can be trusted to give correct answers.

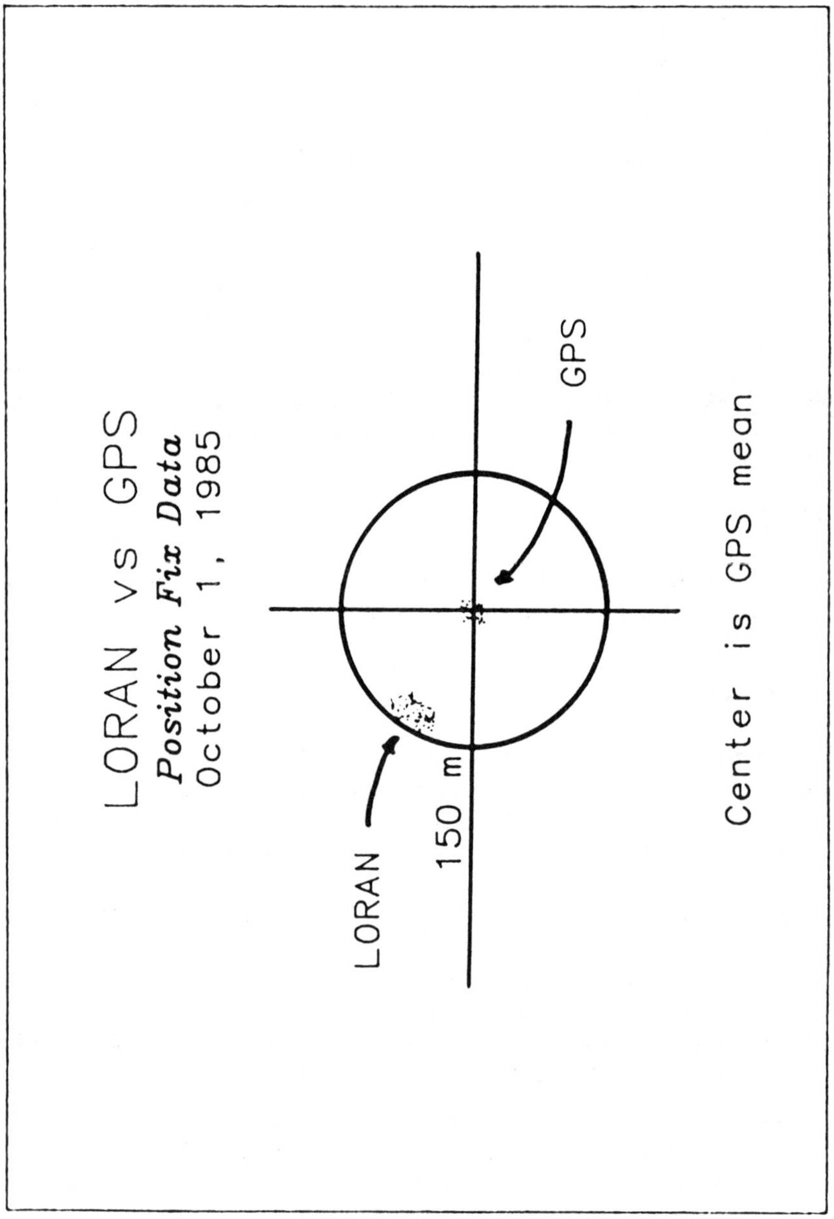


*Fig. 2—Typical LORAN and GPS Position Fixes*

### *LORAN aids GPS*

***Initialization***—From a cold start, the time to first GPS fix can be reduced if an approximate initial position is available. This is particularly true with the present reduced satellite constellation. If the user is in a LORAN coverage area, then the initial position can be provided by the LORAN.

### *Systems aid each other*

There are two ways the systems can aid each other. The first has to do with the fact that one system completely covers time and the other covers space. That is to say, the GPS gives global coverage but not continuous, and the LORAN gives continuous coverage, but not global. Thus the two systems compliment each other in this regard.

The second way they aid each other concerns what we call the extended coverage capability. With some modifications in software, a position fix can be obtained by taking only two satellites and two LORAN transmitters. Thus with one TD and two satellites a position fix can be obtained. This solution extends the coverage area of LORAN and the coverage time of GPS.

## OTHER COMBINATIONS

***GPS/INS***—This combination has several attractive features. If the GPS is not available, the INS can provide a solution, and with GPS, the INS can be continuously calibrated. This is quite attractive in a jamming environment since the INS does not require any external signals. For a marine application, however, this advantage is not pertinent. In addition, this solution is quite expensive, as even low cost INSs cost more than $25,000.

***GPS/Omega***—This solution is also expensive. Even though Omega is global, it does not have the accuracy that LORAN does, and GPS provides the global coverage in any case. Thus this combination is both more expensive and less accurate than the LORAN/GPS.

***GPS/Transit***—The problem with this combination is that the capabilities of both systems overlap. Both systems provide global coverage and both have time gaps in their coverage. What Transit needs most is accurate velocity information, and when GPS could provide this, a GPS solution is available. Thus, a Transit fix is not necessary.

## TYPICAL SCENARIOS

***Coastal User***—For the typical coastal user, the LORAN/GPS combination has many benefits. When GPS is available, much more accuracy is achievable. Since the LORAN ASF corrections will be calibrated by GPS, the LORAN will be more accurate when the GPS is not available. Also, because of the aiding from the GPS, the LORAN will have greatly reduced errors caused by cycle slips and ambiguities.

***Oceanic User***—The oceanic user gets all of the benefits of LORAN when near port at both ends of the journey. In the middle of the trip, when LORAN is not available, GPS can provide accurate periodic navigation. With GPS, set and drift can be computed in the open ocean, and more direct headings and routings can thus be achieved. This is not possible with a system that provides only periodic fixes. Real time navigation as provided by GPS is necessary for this computation. Lastly, the oceanic user can get the extended range capability available only to the user of the LORAN/GPS combination.

## SUMMARY

In summary, we have looked at the two systems, GPS and LORAN, and found that both have some deficiencies, especially in the interim period before all the GPS satellites are launched. More importantly, however, the advantages of one system can compensate for the disadvantages of the other giving the navigator a more reliable and more accurate system than either system alone can provide. In particular, LORAN helps with initialization of GPS and gives continuous coastal coverage. GPS aids LORAN by assisting with cycle selection and ambiguities. It also improves LORAN accuracies by calibrating ASF factors. In addition, it currently gives about twelve hours per day of global coverage.

We have also looked at other combinations and shown them not to be as effective for the marine user as the LORAN/GPS. Until GPS becomes fully operational in late 1988, the best solution appears to be a LORAN/GPS.

## REFERENCES

1. Janiczek, P. M., Ed., *Global Positioning System—Papers published in NAVIGATION,* ION, 1980.
2. Janiczek, P. M., Ed., *Global Positioning System—Papers published in NAVIGATION, VOLUME II,* ION, 1984.
3. Anon., *LORAN-C Users Handbook,* Department of Transportation, U.S. Coast Guard, Comdtinst M16562.3, 1980.
4. Eschenbach, R. F., *GPS, The Coming Revolution in Marine Navigation,* NMEA News, Vol. 12, No. 1, Jan–Feb. 1985.

# Global Positioning System Operational Control System Accuracies

R. BOWEN

Headquarters, Space Division, USAF, E1 Segundo, California

P. L. SWANSON, F. B. WINN, N. W. RHODUS, W. A. FEESS

The Aerospace Corporation, El Segundo, California

*Received January 1985*

## ABSTRACT

In early 1985, an improved ground Operational Control System (OCS) will maintain the navigation service. Primary among the OCS improvements over past GPS navigation systems is a global network of ground antennas (to upload satellite navigation data) and tracking/monitor stations. In addition, a refined Kalman filter will continuously estimate the Global Positioning System (GPS) satellite ephemerides and clock phases and frequencies. Although the user-GPS interface remains unaltered, the accuracy of the GPS navigation service is expected to improve by a factor of three.

Using real and simulated GPS pseudo range radiometric tracking data, The Aerospace Corporation has completed a detailed error analysis which shows that the satellite clock noise contributes more than 90 percent of the total satellite-to-user pseudo range error. If the number of OCS uploads is increased to three per day (as planned), then the accuracy of the navigation service is also expected to improve by nearly a factor of three because the clock-noise contribution to the range error increases linearly with time. Also, this analysis shows the consequence of satellite ephermeris uncertainties in the GPS navigation application.

## INTRODUCTION

The Initial Control System (ICS) is about to be retired. Nearly 5 years of GPS navigation service to the Army Proving Ground at Yuma, Arizona has shown the potential of GPS as a navigation system. The ICS is a ground data reduction and orbit/clock estimation and prediction system—a system that uploads those predictions into GPS Block I Space vehicle (SV) memories for later spacecraft broadcast. In recent years, the ICS has been required to provide the Army Proving Ground with satellite-broadcast pseudo range accurate to 4.3 m ($1\sigma$) for the 2-hr period immediately following satellite state prediction.

Although the GPS-Yuma navigation support is statistically superior to specification requirements (<5 m ($1\sigma$) at 24 hr), there are occasions when ground and on-orbit hardware failures, natural phenomena, or operational difficulties have perturbed individual SV performance on time scales as long as days. On such occasions, the ICS has isolated the offending satellite from the operational constellation. It is the stability, consistency, and accuracy of the ICS navigation service that is the focus of this paper. Stability and consistency are directly observable in the repetitive performance of the ICS. Accuracy, however, has to be indirectly inferred from Kalman filter adjustments to the assumed models. If the data is used for which the truth is known, (i.e., if simulated data is used, the assumed models are not challenged. However, the ability of the filter to follow/track the SV force model dynamics (both deterministic and statistical) and to follow the SV clock phase, frequency, and frequency rate (also both deterministic and statistical) can be shown.

This paper examines recent ICS operational data and shows that the prime limitation to GPS pseudo range (PR) prediction accuracy is SV clock noise. Simulations are used to demonstrate how a Kalman filter of GPS baseline design responds to SV clock noise. Since, the ICS is to be replaced by the OCS prior to March 1985, both OCS and ICS performances are investigated.

A brief description includes the fact that the ICS tracks space vehicles (SV) with only four monitor stations (MS): Vandenberg Air Force Base, Hawaii, Guam, and Alaska (Figure 1). All are located in the North Pacific basin. In addition, the ICS operates with a single ground antenna (GA) located at the Vandenberg Air Force Base. With the single ground antenna, the ICS generally uploads orbit predictions and clock predictions to each satellite once per day. For elevation angles above 15 deg, the ICS monitor stations (MS) track the SVs about 50 percent of the time.

The OCS will track satellites using five MSs (1, Colorado Springs; 2, Ascen-

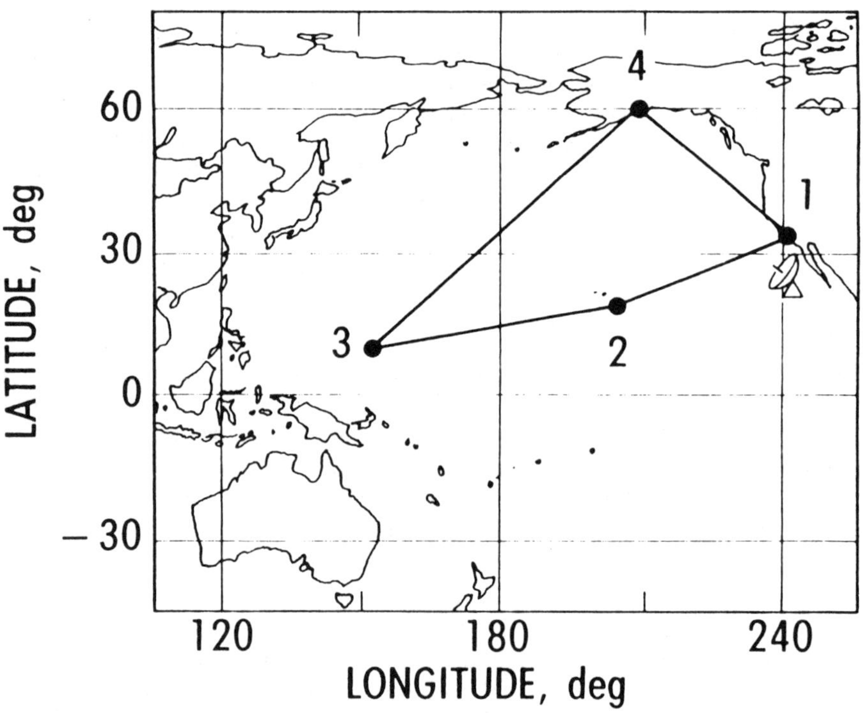


*Fig. 1—ICS monitor station and upload antenna locations*

sion Island; 3, Diego Garcia; 4, Kwajalein; 5, Hawaii). The OCS monitor stations are globally dispersed in longitude such that the maximum longitude separation is less than 90 deg (Figure 2). The OCS MSs can track the average SV over 80 percent of the time. Ground antennas are co-located with the MSs at Ascension, Diego Garcia, and Kwajalein (Figure 2), which allows uploading of GPS satellites during most of each GPS orbit. In short, the OCS will exploit nearly continuous two-way communications to monitor/track and upload all GPS satellites. The current plan is to upload all SVs three times per day.

### *Initial Control System (ICS)*

The ICS still supports GPS navigation applications at the Army Proving Ground in Yuma, Arizona, and there are other GPS routine users that have gone into operation. At the time of this writing, the ICS uploads six operational Block I GPS satellites. The ICS tracks and uploads navigation data while the SV constellation is over the Pacific basin (Figure 1). Tracking data, radiometric pseudo range (PR), is transferred from the four Pacific MSs to the ICS Master Control Station (MCS) at Vandenberg AFB. There the data is processed by a Kalman filter designed to estimate the trajectory and clock phase, frequency, and aging of each SV, as well as other ground or MS states. Once the SV orbit and clock states have been estimated, the ICS predicts the SV orbit/clock states for the next 26 hours and uploads that prediction into the memory of each SV. In the subsequent 24 hours, the predictions are retransmitted by each SV so that Yuma and other users have "knowledge" of the SV orbits and clocks. Each day the ICS repeats the process.

A conceptual definition of pseudo range takes into account the SV operation. The SV transmits a clock code from which an equipped user can note the detection of the SV clock code modulation relative to his reference clock. A modulation on the SV carrier specifies to the user the SV ephemeris and clock phase and frequency. With four equations in four unknowns, the user can solve for location and clock offset. The four equations have the form:

$$\mathrm{PR} = \left\| \begin{bmatrix} X - x \\ Y - y \\ Z - z \end{bmatrix} \right\| + [\text{SV clock time} - \text{user clock time}]\, C \qquad (1)$$

where PR is the satellite-to-user pseudo range,

$\begin{bmatrix} X \\ Y \\ Z \end{bmatrix}$ is the SV cartesian coordinates in World Geodetic Survey, 1972, earth-fixed coordinates,

$\begin{bmatrix} x \\ y \\ z \end{bmatrix}$ is the assumed position of the user in WGS 1972 coordinates,

and C is the velocity of light.

Obviously, the strength of the user's solution is dependent upon the accuracy of the SV-transmitted "knowledge."

The user range error (URE) is the projection of the SV ephemeris position

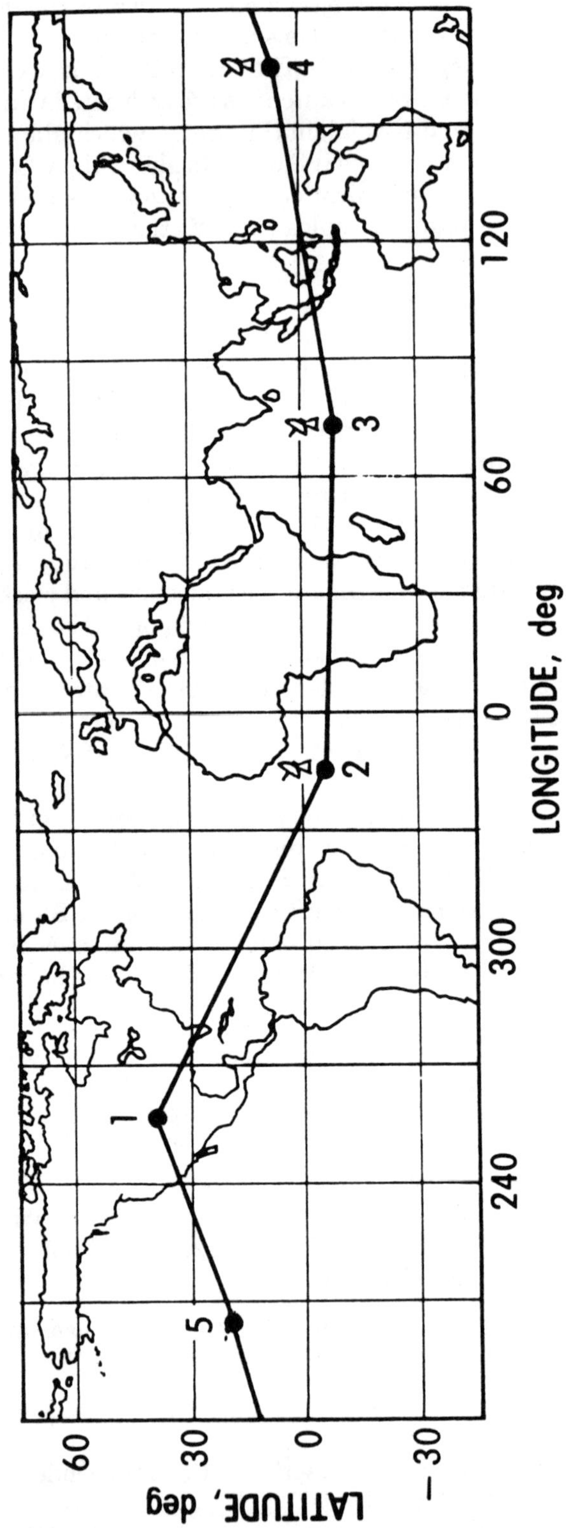


*Fig. 2—OCS monitor station and upload antenna locations*

error onto the SV-to-user line-of-sight, plus the SV clock phase prediction error, multiplied by the speed of light.

$$\text{URE} = \delta\text{PR} = \frac{\begin{bmatrix}\Delta X \\ \Delta Y \\ \Delta Z\end{bmatrix} \cdot \begin{bmatrix}X - x \\ Y - y \\ Z - z\end{bmatrix}}{\left\| \begin{bmatrix}X - x \\ Y - y \\ Z - z\end{bmatrix} \right\|} + (\text{SV clock phase error})\ C \qquad (2)$$

Hence, the URE provides a measure of the ability of the ICS to predict SV orbits and clocks. An example of URE growth as a function of prediction interval up to 10 days is shown in Figure 3. The $\Delta X$, $\Delta Y$, $\Delta Z$, and SV clock phase errors of (2) were calculated by differencing the predicted SV states from those estimated in real time. Both predictions and estimation are ICS products. The URE is factored into its ephemeris and clock error contributions. The SV clock prediction error dominates. Indeed, that impression is reinforced by Table 1.

For this arbitrary sample, the differences between the ephemeris predictions and real time Kalman filter estimates of the trajectory when projected onto the SV-to-Yuma line-of-sight never exceeded 70 m and had an rms difference of 38.7 m. The clock predictions differed from estimates by a maximum of 318 m and had an rms difference of 118.4 m. The clock prediction errors, on average, exceed the ephemeris prediction errors by nearly three times. It is to be noted that the ephemeris prediction error is of orbit and diurnal period while the clock prediction error growth is quadratic in this instance.

For prediction times of one day or less, the SV clock appears still the dominant error source in that the predicted minus estimated SV clock phase difference

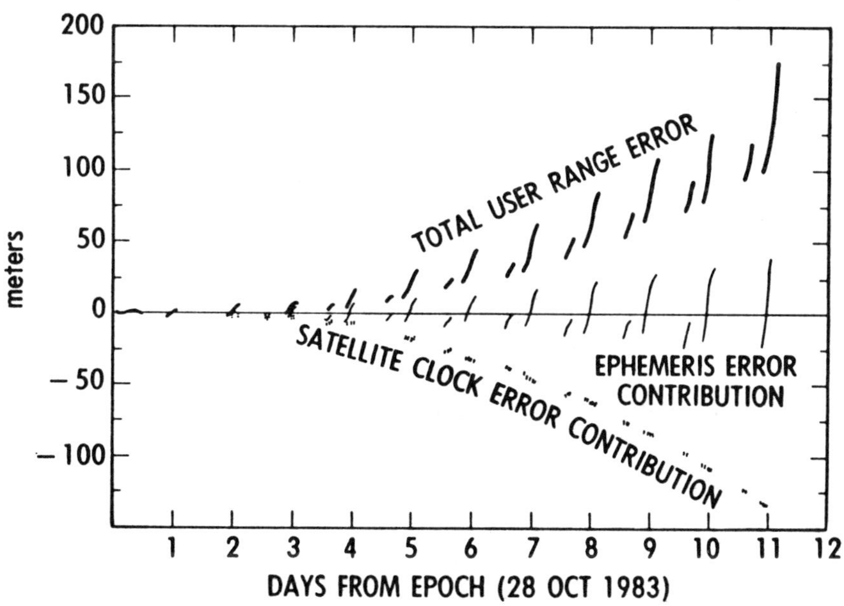


*Fig. 3—Pseudo range error growth*

Table 1.—Components of User Range Errors

| Upload Date | Satellite | 10 Days, User Range Error, meters | Error Contribution meters | |
|---|---|---|---|---|
| | | | Ephemeris | Clock |
| 18 Sept 83 | | | | |
| | 5 | 318.31 | 14.31 | 318.00 |
| | 3 | 101.67 | 9.02 | 101.28 |
| | 4 | 60.86 | 60.39 | 12.88 |
| | 6 | 162.29 | 10.99 | 161.93 |
| | 8 | — | — | — |
| 30 Sept 83 | | | | |
| | 5 | 49.37 | 48.76 | 11.44 |
| | 3 | 77.02 | 24.30 | 73.21 |
| | 4 | 73.26 | 68.50 | 28.57 |
| | 6 | 51.17 | 11.06 | 50.00 |
| | 8 | 133.22 | 50.61 | 123.35 |
| 1 Oct 83 | | | | |
| | 5 | 35.33 | 23.45 | 26.74 |
| | 3 | 93.68 | 34.48 | 87.31 |
| | 4 | 16.18 | 6.64 | 14.80 |
| | 6 | 62.00 | 13.54 | 60.55 |
| | 8 | 96.65 | 36.79 | 89.60 |
| 15 Oct 83 | | | | |
| | 5 | 154.46 | 50.08 | 146.37 |
| | 3 | 33.69 | 17.67 | 28.85 |
| | 4 | 121.04 | 67.72 | 101.01 |
| | 6 | 279.32 | 69.38 | 270.84 |
| | 8 | 52.43 | 26.54 | 45.46 |
| 28 Oct 83 | | | | |
| | 5 | 40.82 | 13.82 | 38.48 |
| | 3 | 122.57 | 47.43 | 113.33 |
| | 4 | 21.34 | 11.93 | 17.82 |
| | 6 | — | — | — |
| | 8 | 67.23 | 34.67 | 57.91 |

statistically accounts for a full 90 percent of the URE. Examples of URE growth over 24 hr are shown in Figure 4. The URE growth is factored into clock and ephemeris component error growths.

An extensive sample of 47 days (day 304 to 351, 1984), supports the evidence that the URE is driven by clock prediction errors (Table 2). This 47-day sample of ICS UREs involves five SVs (Navstars 3, 4, 8, 9, and 10). The URE sample rms is 4.5 m; the rms of the SV clock error components is 4.1 m; and the rms of the SV ephemeris error component is approximately 2.0 m (Table 2). For this sample that excludes Navstar 6, which had a marginal rubidium clock performance during this time, the SV clock phase 24-hr prediction accounts for approximately 90 percent of the URE.

If Navstar 6 data is added to the sample, and perhaps it should be since the clock was judged to have a flicker noise level below 5 pp $10^{13}$ (the specification), the system rms URE is 5.8 m at the 24-hr prediction point. Although the

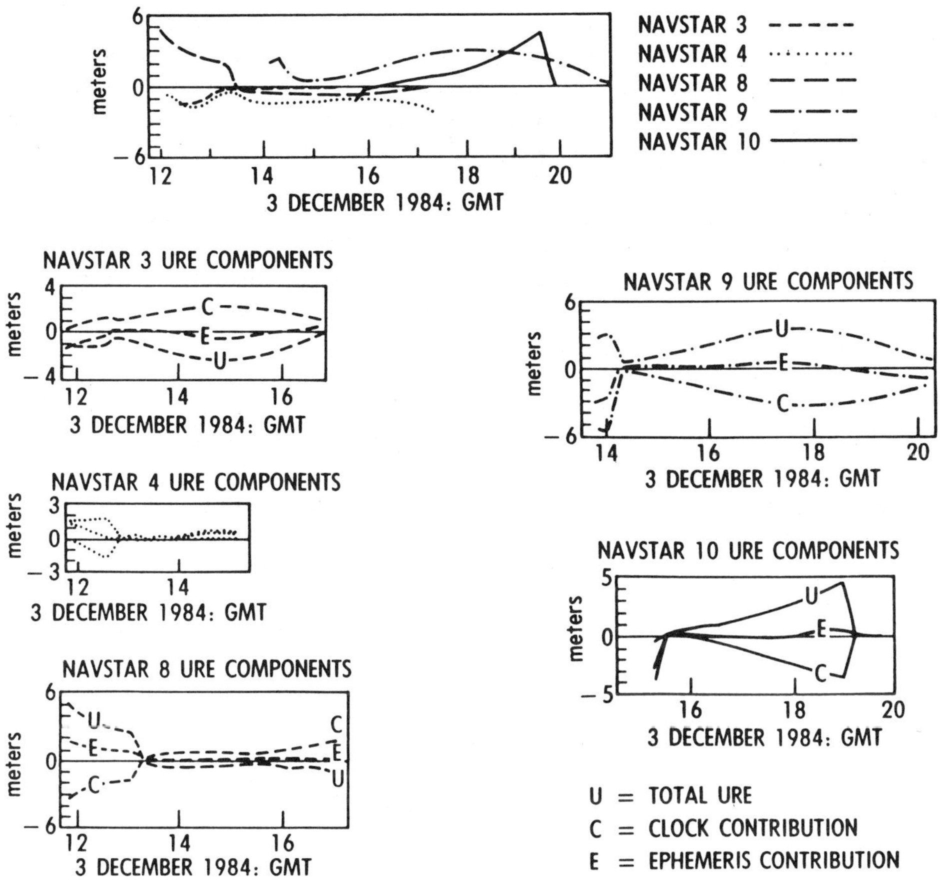


*Fig. 4—ICS computed user range errors for Yuma*

Table 2.—User Range Error Statistics

| | Navstar Satellite Number | | | | | | | |
|---|---|---|---|---|---|---|---|---|
| | 3 | 4 | 6 | 8 | 9 | 10 | Combined | |
| Total URE, meters | 4.0 | 4.4 | (9.7) | 3.9 | 5.5 | 5.3 | 4.7 | (5.8) |
| SV Clock Error, meters | 3.3 | 4.1 | (9.6) | 3.6 | 4.9 | 4.4 | 4.1 | (5.4) |
| Ephemeris Error, meters | 2.3 | 1.0 | (1.3) | 1.6 | 2.2 | 3.1 | 2.1 | (2.0) |

Navstar 6 clock performance was within specification, its consequence to the November 1984 statistics is degrading. The other SV clocks are a factor of two or three times more stable.

With Navstar 6 data excluded, a comparison of the URE performance in the

November 1984 performance period (day 304 to 351, 1984) with that from a July 1984 period (190:84 to 240:84) shows similar statistics: 73 percent of the UREs of the November uploads were below 4 m at 24 hours and 68 percent of the July 1984 uploads were below 4 m (Figure 5 and Figure 6). As stated, the November rms error of the user range was 4.5 m, while the July 1984 50-day period produced a 4.2 m (1$\sigma$) URE.

Although there is comfort in the better than 5 m (1$\sigma$) URE performance of the ICS (and that performance is realized using SV clock and ephemeris predictions that are generally 24-hours old), there are occasions when prediction accuracy falters to the extent that ICS operators perform contingency uploads. Often the contingency uploads are preventative; preventative to the extent that Kalman estimates of the SV clock and ephemeris appear to be diverging from what was uploaded into the SV. To wait until the filter-upload potential problem is confirmed or dispelled would mean waiting until after the SV was no longer in view of the single ICS GA. Contigency uploads are employed for less than 15 percent of the SV tracking passes. The need for contingency uploads is not fully understood.

There are indications that GPS models of the SV forces and clocks could be improved.

1. It is known that the SVs radiate tens of Watts of thermal energy and that the wattage varies with spacecraft equipment duty cycles and the SV-

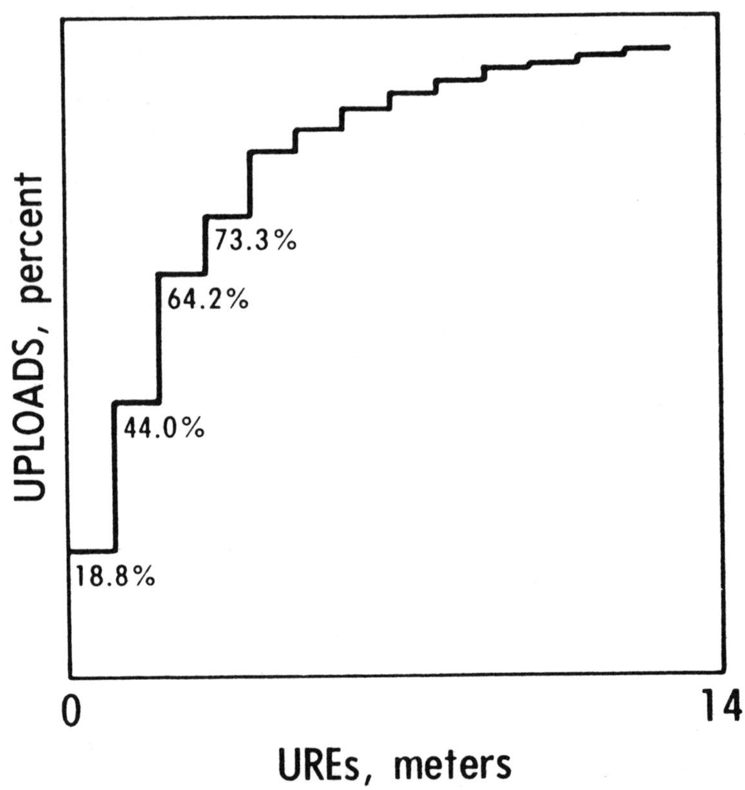


*Fig. 5—Upload distribution of user range errors (Nov. 1984)*

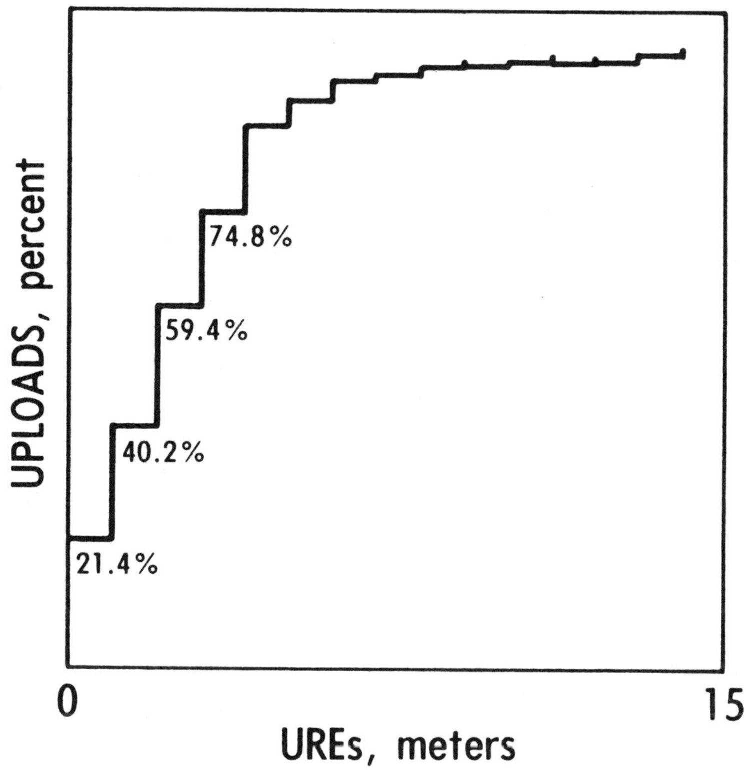


*Fig. 6—Upload distribution of user range errors (Jul. 1984)*

body orientation with respect to the sun. GPS employs no model for thermal radiation.

2. The solar radiation force model for GPS satellites also requires augmentation in that the Kalman filter not only solves for SV position and velocity but also estimates two unmodeled components of acceleration: one component along the SV, body-fixed, Y-axis and a second component assumed normal to the SV polar panel surfaces.
3. Frequently, when SVs enter or exit earth shadow or lunar shadow, SV clocks experience thermal variation and frequency change. Occasionally, there is evidence that SVs have experienced unmodeled forces that change the semi-major axis of the orbit by meters.

The GPS Joint Program Office is directing a technical effort to eliminate or lessen the need for contingency uploads that involves IBM, Rockwell, The Aerospace Corporation, the Defense Mapping Agency, the Naval Surface Weapons Center, the Naval Research Laboratory, the U.S. Naval Observatory, and the National Bureau of Standards.

Thus, despite an ICS statistical performance considerably better than requirements, there is the expectation that the OCS will benefit operationally by refining the SV force and clock models.

### *Operational Control System (OCS)*

By March 1985, the Operational Control System becomes the controller of the GPS. It will estimate satellite trajectories and clock phase, frequency, and aging from the pseudo range and accumulated-delta-range (ADR) data reductions using a Kalman filter, much as the ICS did. However, OCS MSs will track or monitor SVs 83 percent of the time and OCS GAs will upload SVs nominally three times per day. The nearly continuous tracking of SV radiometric data will allow the OCS filter to maintain a near steady-state "knowledge" of the SV clock states and ephemeris states. Eight-hour uploads of clock and ephemeris predictions means that the excursions of the SV clocks will be three times shorter than those characteristic of the ICS operations.

As a means to test the emerging OCS, the Joint Program Office arranged for The Aerospace Corporation to simulate tracking data that incorporated the OCS deterministic models of SV dynamics and clock behavior, tracking data that also modeled statistical uncertainties consistent with the dynamical noise "contractually assumed" for the Rockwell SV force model (unmodeled acceleration set at $10^{-9}$ m/sec$^2$ ($3\sigma$)), and statistical noise "contractually assumed" for the Rockwell SV clock (Figure 7).

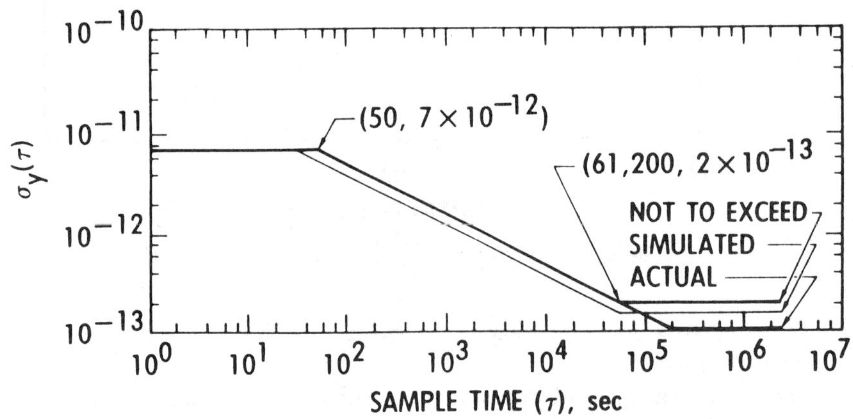


*Fig. 7—SV clock Allan variances*

A not-to-exceed Allan variance is included in Figure 7 that shows the frequency modulation white phase noise and flicker floor noise levels to be equal at 61,200 sec. For the purposes of this simulation, the phase and flicker noises were chosen such that the simulated noise does not exceed the specified bound. This was achieved by scaling the specification Allan variance bound on phase and flicker levels by $\sqrt{2}$ to define the asymptotic Allan variance curve. The specification Allan bounds, the asymptotic Allan variance used in this simulation, and the SV cesium clock variance are shown in Figure 7.

The Aerospace simulation assumes the full GPS 18-satellite constellation (Figure 8). The simulation was run for 30 days.

White-noise jerk and white-noise accelerations were added to the SV accelerations integrated to define the orbits specified in Figure 8. In all, 50 perturbed trajectories were generated such that the rms of 16-hour and 14-day UREs approximated 1 m and 110 m, respectively. Actual UREs were 1.01131 m and 114.5036 m, respectively. The root-mean-squares of the UREs were computed from the formula

$$\text{URE} = [(\text{radial perturbation})^2 + 0.0192\,(\text{in track perturbation})^2 + 0.0192\,(\text{cross-track perturbation})^2]^{1/2}$$

From these 50 samples of perturbed trajectories, 18 were chosen at random for this GPS tracking data simulation. The 16-hour and 14-day UREs (root-mean-square) associated with these 18 selected orbits are shown in Table 3. One sample of trajectory perturbation is shown in Figure 9.

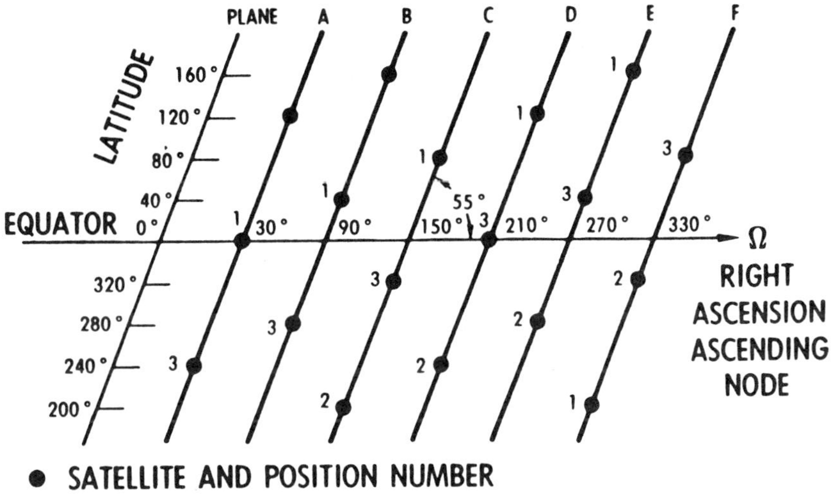


*Fig. 8—OCS baseline satellite constellation*

Table 3.—Assumed Satellite Trajectory RMS Errors

| Plane | Satellite | 10-hr URE meters | 14-day URE meters |
|---|---|---|---|
| A | 1 | 0.9624 | 148.7778 |
| A | 2 | 1.1261 | 88.2452 |
| A | 3 | 1.0776 | 59.3367 |
| B | 1 | 0.9851 | 118.5011 |
| B | 2 | 1.0613 | 123.8261 |
| B | 3 | 1.0027 | 100.1853 |
| C | 1 | 1.0951 | 156.7645 |
| C | 2 | 0.9680 | 88.9386 |
| C | 3 | 1.0615 | 123.0544 |
| D | 1 | 1.0067 | 88.1200 |
| D | 2 | 1.0428 | 100.8753 |
| D | 3 | 1.0401 | 98.7732 |
| E | 1 | 0.9912 | 152.3074 |
| E | 2 | 0.9425 | 77.9240 |
| E | 3 | 1.0235 | 164.9850 |
| F | 1 | 0.9156 | 85.0566 |
| F | 2 | 1.0377 | 93.6489 |
| F | 3 | 0.9498 | 85.5475 |
| Total RMS | | 1.0176 | 112.5445 |

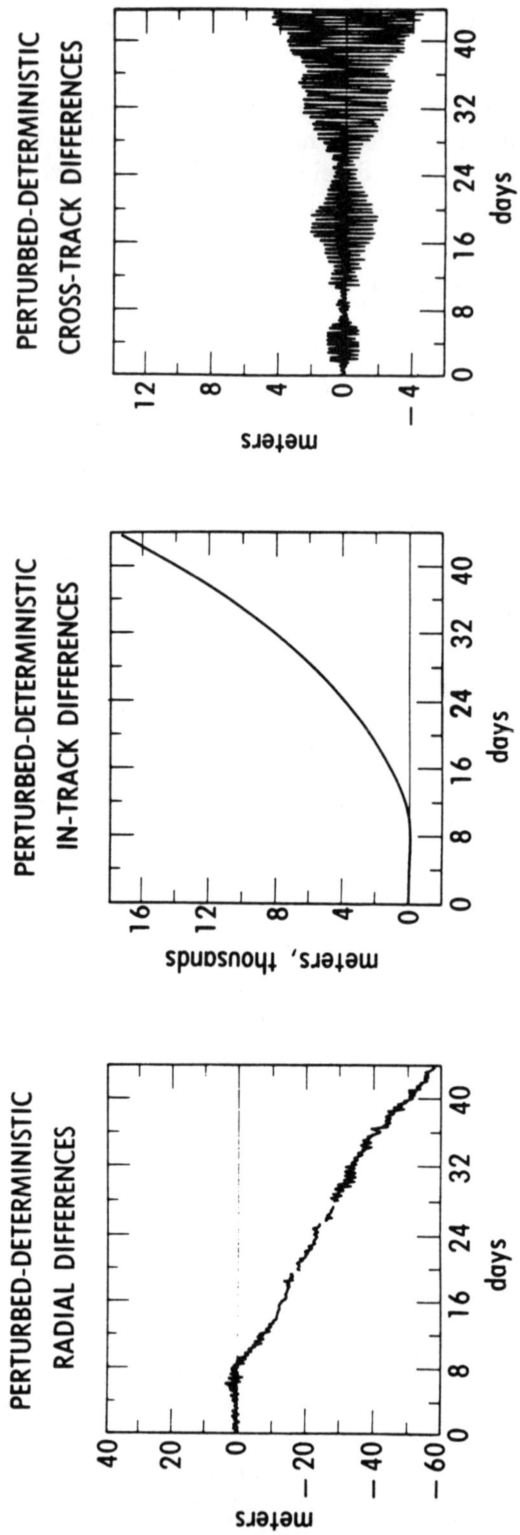


*Fig. 9—Example of radial, in-track and cross-track random walks*

The trajectory perturbations imposed on the simulation (Table 3, Figure 9) are primarily in-track and amount to approximately 196 m (1σ) per week. Actual trajectory perturbations observed in ICS operations are approximately 50 m/week. Hence, the "contractually assumed" is considerably worse than the real world.

Comparisons of inferred clock stability with that "assumed" also implies that actual SV clocks are $\sqrt{2}$ times more stable than assumed (Figure 7) in this simulation. A statistical overview of the assumed SV clock noise is provided in Table 4. For convenience, the rms of the integrated phase error is expressed in terms of URE.

The principal understanding to emerge from the analysis of this simulated data is that a Kalman filter of OCS design can perform 10-hour predictions of SV orbit and clock states with sufficient accuracy to support 6 m (1σ) UREs. One example of UREs is pictured in Figure 10. The UREs are computed using (2) where $\triangle X$, $\triangle Y$, and $\triangle Z$ is the difference between ephemeris prediction and estimation. For the clock phase error, it is the difference between what was predicted and what is estimated. The results of exercising (2) every 15 min of the 30-day simulation to compute 3-hour, 6-hour, 10-hour, 1-day, 3-day, 7-day and 14-day UREs as well are shown in Figure 10.

Figure 11 is much like Figure 10. However, $\triangle X$, $\triangle Y$, $\triangle Z$ and clock phase error, as the difference between prediction and the truth, are shown in Figure 11.

The data in Figure 11 and Figure 10 show 10-hour UREs for the entire 30-day simulation that are quite similar. This is consistent with the quality of the

Table 4.—Space Vehicle Clock Random Walk Contributions to 10-hr/14-day UREs

| SV Clock | | RMS Error | |
|---|---|---|---|
| Plane | Satellite | 10-hr, meters | 14-day, meters |
| A | 1 | 4.9 | 107 |
| A | 2 | 3.8 | 69 |
| A | 3 | 5.1 | 114 |
| B | 1 | 7.9 | 260 |
| B | 2 | 4.8 | 133 |
| B | 3 | 5.1 | 138 |
| C | 1 | 5.0 | 111 |
| C | 2 | 5.1 | 143 |
| C | 3 | 4.8 | 114 |
| D | 1 | 3.8 | 70 |
| D | 2 | 3.9 | 65 |
| D | 3 | 3.4 | 26 |
| E | 1 | 4.3 | 60 |
| E | 2 | 4.2 | 80 |
| E | 3 | 5.3 | 141 |
| F | 1 | 4.3 | 94 |
| F | 2 | 5.6 | 155 |
| F | 3 | 4.0 | 73 |
| Combined RMS | | 4.7 | 126 |

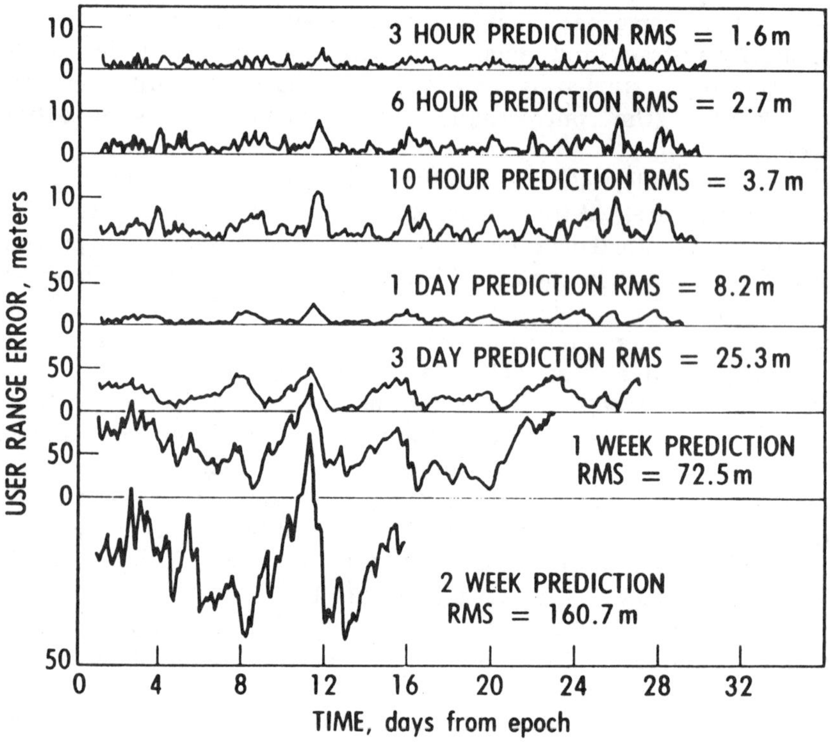


*Fig. 10—URE relative to Kalman filter estimate*

filter running estimates of the orbits and clocks. The rms difference between filter running estimates and the truth in terms of URE is 1.4 m, with SV clock estimations again contributing 90 percent of the error.

The rms difference between UREs relative to the filter estimates and UREs relative to truth orbits and clocks is 0.6 m. The URE statistics of the total simulation are presented in Tables 5 and 6. The UREs relative to the filter estimates are presented in Table 5. The data of Table 6 are relative to the truth orbits and truth clocks. The 10-hour URE is approximately 4.3 m ($1\sigma$) relative to truth and 3.7 m ($1\sigma$) relative to filter estimation.

As stated above, a second finding to emerge from the simulation is that SV clock noise is the prime contributor to UREs. The error of the SV clock phase prediction contributes 1.33 m ($1\sigma$) to the UREs without prediction and contributes 4.2 m ($1\sigma$) to the UREs associated with 10-hr predictions. Over 24 hrs the rms of the URE is linear. Over 24 hrs the SV clock error rms is also linear.

The ICS operational filter performance with real data supports the same conclusion: 90 percent of the total URE is the clock error contribution and the growth is linear. For specification SV cesium clocks (Figure 7) assumed in this simulation, the clock phase excursions accrue to 4.7 m ($1\sigma$) URE in 10-hr (Table 4), while ephemeris perturbations as simulated contribute another meter ($1\sigma$).

It is interesting to note that the filter managed to absorb some of the clock/ephemeris random walk into the ephemeris and clock state estimations. The data of Table 6 shows that the 10-hour URE has an rms of only 4.3 m ($1\sigma$) and

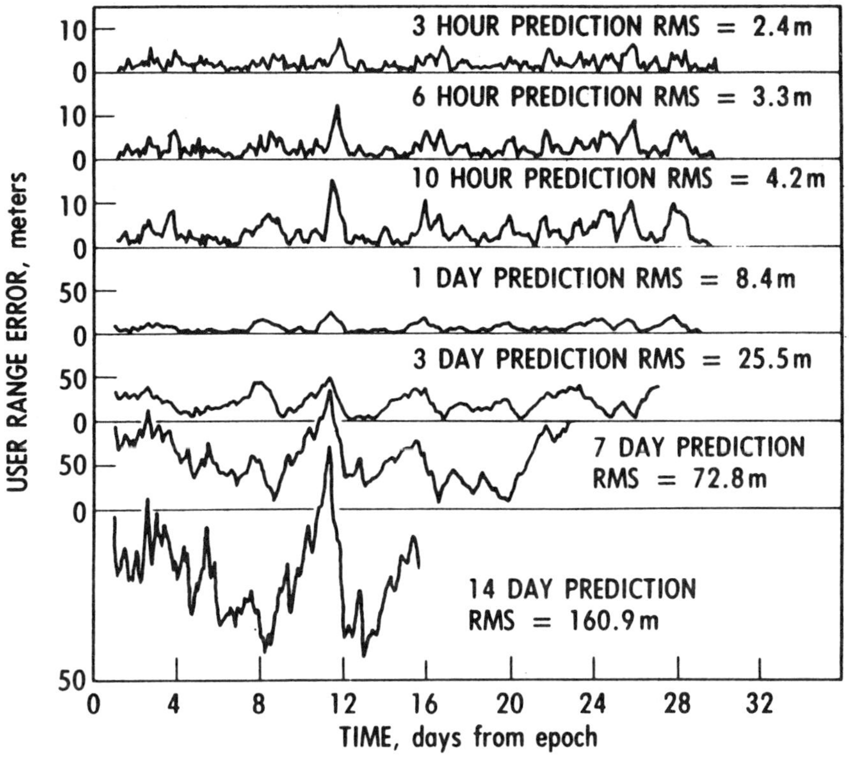


*Fig. 11—URE relative to simulated true*

that 95 percent of the URE is the consequence of accumulated clock noise. The clocks were simulated to experience 4.7 m (1σ) phase drifts in 10 hours (Table 4). When the SV clock flicker is reduced to observed operational levels (1 pp $10^{13}$), the URE at 10 hours is 2.3 m rms.

## CONCLUSIONS

The Initial Control System (ICS) has provided the Army Proving Ground in Yuma, Arizona, with better than 5 m (1σ) satellite-to-user ranging. The 5 m (1σ) accuracy is a realized performance that is two times better than the system design accuracy. This enhanced performance is primarily due to Rockwell satellite clock stabilities that exceed contractual specification by a factor of almost two.

In 1985, the Operational Control System (OCS) is scheduled to replace the ICS. Based on ICS performance, the OCS should meet the 6 m (1σ) user range accuracy requirement with ease. In addition, it is shown that the OCS sensitivity to space vehicle clock instability is three times less than that of the ICS. This is because the OCS will upload satellites three times more frequently than the ICS. However, even with three uploads per day, satellite clock predictability continues to limit GPS user range accuracy to 3 m (1σ) 10 hours after satellite ephemeris and clock phase prediction.

Table 5.—URE Performance Statistics (Filter Prediction—Filter Estimate)

| Plane | SV Number | Prediction Interval, meters | | | | | | |
|---|---|---|---|---|---|---|---|---|
| | | 3-hr | 6-hr | 10-hr | 24-hr | 3-day | 7-day | 14-day |
| A | 1 | 1.6 | 2.6 | 3.6 | 8.1 | 28.6 | 90.3 | 302.8 |
| A | 2 | 1.6 | 2.6 | 3.5 | 7.5 | 19.2 | 61.4 | 111.6 |
| A | 3 | 1.6 | 2.7 | 3.7 | 8.2 | 25.3 | 72.5 | 160.7 |
| B | 1 | 1.5 | 2.4 | 3.4 | 7.3 | 25.0 | 75.2 | 251.0 |
| B | 2 | 1.5 | 2.5 | 3.4 | 7.3 | 23.2 | 56.8 | 164.7 |
| B | 3 | 1.6 | 2.7 | 3.8 | 8.4 | 20.6 | 45.3 | 108.8 |
| C | 1 | 2.2 | 3.3 | 4.3 | 9.0 | 25.5 | 68.9 | 193.4 |
| C | 2 | 1.5 | 2.5 | 3.4 | 6.9 | 20.3 | 69.9 | 167.0 |
| C | 3 | 1.6 | 2.7 | 3.7 | 8.2 | 25.1 | 66.3 | 193.4 |
| D | 1 | 1.5 | 2.4 | 3.2 | 6.8 | 18.0 | 56.9 | 293.7 |
| D | 2 | 1.6 | 2.7 | 3.7 | 7.7 | 25.5 | 75.6 | 165.3 |
| D | 3 | 1.4 | 2.4 | 3.3 | 7.7 | 23.9 | 84.8 | 117.2 |
| E | 1 | 1.8 | 3.2 | 4.7 | 10.3 | 29.8 | 83.1 | 233.8 |
| E | 2 | 1.6 | 2.7 | 3.7 | 8.5 | 26.3 | 64.9 | 186.7 |
| E | 3 | 1.9 | 3.1 | 4.4 | 10.0 | 30.7 | 80.4 | 206.4 |
| F | 1 | 1.5 | 2.5 | 3.4 | 7.9 | 23.7 | 73.7 | 198.6 |
| F | 2 | 1.4 | 2.4 | 3.2 | 6.9 | 21.0 | 71.7 | 182.1 |
| F | 3 | 1.6 | 2.7 | 3.9 | 8.4 | 21.3 | 67.6 | 183.2 |
| Total RMS | | 1.6 | 2.5 | 3.7 | 8.1 | 24.3 | 71.0 | 197.2 |

Table 6.—URE Performance Statistics (Filter Prediction—Truth)

| Plane | SV Number | Prediction Interval, meters | | | | | | | |
|---|---|---|---|---|---|---|---|---|---|
| | | 0-hr | 3-hr | 6-hr | 10-hr | 24-hr | 3-day | 7-day | 14-day |
| A | 1 | 1.5 | 2.4 | 3.4 | 4.4 | 8.6 | 28.7 | 90.2 | 302.4 |
| A | 2 | 1.4 | 2.4 | 3.3 | 4.1 | 7.6 | 19.1 | 62.1 | 111.6 |
| A | 3 | 1.3 | 2.4 | 3.3 | 4.2 | 8.4 | 25.5 | 72.8 | 160.9 |
| B | 1 | 1.4 | 2.2 | 3.1 | 4.0 | 7.7 | 25.3 | 75.2 | 251.6 |
| B | 2 | 1.4 | 2.3 | 3.2 | 4.0 | 7.7 | 25.3 | 56.8 | 165.3 |
| B | 3 | 1.3 | 2.3 | 3.4 | 4.4 | 8.7 | 20.5 | 45.3 | 109.0 |
| C | 1 | 1.7 | 2.8 | 3.7 | 4.7 | 9.0 | 25.4 | 68.9 | 193.7 |
| C | 2 | 1.1 | 2.1 | 3.0 | 3.9 | 7.0 | 20.3 | 70.3 | 167.7 |
| C | 3 | 1.4 | 2.4 | 3.4 | 4.3 | 8.7 | 25.1 | 66.8 | 193.9 |
| D | 1 | 1.2 | 2.3 | 2.8 | 3.6 | 6.9 | 18.0 | 57.1 | 293.8 |
| D | 2 | 1.4 | 2.4 | 3.4 | 4.3 | 8.1 | 25.7 | 75.4 | 165.5 |
| D | 3 | 1.4 | 2.3 | 3.1 | 4.0 | 8.0 | 23.9 | 85.1 | 116.8 |
| E | 1 | 1.6 | 2.7 | 3.9 | 5.2 | 10.7 | 30.1 | 83.6 | 233.3 |
| E | 2 | 1.2 | 2.2 | 3.2 | 4.1 | 8.7 | 28.1 | 64.6 | 186.6 |
| E | 3 | 1.6 | 2.7 | 3.9 | 5.2 | 10.7 | 30.8 | 80.0 | 206.2 |
| F | 1 | 1.1 | 2.1 | 3.0 | 3.9 | 7.9 | 23.6 | 73.9 | 198.8 |
| F | 2 | 1.3 | 2.2 | 3.1 | 3.8 | 7.2 | 21.0 | 71.7 | 182.1 |
| F | 3 | 1.4 | 2.4 | 3.5 | 4.6 | 8.6 | 21.5 | 67.9 | 183.5 |
| Total RMS | | 1.4 | 2.4 | 3.2 | 4.3 | 8.4 | 24.6 | 71.2 | 197.3 |

ACKNOWLEDGEMENTS

Dr. H. Beisner of the IBM Corporation developed the simulated clock model used in this analysis. Mr. S. Francisco of IBM, Mr. J. Carr and Mr. E. Swift of the Naval Surface Weapons Center, and Dr. Wm. Wooden and Mr. B. Merritt of the Defense Mapping Agency contributed significantly to this effort through a comparative calibration of the IBM, Aerospace, and NSWC GPS orbit and radiometric data systems.

# GPS Integrity Channel*

RONALD BRAFF, CURTIS SHIVELY
The MITRE Corporation, McLean, VA
*Received January 1986*

## ABSTRACT

This paper documents a system description and operational concept for ensuring the integrity of the Global Positioning System (GPS) signals-in-space for use in the National Airspace System. The concept employs a ground monitoring network, L-band transponders placed onboard already planned geostationary communication satellites to broadcast integrity data, and an integrated receiver containing both the GPS and the GPS Integrity Channel (GIC). The three, aforementioned system segments are described. The feasibility of employing a network of only four ground monitors to cover CONUS is analyzed. A discussion concerning implementation of such a system is also presented.

## INTRODUCTION

This paper documents an investigation of an approach for ensuring the integrity of the Global Positioning System (GPS). This is required so that GPS can be used as a fully capable supplemental navigation system in the National Airspace System (NAS) when it is implemented by the Department of Defense in the 1988–1989 time frame.

### *Background*

GPS will consist of a constellation of 21 satellites (18 operating satellites plus 3 active spare satellites on orbit) distributed in 6 inclined orbital planes. At least 4 satellites are required for unaided position determination. The period of the satellites will be 12 hours, so the satellite geometry varies as a function of time. GPS will provide accurate position and time determination. A concise description of GPS is given in Reference 1.

The FAA has defined the operational requirements to establish GPS as a fully capable supplemental air navigation system to be used in the National Airspace System (NAS) in the 1990s. A supplemental air navigation system is an approved navigation system that can be used in controlled airspace in conjunction with a *sole means air navigation system.*** A fully capable supplemental system can also be used under certain conditions where there is no sole

*This paper is based on navigation system analysis studies performed by The MITRE Corporation for the FAA. The contents of this paper reflect the views of the authors who are responsible for the accuracy, analyses and suggestions presented herein, and do not necessarily reflect the official views or policy of the FAA.

**A *sole means air navigation system* is an approved system that can be used for specific phases of air navigation in controlled airspace without the need for any other navigation system.

means navigation system coverage. The use of GPS where there is no sole means navigation coverage is of most interest. Such applications in the NAS include nonprecision approaches at all airports where there is no VOR/DME coverage, low altitude coverage (especially for helicopters), and off-shore coverage.

### *Nonprecision Approach Operational Requirements*

A nonprecision approach is an instrument approach that requires only horizontal guidance from a radio-navigation system. Its operational requirements are the most stringent ones for using GPS in the NAS. The two major operational requirements set down by the FAA for GPS to be a fully capable supplemental navigation system for nonprecision approaches are the horizontal position determination accuracy requirement of 100 m for the GPS signals-in-space (95 percent error)[1] and a GPS malfunction warning response time of 10 seconds to guarantee the signals' integrity. The 100 m accuracy requirement will be met by GPS. The 10 seconds malfunction response time is necessary to provide the pilot with enough time to safely execute a missed approach maneuver. For instance, the VOR/DME automatically monitors itself and terminates transmission if a performance parameter is out of tolerance for more than 8 seconds.

The 10 second malfunction warning response time includes the time to detect a malfunction, the time to communicate the command to the GPS receiver to drop the malfunctioning signal from the position determination process, and to present the flag alarm to the pilot if there are insufficient remaining GPS signals being tracked for position determination. If a GPS receiver is tracking signals from 4 GPS satellites and one of these is determined to be out of tolerance, then the receiver must automatically drop the satellite from the position determination calculation and activate the warning flag displayed to the pilot. With an aided GPS receiver implementation (e.g., allows its clock to coast or can use a barometric altimeter input), it may be possible to continue the approach.

The Block II GPS satellites to be deployed by 1988–89 will have an integral monitoring capability onboard the satellites and the signals-in-space will be monitored on the ground by the GPS Control Segment. However, the FAA has determined that these measures do not fully guarantee the integrity of the GPS signals-in-space to the same level as in the VOR/DME system. The main concern is that the response time of the GPS Control Segment, which is the main guarantor of integrity, is too long since a typical response time is about 5 to 15 minutes. Therefore, the FAA is exploring alternative means for providing the integrity function.

### *Purpose and Scope of Paper*

The purpose of this paper is to present a proposed satellite based method for guaranteeing the integrity of GPS signals-in-space that meets the FAA 10 second response time requirement for nonprecision approaches. First, the method is described in some technical and operational detail. Then, the problem of monitoring GPS integrity with a few ground stations is discussed. The final section contains a discussion concerning the implementation of such a system.

## DESCRIPTION OF GPS INTEGRITY CHANNEL (GIC)

### Overview

The GIC concept provides GPS signals-in-space monitoring coverage for the conterminous U.S. with field Monitoring Stations located at the 4 corners of the U.S.,[2] as illustrated in Figure 1. The Monitoring Stations are located at the Miami, Boston, Los Angeles, and Seattle Air Route Traffic Control Centers (ARTCC). Any two of these monitoring sites serve as Integrity Control Nodes (Los Angeles and Miami are illustrated in Figure 1). Each Integrity Control Node receives the data from each of the Monitoring Stations, processes and formats the data into continuous integrity messages suitable for modulating a geostationary satellite feeder link as illustrated in Figure 1. Each Integrity Control Node feeds a separate geostationary satellite for the purpose of redundancy. The GIC concept was originally outlined in Reference 3.

In this concept, the geostationary satellites that provide the facility for broadcasting the integrity messages are commercial or government satellites that are already planned for the late 1980 and/or early 1990's time frame. An L-band integrity transponder package is carried onboard each satellite through a financial arrangement between the FAA and the satellite supplier.

The down link integrity broadcast is a narrow band signal located in the 1545–1559 MHz aeronautical mobile-satellite band, which is near the GPS frequency of 1575 MHz. This allows the GIC receiver to share the antenna and front end of the GPS receiver. Thus, to the user, the integration of GPS and GIC appears to be a single navigation system with an integrity that meets the FAA's requirements. A patent, assigned to NASA, is on record for an integrated GPS and separate data channel receiver for signal from a geostationary satellite[4]. In this case, the purpose of the data channel was to provide GPS aiding information for inexpensive receivers, such as identifying the satellites in view having the best geometry and their coordinates.

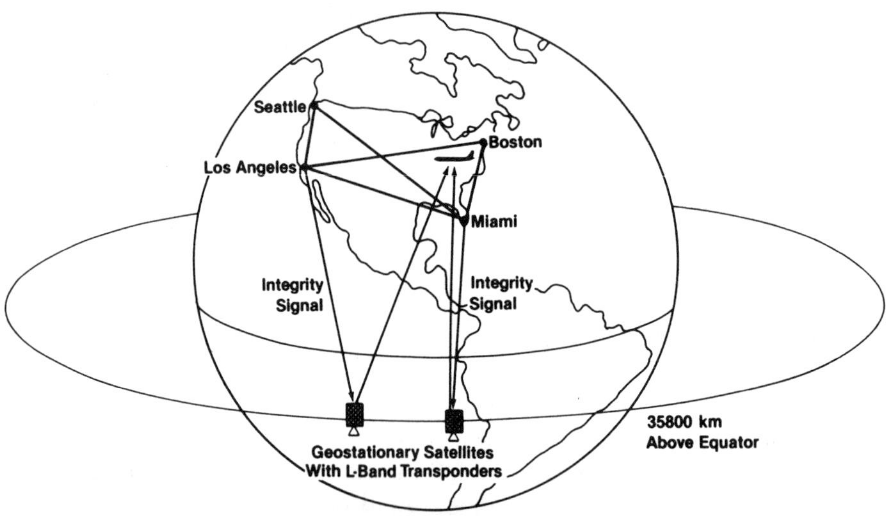


*Fig. 1—GPS Integrity Channel (GIC) Network*

### *Integrity Message*

An example GIC integrity message is illustrated in Table 1. For each GPS satellite in view of at least one of the four Monitoring Stations, an integrity message is generated at each Integrity Control Node. The satellite identification consists of five bits. The state of integrity for each phase of flight is given so that if a signal-in-space exceeds the tighter alarm limits for nonprecision approach it still may be useful for the other phases of flight.

Before a particular satellite can be used for a given phase of flight, the integrity channel of the receiver must continuously decide that the integrity message for that phase of flight is positive. For the case of the nonprecision approach, the complete 8-bit subframe for each tracked satellite must indicate the correct satellite identification and positive integrity for each of the three phases of flight according to the decision criterion described below.

For the purpose of analysis, it is assumed that a complete integrity message frame will contain information for up to 12 satellites at a time, since this is an upper bound on visible satellites for a 24 satellite GPS constellation. For a larger GPS constellation, the message size would have to be appropriately increased. It is further assumed that each integrity message frame contains a 16-bit preamble to indicate the start of the frame and possibly additional data. The total integrity message frame contains 112 data bits as illustrated in Table 1. This total message is updated and transmitted once per second.

A realistic maximum bit error rate (BER) is assumed to be $1 \times 10^{-5}$ for the purpose of minimizing the required satellite radiated power and, thus, cost. An alarm is set in the receiver for a GPS satellite if two out of three integrity messages for a particular satellite indicate a signal-in-space malfunction. That is, after the integrity channel first detects a malfunction bit, if either of the next two frames indicate a malfunction, then the alarm is set in the GPS receiver. In the alarm state, the GPS receiver does not use the indicated satellite for position determination. If there are only three remaining satellites in track, the flag alarm is generated for GPS. With an aided GPS receiver implemen

Table 1—GIC Message Contents and Parameters

| Function | | | Bit Length |
|---|---|---|---|
| Satellite Identification | | | 5 |
| Nonprecision Approach Integrity Indication | | | 1 |
| Terminal Area Integrity Indication | | | 1 |
| En Route Integrity Indication | | | 1 |
| 8 Bits × 12 Satellites | = | | 96 Bits |
| Preamble | = | | 16 |
| Total Data Bits Per Message Frame | = | | 112 |
| Maximum Bit Error Rate (BER) | = | $1 \times 10^{-5}$ | |
| False Alarm Time Due to BER | = | Every 5400 Hrs. of Flight Time | |
| Probability of Missed Alarm Due to BER | = | $3 \times 10^{-10}$ | |

tation (e.g., allows its clock to coast or uses a barometric altimeter input), it may be possible to continue to navigate with GPS. Since an integrity message frame is transmitted once per second, three seconds are allocated for the GIC receiver's decision.

The two out of three criterion, with a BER $= 1 \times 10^{-5}$, yields a mean time between false alarms of approximately 5400 hrs of flight time.[5] The probability of missing a malfunction indication within 3 seconds, due to bit errors, is approximately $3 \times 10^{-10}$.

### *Space Segment*

The GIC space segment consists of L-band down link transponders placed onboard two planned commercial or government geostationary satellites. Only the feasibility of using planned commercial domestic communication satellites has been investigated to date. Several major commercial communication satellite service suppliers were contacted for their opinion of the feasibility of using at least one of their planned satellites in the 1990's time frame for the GIC space segment. The consensus is that it is technically feasible to place an L-band transponder onboard commercial communication satellites. The required effective isotropic radiated power (EIRP) of the L-band down link (22 dBW, to be derived later in this subsection) would require a small fraction of the available power onboard the satellite. The basic approaches for implementing the L-band down link capability are adding an L-band antenna or feeding an existing C-band or Ku-band reflector. From the survey, it appears that adding an L-band antenna is the preferred approach for ease of implementation. The antenna would necessarily have to be small to fit the satellite; therefore, it would provide earth coverage. Figure 2 is a block diagram, provided by Western Union, illustrating how an L-band transponder could be implemented on a commercial communication satellite.

The link budget for the L-band down link is estimated in Table 2. The assumed differentially coherent phase shift keying (DPSK) modulation, the

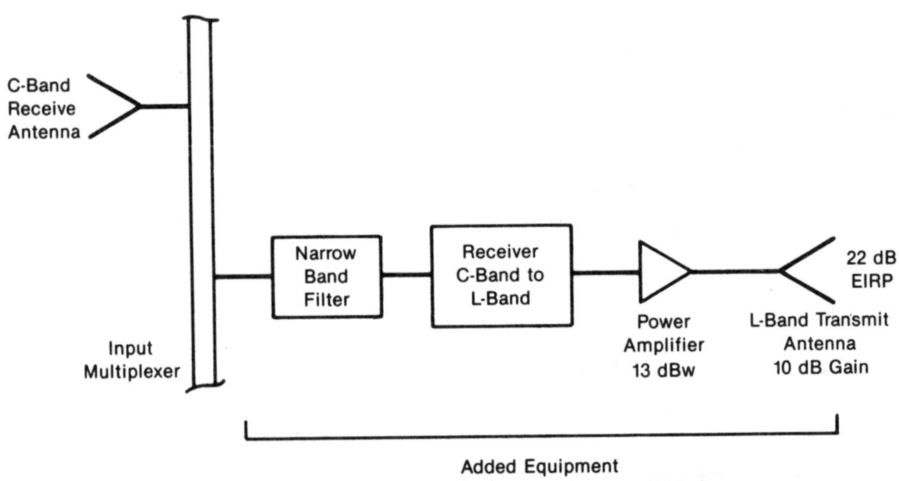


*Fig. 2—Integrity Message Repeater Block Diagram (Courtesy of Western Union and Hughes)*

Table 2—L-Band Down Link Budget (Frequency = 1559 MHz)

| | |
|---|---|
| Satellite EIRP | 22 dBW |
| Free Space Loss | 188.6 dB |
| Propagation Margin | 5.0 dB |
| Aircraft G/T (note 1) | −27.0 dB/°K |
| Boltzmann's Constant | −228.6 dBJ/°K |
| C/No (note 2) | 30.0 dBHz |
| Data Rate | 112 bps |
| Uncoded $E_b/N_0$ (note 3) | 9.5 dB |
| Coding Improvement (note 4) | 3.5 dB |
| Receiver Implementation Margin | 2.5 dB |
| Calculated $E_b/N_0$ | 10.5 dB |
| Required $E_b/N_0$ (BER = $10^{-5}$) | 10.4 dB |
| Symbol Rate | 224 bps |

[1]The ratio of aircraft antenna gain to receiver system noise temperature (G/T) assumes a 0 dB antenna gain and receiver system (antenna plus receiver) noise temperature of 500 °K.

[2]$C/N_0$ is the ratio of received carrier power to receiver system noise power density.

[3]$E_b/N_0$ is the ratio of received energy per bit to the receiver system noise power density. The BER depends upon $E_b/N_0$.

[4]A rate 1/2 error correcting convolutional code is assumed as described in Reference 6 resulting in the transmission of twice as many bits.

assumed convolutional error correcting code and the parameter values in Table 2 are based on the MITRE developed aeronautical data link described in.[6]

This power budget is considered conservative. However, the false alarm rate is very sensitive to a slight reduction from the signal-to-noise threshold for BER = $1 \times 10^{-5}$. For example, if the assumed 5.0 dB propagation margin (e.g., fading) is too small, then the satellite EIRP would also have to be increased accordingly. The satellite EIRP may also have to be slightly increased to compensate for noise in the up link. The up link power budget has not been estimated at this time because it is not known whether the earth station would be located at the FAA Integrity Control Nodes or whether the integrity messages would have to be transmitted through the service supplier's earth station.

### *User Segment*

The GIC User Segment receives and processes the integrity messages. It shares the same antenna and front end of the GPS receiver as illustrated in Figure 3. In addition, the GPS clock and power supply are shared. Both the GPS and GIC signals are converted to IF frequencies through the same mixer. A detailed block diagram of an integrated GPS/GIC receiver front end has been configured by Avantek, Inc. of Santa Clara, California, and provided to illustrate the feasibility of integration with off-the-shelf components. Figure 3 is based on the Avantek concept. The receiver is required to have the capability to switch to the standby integrity broadcast from the other geostationary satellite in the event that one of the satellites has a failure.

The noise figure of such a receiver is estimated to be 2.3 dB, which is equivalent to a noise temperature of 202°K. A receiver system noise temperature of 500°K is assumed in the down link budget (Table 2). Thus, the aircraft equivalent antenna temperature, due to background noise, needs to be less than 300°K. It is expected that the antenna temperature should be around 225°K,

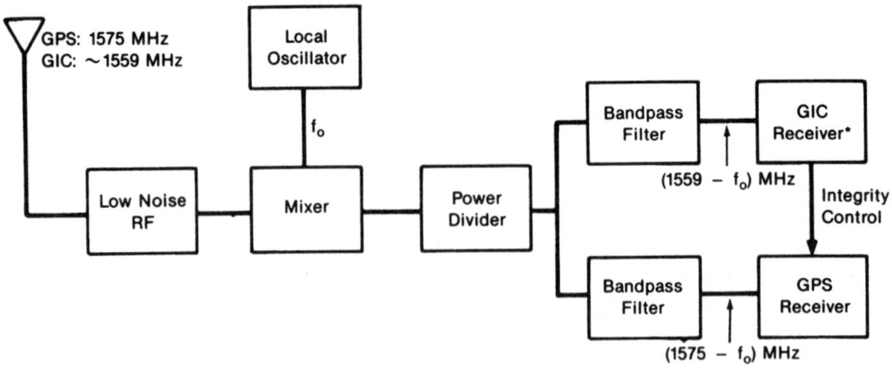


*The GIC Receiver Has the Capability to Filter and Select 1 of the 2 Integrity Broadcasts and Automatically Switch to the Other Broadcast in the Event of a Failure of the Main Broadcast in Use.

*Fig. 3—GPS/GIC Receiver Concept*

thus a 500°K receiver system noise temperature is a reasonable assumption. The aircraft antenna is envisioned to be a microstrip antenna with hemispherical coverage. For the purpose of developing the down link power budget, the antenna is assumed to have 0 dB gain due to cases where there are very low relative elevation angles between the aircraft antenna and the geostationary satellite.

### *Ground Segment*

The GIC Ground Segment monitors the GPS navigation signals from all visible GPS satellites, makes decisions on the integrity of GPS satellites, uplinks the integrity messages, and coordinates with the GPS Master Control Station. Figure 4 is a block diagram of an operational concept for the monitoring process. Each Monitoring Station measures the ranging accuracy of each satellite in view based on a precision clock, the GPS ephemerides supplied by the Integrity Control Nodes, and the known location of its receiver antenna. Each station has the capability to format and send messages to each of the two Integrity Control Nodes. The messages contain the Monitoring Station's assessment of the accuracy and a copy of the received 50 bps navigation message for each GPS satellite in view.

The two Integrity Control Nodes process the data received from each Monitoring Station and make a decision on the integrity of each satellite that has been monitored. Figure 5 is a block diagram of an operational concept for an Integrity Control Node with its collocated Monitoring Station. The Integrity Control Node estimates whether the accuracy of a GPS satellite is sufficient for each phase of flight, based on the accuracy assessment inputs available from the four Monitoring Stations and the GPS Master Control Station at Colorado Springs. The Integrity Control Nodes also coordinate any error trends they detect with the GPS Master Control Station. Similarly, the GPS navigation messages received from the Monitoring Stations are compared with the true messages as received from the GPS Master Control Station. Any errors detected

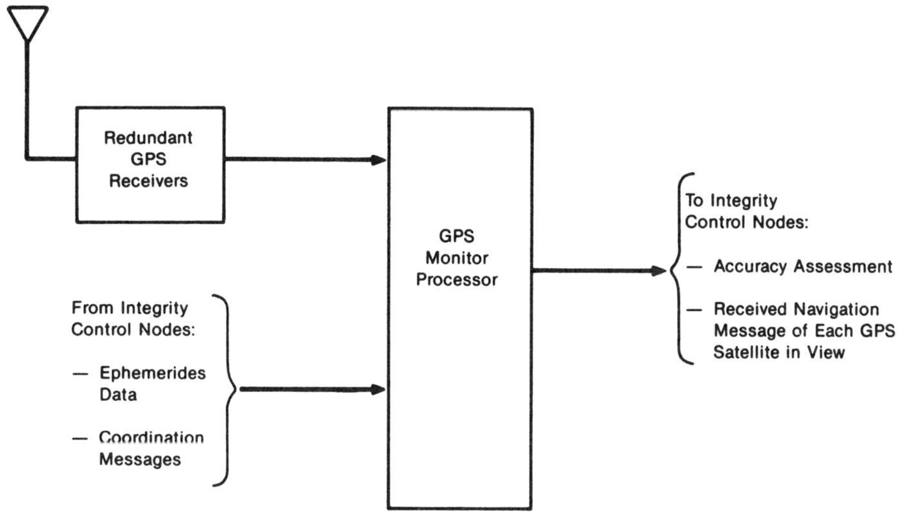


*Fig. 4—GPS Monitoring Station Concept (Located at Miami, Los Angeles, Boston and Seattle ARTCCs)*

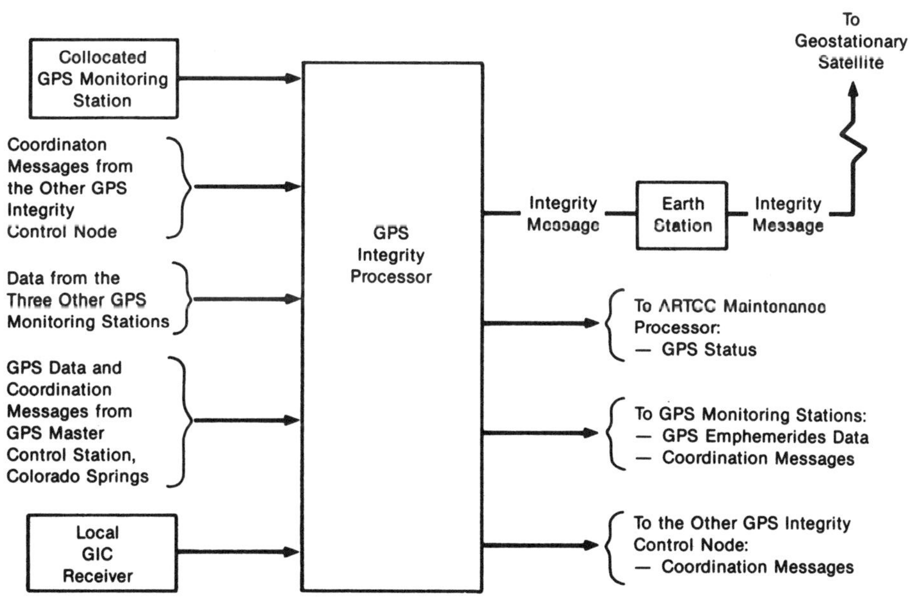


*Fig. 5—GPS Integrity Control Node Concept (Located at two ARTCCs)*

in the GPS navigation message that could impact the safety of flight in the NAS are flagged (e.g., error in satellite ephemeris data). The Integrity Control Node monitors the GIC satellite down link to ensure its integrity. It also interfaces with the ARTCC's Maintenance Processor Subsystem to provide GPS status to the FAA air traffic control facilities, nationwide.

The Integrity Control Nodes have the capability to make the decision as to whether a GPS satellite is suitable for navigation in the NAS. For the FAA's 10 second integrity response time requirement, the decision time has to be about 6 seconds to account for the user receiver's 3 second decision time, communication delays in receiving data from the Monitoring Stations and the 0.25 second satellite propagation delay. The decision process also needs to be exact since an alarm decision will simultaneously impact the navigation capability of many users. Therefore, the decision process needs to be automated to meet this level of performance. As with the GIC broadcast, the challenge will be the elimination of false alarms.

The ground communication network interconnecting the GIC Ground Segment utilizes the FAA's NADIN 2 Packet Data Switching Network, which will provide highly reliable interfacility connectivity. The connectivity to Colorado Springs could be established through the Denver ARTCC.

The GPS Integrity Processor composes and formats each integrity data subframe for each GPS satellite and the frame preamble. It continuously passes the integrity information to the earth station where it is encoded and transmitted to the satellite. The FAA is responsible for providing the integrity messages. The satellite service supplier is responsible for the earth station and satellite links. It would be preferable to have earth stations collocated with the Integrity Control Nodes. However, this could be difficult if commercial satellites are used. If the satellite service supplier's earth station has to be used, the communications link from the Integrity Control Node will have to be highly reliable with very little network delay.

The Earth Station and GPS Integrity Control Node coordinate the accuracy of the down link integrity messages to ensure the integrity of the GIC down link to the GIC User Segment; i.e., the User Segment sees a bit error rate that is no greater than the maximum value (e.g., $1 \times 10^{-5}$), given that the user's receiver is operating within specifications.

## EVALUATION OF SATELLITE INTEGRITY BY THE GIC GROUND SEGMENT

In order to assess the integrity of a particular GPS satellite, an Integrity Control Node must gauge how errors from that satellite will affect user position estimates and compare the resulting accuracy to the level required for various phases of flight. This effect could be measured directly at a Monitoring Station by comparing position solutions involving the satellite in question to the known surveyed location of the Monitoring Station. However, as will be illustrated, the impact of a satellite error on user position accuracy varies somewhat with increasing geographic separation from the Monitoring Station, due to differences in satellite geometry seen at the Monitoring Station and user locations. As a result, it does not appear feasible to assess GPS satellite integrity over CONUS based on position accuracy measured directly by only a few Monitoring Stations, such as the four proposed in this paper (Boston, Miami, Seattle, and Los Angeles). Therefore, the following discussion will focus on extrapolating satellite ranging errors measured at the Monitoring Station to assess the integrity of individual GPS satellites.

### *Review of GPS Position Determination Error Theory*

Errors of GPS derived position coordinates x, y, and z and time t are characterized by the covariance matrix P

$$\mathbf{P} = \begin{bmatrix} \sigma_x^2 & — & — & — \\ — & \sigma_y^2 & — & — \\ — & — & \sigma_z^2 & — \\ — & — & — & \sigma_t^2 \end{bmatrix} \tag{1}$$

These errors are minimized (*a priori*) by employing all visible satellites in the solution. However, some aviation users may not be able to afford a receiver that tracks all satellites in view. It is therefore assumed in this analysis that a user at any particular place employs the four visible satellites with best satellite-user geometry. Under this assumption, P is related to the covariance of the satellite ranging errors R by

$$\mathbf{P} = (\mathbf{G}^{-1})\,\mathbf{R}\,(\mathbf{G}^{-1})^T \tag{2}$$

where G is the satellite-user geometry matrix that translates changes in user position and time to corresponding changes in ranges to the four satellites.[7] If the satellite ranging errors are independent and the variance of the error on satellite i is $\sigma_i^2$, the satellite error covariance becomes the diagonal matrix

$$\mathbf{R} = \begin{bmatrix} \sigma_1^2 & 0 & 0 & 0 \\ 0 & \sigma_2^2 & 0 & 0 \\ 0 & 0 & \sigma_3^2 & 0 \\ 0 & 0 & 0 & \sigma_4^2 \end{bmatrix} \tag{3}$$

Of particular interest to civil aviation is the 2 drms horizontal position error given by the x and y terms of the covariance matrix

$$2\text{ drms} = 2\sqrt{\sigma_x^2 + \sigma_y^2} \tag{4}$$

Under the assumption that the satellite ranging errors have equal variances $\sigma_r^2$ the user position error covariance becomes

$$\mathbf{P} = \sigma_r^2\,(\mathbf{G}^{-1})\,(\mathbf{G}^{-1})^T \tag{5}$$

and the corresponding 2 drms error is

$$2\text{ drms} = 2\,\sigma_r\,\text{HDOP} \tag{6}$$

Where HDOP is the familiar horizontal dilution of precision coefficient, which depends only on the satellite-user geometry.

### *Abnormal Satellite Ranging Errors*

The impact on position error caused by abnormally large ranging errors on one of the four satellites can be predicted by using unequal variances in (2) above. Let it be assumed that satellite i has an abnormal ranging error with variance $\sigma_a^2$ and that the other three satellites have a baseline ranging error with variance $\sigma_b^2$. Then, it can easily be shown that the corresponding 2 drms position error can be expressed by

$$2 \text{ drms} = 2\,\sigma_b\,\text{HDOPAE}\,(i,\ \sigma_a/\sigma_b) \tag{7}$$

where, for the abnormal error on a given satellite i, HDOPAE depends only on the ratio $\sigma_a/\sigma_b$.

To illustrate the impact of an abnormal satellite ranging error, it is useful to consider the GPS satellite visibility condition that would be present over CONUS at some particular time. Figure 6 shows the number of satellites visible above 10° elevation at various places in CONUS at 11.0 hours GMT. The GPS constellation is assumed to consist of 18 satellites on six orbits, with three additional active spares on orbit. The situation shown in the figure is of particular interest because only four satellites are visible (above 10 degrees) in a region of south central U.S. At Dallas, Texas, for example, such a visibility condition persists for about an hour. The evaluation of satellite integrity is most critical when only four are visible, since that is the minimum number needed to estimate position and time from GPS alone.

Figure 7 shows HDOPAE as a function of the ratio $\sigma_a/\sigma_b$ for the four satellites visible, at Dallas, at 11.0 hours GMT. It is interesting to note that the impact from an abnormal ranging error of given size varies somewhat depending on the particular satellite affected. For example, if the abnormal error is on satellite 6 in orbit 2, the impact is only about one half the effect if the error is on any one of the other three satellites. Note, however, that the curves for indi-

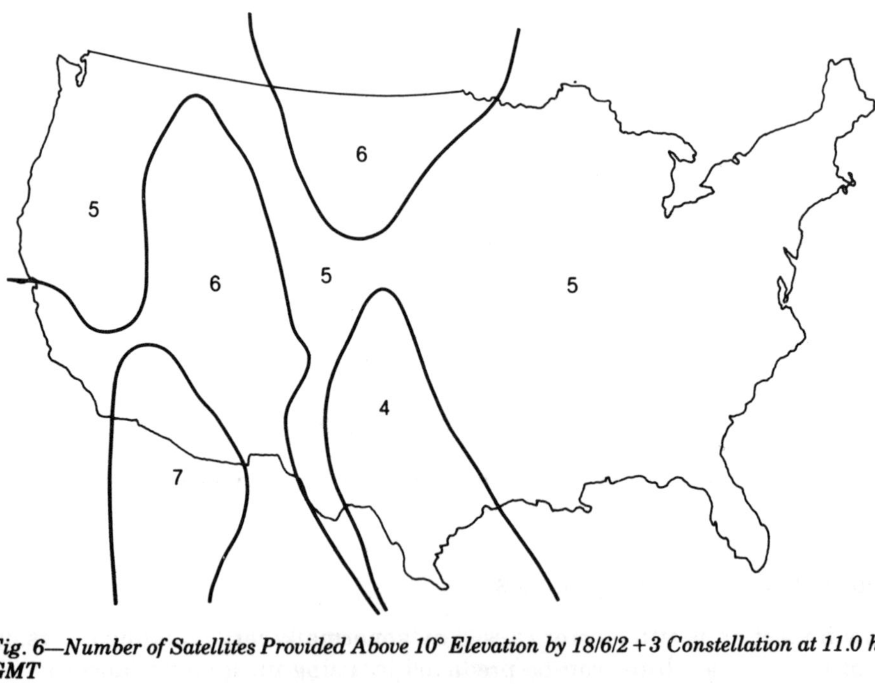


*Fig. 6—Number of Satellites Provided Above 10° Elevation by 18/6/2 + 3 Constellation at 11.0 hrs GMT*

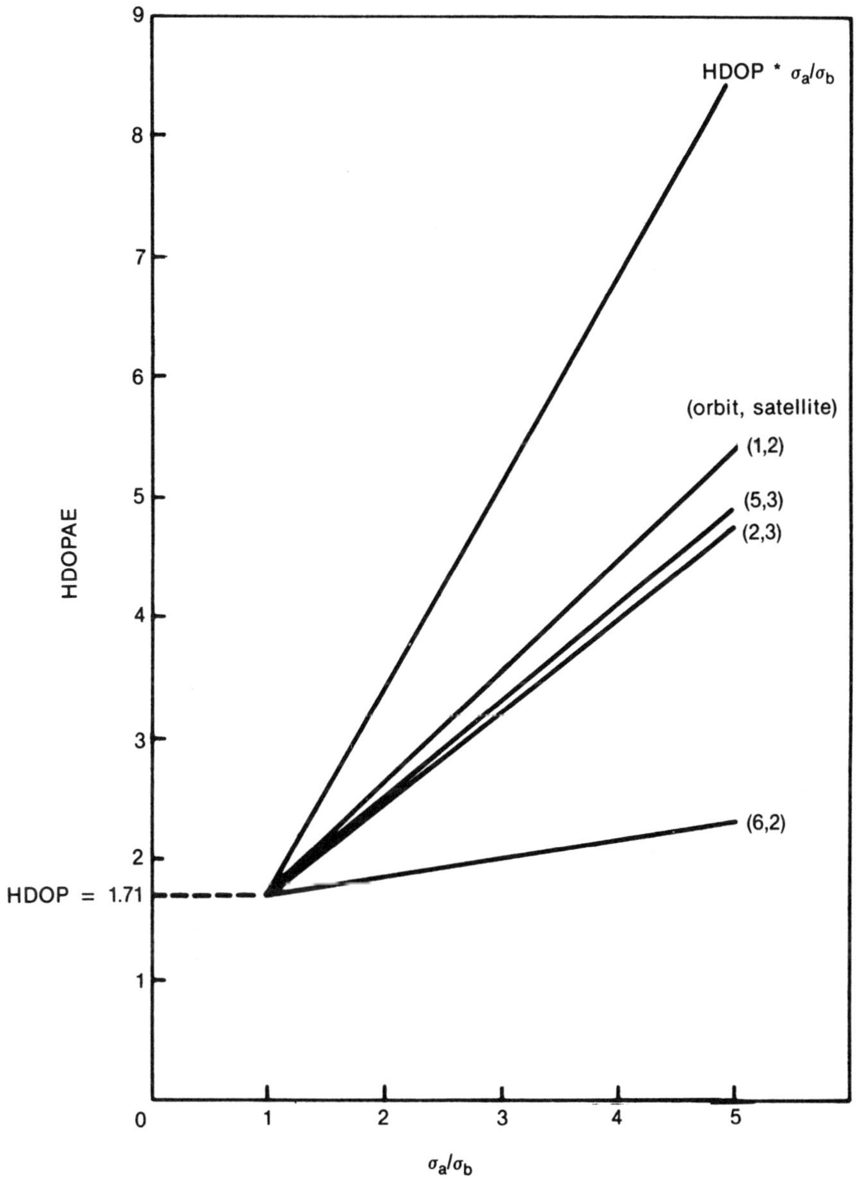


*Fig. 7—Position Accuracy Degradation from Abnormal Satellite Ranging Error at Dallas (11.0 hrs GMT)*

vidual satellites are all bounded by the curve labeled "HDOP*$\sigma_a/\sigma_b$" which corresponds to the abnormal error being present on all satellites at once.

It is important to consider how the position error caused by an abnormal ranging error from a particular satellite varies over the geographic region where the same satellites are visible. Figure 8 illustrates this effect, in the region around Dallas, at 11.0 hours GMT. The figure shows HDOPAE for a ranging error on satellite 3 in orbit 5 that is twice the baseline ranging error on the other three satellites in view. Note that for the situation shown, the

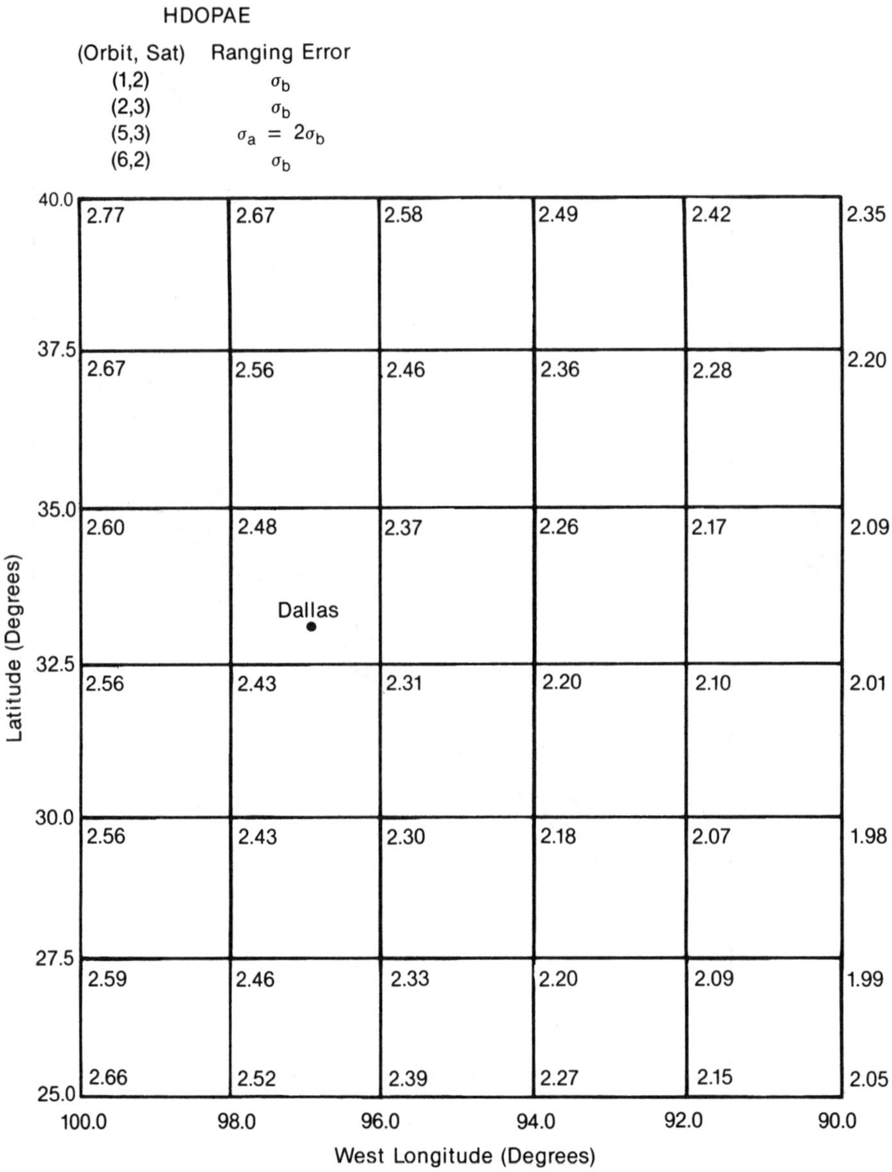


*Fig. 8—Example of Regional Position Accuracy Degradation from Abnormal Satellite Ranging Error*

position error variation over the region for a given satellite (Figure 8) is relatively small compared to the variation if the abnormal error is on different satellites at a particular place (Figure 7). However, abnormal satellite ranging error can impact position accuracy over a wide area when even more than four satellites are visible. Table 3 illustrates this effect for six widely scattered locations in CONUS (Boston, Miami, Chicago, Dallas, Seattle, and Los Angeles). The table shows HDOPAE assuming that satellite 3 in orbit 5 has a ranging error that is twice the normal value. This particular satellite was chosen

Table 3—Examples of Nationwide Position Accuracy Degradation from Abnormal Satellite Ranging Error

| | No. of Satellites Visible Above 10° at 11.0 Hours GMT | Best Set of 4 Satellites (Orbit, Sat) | HDOP[1] | HDOPAE[2] |
|---|---|---|---|---|
| Boston | 5 | (1,2) (3,2) (5,3) (6,2) | 1.64 | 1.89 |
| Chicago | 5 | (1,2) (2,3) (3,2) (5,3) | 1.68 | 2.50 |
| Dallas | 4 | (1,2) (2,3) (5,3) (6,2) | 1.71 | 2.37 |
| Los Angeles | 6 | (1,2) (4,2) (5,3) (6,1) | 1.60 | 1.90 |
| Miami | 5 | (1,2) (2,3) (5,3) (6,2) | 1.75 | 1.80 |
| Seattle | 5 | (1,2) (2,3) (5,3) (6,1) | 1.66 | 2.03 |

[1]Best Set of 4 Satellites
[2]For Abnormal Ranging Error with $\sigma_a - 2\sigma_b$ on Satellite 3 in Orbit 5

because it is in the best set of four at all six locations at 11.0 hours GMT. The baseline HDOP at each location (assuming equal satellite ranging errors) is also shown for comparison. Note that although the baseline HDOP and, therefore, position accuracy is nearly the same at all locations, the impact of the abnormal satellite error varies by about 50 percent.

### *Minimum Resolution Method for Assessing Integrity*

An elementary method similar to one being used by the DoD in the GPS Control Segment[8], could be employed by the GIC ground segment for assessing the impact of abnormal satellite ranging errors on user position accuracy at a particular location as follows:

1. Estimate the covariance R by substituting the satellite ranging errors observed at the Monitoring Stations into (3). For simplicity, both noise and random bias errors are lumped together in this analysis to illustrate the long term effect in a statistical sense.
2. Compute the corresponding user position covariance P by evaluating (2) at sample user locations.
3. Compare the resulting 2 drms horizontal position error estimate as given by (4) to the level acceptable for a particular phase of flight.

To gauge the effect of abnormal ranging errors on a particular satellite the calculation would be performed at a representative set of user locations over all CONUS where the questionable satellite appears in the best set of four. If, at any such location, the resulting accuracy is predicted for a given period of time (e.g., 6 seconds) to be less than that required for a particular phase of flight, the satellite would be declared to have the corresponding lack of integrity. In this manner, the decision about a satellite's integrity would be based on the worst case impact of an abnormal error, to assure that no user in CONUS would have insufficient accuracy.

### *Illustration of Method*

In order to illustrate the performance of the above method, an assumption must first be made regarding the baseline satellite ranging error $\sigma_b^2$. The DoD has proposed to provide civil users with a GPS positioning accuracy (Standard Positioning Service) of 100 meters 2-drms, 95 percent of the time[9]. For the 18 satellite constellation with three active spares, the 95th percentile value of HDOP in CONUS achieved by using the best set of four satellites is about 2.2.[10] From this, we may infer that the baseline ranging error is of such a size that $2(\sigma_b)2.2 = 100$ meters. Thus, if one satellite has an abnormal ranging error (Equation 7), 100 meters 2-drms error will be exceeded for values of HDOPAE exceeding 2.2.

The six locations cited in Table 3 may be used to illustrate the integrity decision that might be performed by the GIC regarding a particular satellite. Recall that the table shows values of HDOPAE assuming twice the normal ranging error on satellite 3 in orbit 5 at 11.0 hours GMT. Notice that the resulting HDOPAE is greater than 2.2 at Chicago and Seattle. Thus, an integrity decision based on predicting the worst impact of ranging error over CONUS would declare the integrity of satellite 3 in orbit 5 inadequate for nonprecision approach guidance in this case. However, users where HDOPAE was less than 2.2 (for example at Miami, Los Angeles, and Boston) would still have better than 100 meters 2-drms accuracy. In this case, the advice against employing a particular satellite might be unnecessary for some users.

Such a conservative method of assessing satellite integrity might be acceptable depending on how frequently abnormal satellite ranging errors occur and how large a user population is affected by integrity decisions. During the early years of GPS availability when the use population is relatively small, the number of users impacted unnecessarily by a conservative integrity decision may be acceptable. However, as the GPS user population grows, the number of users unnecessarily advised not to use a particular satellite may become too large.

A possible remedy for this problem is for the GIC to make integrity decisions at higher geographic resolution. The GIC would evaluate the impact of satellite ranging error on position accuracy for various zones of CONUS and broadcast the zonal integrity to users. To minimize the corresponding increase in the GIC broadcast channel data rate, details of the zonal integrity would be included in the message only if the satellite did not have sufficient integrity everywhere over CONUS. Further study is needed to determine the number of such zones that might be required, especially to ensure coverage of areas where position error is particularly sensitive to satellite ranging error.

## IMPLEMENTATION DISCUSSION

In this section, only those implementation considerations involving the feasibility of the GIC concept are discussed; user and Government institutional issues are considered too premature for discussion. The implementation of each of the three segments is discussed in turn.

### *Space Segment Implementation*

As previously discussed, it is technically feasible to place the special L-band transponders onboard planned commercial geostationary communication satellites that will provide coverage of the U.S. An arrangement to use commercial satellites has the advantage of providing high certainty that the GIC space segment will be on orbit when planned. There are several planned replacements of commercial communication satellites in the late 1980s and 1990s. On the other hand, from both a technical and economic viewpoint, a planned UHF or L-band mobile satellite would be the best choice, such as the current mobile satellite system under review by the Federal Communications Commission (FCC) where there are 12 applications[11]. However, there is also potential for greater schedule risk since there are no firm plans for a mobile satellite launch date. Another potential source for the L-band transponder is planned Government communication satellites (civil and military). If it is desired to implement the GPS Integrity Channel, arrangements for two suitable satellites would have to be made about 3–5 years prior to the scheduled launch to provide sufficient time to develop the suitable L-band transponder and to obtain the required frequency allocation. Hopefully, frequency allocation would not be a problem in the contested 1545–1559 MHz band since the required bandwidth is very small (on the order of 10 kHz to account for the Doppler effect due to aircraft motion) and the resulting benefit to aviation is potentially very high. However, frequency allocation efforts need to start, as soon as possible, because of the potential long lead time.

### User Segment Implementation

As previously discussed, it is technically feasible to develop an integrated GPS/GIC receiver. The minimum operational performance standard (MOPS) for the receiver would have to be determined by the Radio Technical Commission for Aeronautics (RTCA). Another consideration is what assurance the user would have that the FAA would continue supporting GIC if it is decided to implement a satellite based communication, navigation and surveillance (CNS) system. This should create no issue since GIC can be retained as long as needed because it would involve little cost in money and bandwidth to implement it along with an L-band CNS system. Since an addition to conventional GPS receivers is required, the earlier the decision is made, the greater the number of fully capable GPS users there will be when GPS is operational.

### Ground Segment Implementation

The real time decisions, by the GPS Integrity Control Nodes, concerning the integrity of the GPS signals-in-space have to be made such that there is a very low false alarm rate since GPS will be simultaneously used by many aircraft. The decision time is about 6 seconds to meet the FAA's 10 second integrity response time (3 seconds for user receiver decision and 1 second for communication delays). It is expected that in many cases the rate of error growth will be such that soft alarm thresholds can be defined to enable the tracking of errors for a period of time, prior to hard alarm, to yield higher confidence in the broadcast of an alarm to the user. The development of the FAA's ground monitoring system can use the experience of the DoD GPS monitoring system. A long lead time may be needed to develop the ground monitoring automated decision algorithms to yield an acceptable response time and false alarm rate. Another consideration is the location of the two earth stations. For the purpose of reliability, it is very advantageous to collocate the earth station with the FAA Integrity Control Node; however, this may not be possible with commercial communication satellites. On the other hand, mobile satellites would more easily permit collocation.

## SUMMARY

The GPS Integrity Channel concept described in this paper has great potential to meet the FAA's integrity requirement for nonprecision approaches. Its timely implementation in the 1990's time frame will require a decision to proceed with it as soon as possible. The system has to be specified, arrangements have to be made to lease transponders on two planned domestic communication satellites, standards have to be developed and coordinated, a prototype system has to be developed and extensively tested, and frequency allocation has to be obtained.

## ACKNOWLEDGMENT

AT&T Communications, Washington, DC; Avantek, Inc., Santa Clara, California; Hughes Communications, Inc., Los Angeles, California; and Western Union Telegraph Company, Upper Saddle River, New Jersey provided information for this study. Mr. Yaroslav Kaminsky and Mr. Samuel Anderson of MITRE provided technical support. Mrs. Betty Keys prepared the manuscript.

## REFERENCES

1. Anon., *Federal Radionavigation Plan,* Department of Defense and Department of Transportation, DoD-4650.4, DOT-TSC-RSPA-84-8, 1984.
2. Shively, C. A., *Visibility of GPS Satellites for Integrity Channel Coverage,* MITRE Memorandum W46-M4239, October 1985.
3. Braff, R. and Bradley, J., *Global Positioning System As A Sole Means for Civil Air Navigation,* IEEE Plans '84 Position Location and Navigation Symposium Record, 1984.
4. Taylor, R. E., Sennott, J. W., *Navigation System and Method,* United States Patent 4,445,118, April 24, 1984.
5. Braff, R., *Derivation of Decision Error Models for GPS Integrity Channel User Receiver,* MITRE Memorandum W46-M4202, August 1985.
6. Anderson, S., *An Experimental Aeronautical Data Link,* presented at the Digital Avionics System Conference, Baltimore, Maryland, December 1984.
7. Bogen, A., *Geometric Performance of the Global Positioning System,* AF Report SAMSO-TR-74-169, June 1974.
8. Private conversation with Mr. Sherman Francisco of IBM Federal Systems Division, Gaithersburg, MD, 1985.
9. Baker, P., *GPS,* a briefing given by the Department of Defense to the RTCA Special Committee—155, Washington, DC, 14 March 1984.
10. Kalafus, R., Vilcans, J., and Knable, N., *Differential Operation of NAVSTAR GPS,* Navigation, Vol. 30, No. 3, pp. 187–204, Fall 1983.
11. Hills, A., *Satellites and Mobile Phones: Planning A Marriage,* IEEE Spectrum, August 1985.

# Terrestrial Evaluation of the GPS Standard Positioning Service

FRANCIS W. MOONEY
U.S. Department of Transportation, TSC, Cambridge, MA
*Received January 1986*

## ABSTRACT

During the period of January through April and also in August 1985, an engineering test team from the Transportation Systems Center (TSC) conducted a static and underway assessment of radionavigation systems in the Boston area. Sponsored by the Research and Special Programs Administration, the object of the tests was to gather data to determine the viability of the Global Positioning System, Standard Positioning Service (SPS), to provide navigation for terrestrial users. Concurrent data were also obtained on the LORAN-C system to provide a current assessment of its capabilities.

Four receivers, two GPS and two LORAN-C, were used in the evaluation. Tests assessed the capabilities of receivers for conducting static survey and underway navigation. Test units were exposed to operating conditions representative of the rural, suburban and urban terrestrial environments. Parameters of interest included navigation accuracy (absolute and repeatable), interference susceptibility and signal availability. GPS results obtained with signals from the current constellation of satellites were compared with those expected from 18 and 21 satellite configurations. This paper presents a discussion of the conduct of the evaluation and a summary of the results.

## INTRODUCTION

Ever since the development of radionavigation services for marine and air navigation, there have been attempts to extend the system's application to land vehicles. Land based, high frequency systems are line-of-sight and are range limited. Low frequency systems provide surface coverage, but propagation velocities are often affected by surface conditions, making accurate modelling difficult. The introduction of the GPS system presents a fresh opportunity to examine whether a wide area radionavigation system can permit accurate position determination in the terrestrial environment without relying on other sensor inputs (e.g., odometer for velocity, compasses for direction, etc.). The Research and Special Programs Administration commissioned the TSC to evaluate the GPS in conditions typical of the terrestrial environment. While not the principal effort, collection of concurrent LORAN-C data permitted comparison of the systems.

Static and underway tests were conducted during January through April 1985. A second GPS receiver, the Magnavox T-Set, was tested during August

1985. A complete set of test results will be presented in a report which will be distributed by the National Technical Information Service (NTIS). This paper presents a discussion of the conduct of the evaluation and a summary of the results.

## POSITION REFERENCE SYSTEM

Anyone who has attempted a terrestrial evaluation of radionavigation systems has learned that local maps are based on the North American Datum-1927, a state plane reference system, relative surveys or any other system which satisfactorily serves the local user, e.g., real estate maps. Accurate navigation evaluation requires establishing positions and control data in a common geodetic reference system. It is easy to lose tens of meters through conversion of one system to another. Area radionavigation systems typically use an earth model established in the World Geodetic System (WGS-72). For this evaluation, static positions were converted to the WGS-72 while underway data remained in NAD-27.

Underway positions were precisely plotted according to a map scale, then overlayed to the map through "best-fit" techniques. Shifts to the position plots were found to be necessary for the LTN-700 which had a position "bias" offset and to accomplish the Additional Secondary Phase correction for LORAN-C. Test routes were selected with sufficient path complexity to ensure that inaccuracies of the navigation solution were apparent by the plotted position going off the highway, roadway or street. Position variations showed that this approach was adequate.

## TEST RECEIVERS

Four receivers, two GPS and two LORAN-C were used during the tests. All were manufactured by different companies and characteristics are summarized in Tables 1 and 2. Typical of commercial products, brochures and equipment manuals don't provide enough information for direct specification comparison. Both GPS receivers were designed for engineering test purposes and are available for commercial purchase. Four versions of software were used in the LTN-

Table 1—GPS Receiver Specifications

| Receiver | **Litton LTN-700** | **Magnavox T-Set** |
|---|---|---|
| Application | Aviation | Marine and Terrestrial |
| Form | Avionics Package | Digital Computer Terminal |
| Type | Single Channel SPS<br>Multiplex(fast sequence) | Dual Channel SPS<br>Sequential |
| Velocity | 600 Meters/Second | 400 Meters/Second |
| Acceleration | 40 Meters/Second$^2$ | 6 Meters/Second$^2$ |
| Time to First Fix | < 3 Minutes | 3–24 Minutes |
| Accuracy | 40 m CEP, 100 m 2drms | 35 m, 2D 90% Probability |
| Geodetic Reference | WGS-72 | Altitude Corrected WGS-72 |
| Satellite Mask Angle | 7.5° | 7.5° |
| RS-232 Data Output<br>Data Update Rate | 9600 Baud<br>Fixed Update | 2400 Baud<br>2 Second Minimum Interval |

Table 2—LORAN-C Receiver Specifications

| Receiver | **Megapulse Accufix 600** | **Digital Marine Northstar 7000** |
|---|---|---|
| Application | Survey | General Purpose |
| Stations | Master & 4 Secondaries | Master & 4 Secondaries |
| Time to First Fix | < 5 Minutes | < 3 Minutes |
| Velocity | 60 Knots | 150 Knots |
| Acceleration | Unspecified | Unspecified |
| Accuracy | TD-10 Nanoseconds | TD-10 Nanoseconds |
| | L/L-0.1 Arcsecond | L/L-0.01 Arcminute |
| ASF | Manual | Automatic |
| Geodetic Reference | WGS-72 | WGS-72 |
| RS-232 Data Output | 300–1200 Baud | 300–1200 Baud |
| Data Output Rate | 1 Second Multiples | 1 Second Multiples |

700 and a developmental version was used in the T-Set (complete except for a special data output that was unnecessary for our tests).

The LORAN-C receivers were standard commercial units. The Accufix 600 was a navigator version of the Accufix 500, a high precision, survey grade, marine receiver. The tracking bandwidth of the Accufix 600 was modified to permit underway evaluations up to 60 mph (tests showed that receiver accuracy was not compromised). The Northstar 7000 was a general purpose, high quality receiver which could be used for marine, terrestrial and slow aircraft navigation.

## DATA ACQUISITION AND ANALYSIS EQUIPMENT

Microcomputers are well suited to acquisition and storing of data for this type of test because they are small and can be programmed to deal with many different data formats. While all receivers had RS-232 data outputs, none used any form of "hand shake" protocol. This required the recording equipment to be able to take a message, store it and be ready for additional messages within the report cycle time of the test receiver. The data messages from the GPS receivers were quite long. After experiments with several microcomputers, a Hewlett Packard HP2647A graphics terminal was chosen to record data from the GPS receivers. The HP2647A was slow, but had a large buffer which could absorb the data messages and store them on tape. The LORAN-C receivers were interfaced with a HP86 computer, the standard computer used in the TSC mobile test facility (MTF). All navigation data was post processed at TSC on the HP2647A, the HP86 and also on a HP9845A scientific microcomputer. Position plots were done on a HP9872B plotter. Basic Loran-C and GPS static analysis programs for the HP9845 were provided by the Coast Guard Research and Development Center. Use of these programs saved several months of programming effort.

## DATA ACCUMULATION AND MANAGEMENT

The receivers provided an enormous amount of data in a short period of time. A typical LTN-700 output during a navigation transit would include time, position (latitude, longitude and altitude), four sets of satellite (SV) identification and status (SV number, signal to noise number, altitude, azimuth, health), horizontal dilution of precision (HDOP), velocity (horizontal and vertical) and course. Each traffic transit, typically 5 minutes long, filled a standard HP storage cassette. LORAN-C receivers provided time, position, time differences (TDs), SNRs for each station tracked, course and speed. Actual output data and format varied considerably between units, requiring individual sort and analysis programs. Tapes and discs were numbered sequentially, carefully dated and a log of all test runs was maintained.

## TEST ORGANIZATION

Static laboratory measurements were made to establish performance bases for each receiver, then the equipment was relocated to the mobile test facility (MTF). The TSC MTF is a standard Plymouth Voyager window van, modified by adding a fiberglass top to provide standing headroom. The MTF receives 115 VAC power from a small, attached trailer which houses two, 4kW generators. After receiver relocation to the MTF, satisfactory performance was established through comparative measurements at TSC. When these tests were complete, a series of survey and underway tests were conducted in the Boston area. Supplemental static laboratory tests were conducted with the LTN-700 in April to test the fourth version of software. The T-set was independently tested in August 1985.

## TEST CONDUCT

Until the Block two satellites are launched, the availability of Block 1 GPS signals dictates when test observations can be made. In Spring 1985, Boston saw two multiple satellite passes each day. During the primary satellite pass, various combinations of the six functional satellites (SVs 6,8,9,11,12 and 13) were available for observation. Four satellite navigation fixes with HDOP better than 2.0 were available for about 2 hours of the primary pass, while less accurate positions from other combinations were available for an additional three hours. A secondary, three satellite pass was also available.

Figure 1 shows satellite availability on April 1, 1985. The bulk of the test data was obtained during the months of January through March 1985. Most observations were made between the local hours of 2 to 6 am. This presented optimal conditions for road transit because there was virtually no traffic. During August, when the T-set was evaluated, the best HDOP conditions existed between 6 and 8 pm. This also proved to be satisfactory because the peak of the Boston rush hour was past.

## TEST RESULTS

### *STATIC TESTS*

Performance signatures for all receivers were initially established through static tests. Ground truth measurements were made at five first order sites. Three were in Boston and two were in Bedford, MA. The MTF was positioned within 3 meters of the Boston sites. A sixth site was studied for multi-path purposes and provided additional data for establishing LORAN-C performance.

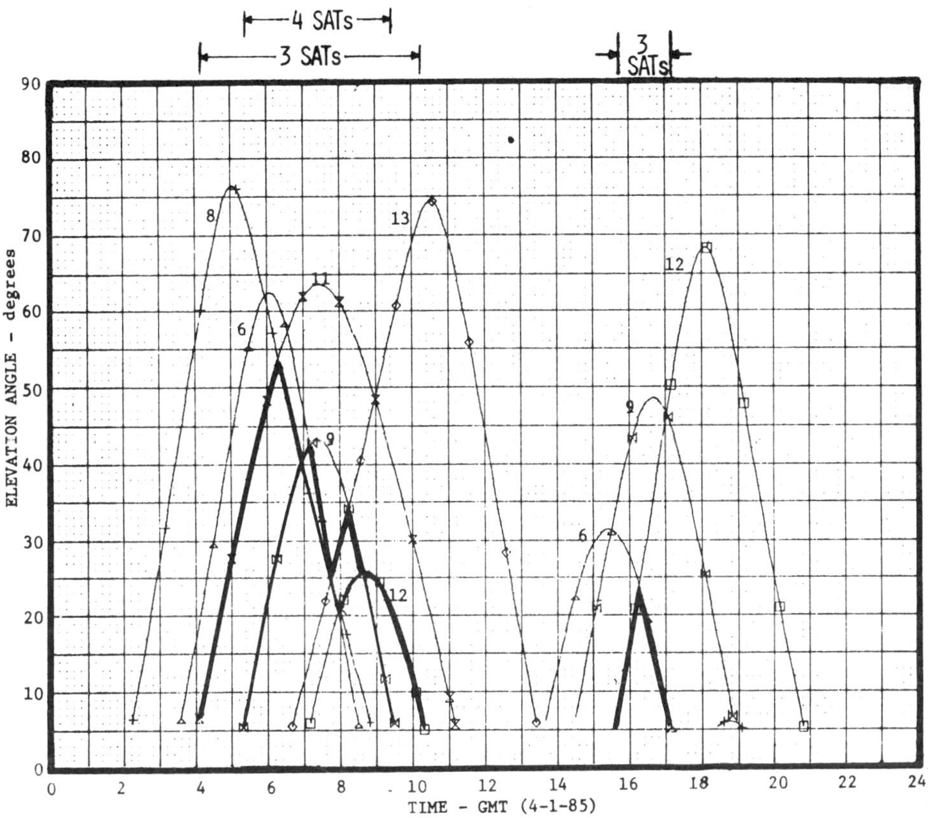


*Fig. 1—Satellite Visibility in Boston*

The LTN-700 showed an average position offset of 0.48 arcsecond in latitude and 1.49 arcseconds in longitude. A fourth software revision reduced the offests to 0.48 arcsecond in latitude (14.8 meters) and 0.44 arcsecond in longitude (10.1 meters). The revised software shifted positions, but did not change scatter plots. An overall position error of 17.9 meters was realized. This performance is comfortably within equipment specifications. Tests with the T-Set at two of the sites produced better results; offset of 0.07 arcsecond in latitude (2.1 meters) and 0.16 arcsecond in longitude (3.7 meters), with an overall offset of 4.3 meters. Scatter plots for each of the receivers are shown in Figure 2.

## LORAN-C

The gradients and crossing angles of LORAN-C signals in the Boston area are excellent. Scatter plots, some taken over a 5 hour period, showed variance of less than 50 ns. The two receivers normally agreed within 50 ns for all TDs. Typical scatter plots are shown for each receiver in Figure 3.

While the TDs agreed, the latitude/longitude outputs from the receivers did not. This was due to the fact that the Northstar 7000 uses preprogrammed, automatic additional secondary phase factor (ASF) corrections while the Accufix 600 uses an operator initiated correction which is added to all fixes once

**LITTON**

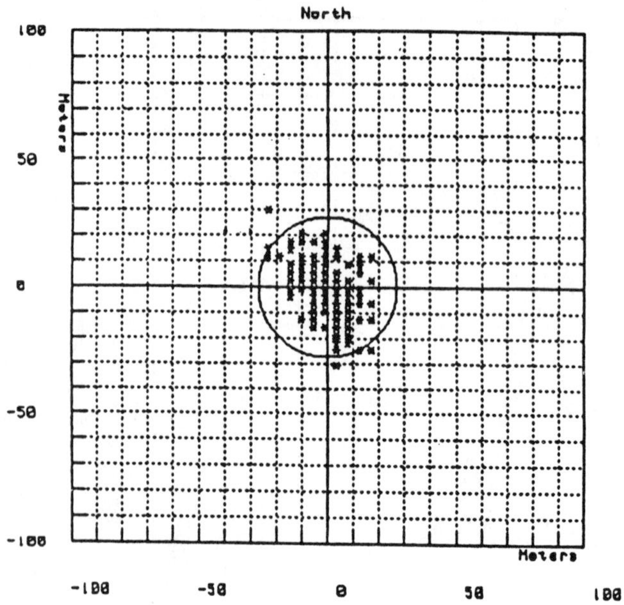


**T-SET**

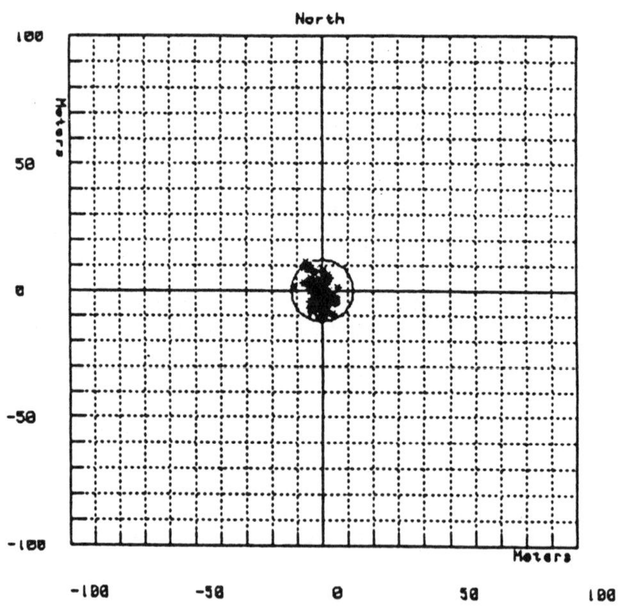


*Fig. 2—Representative GPS Scatter Plots at Artery #12*

**ACCUFIX**

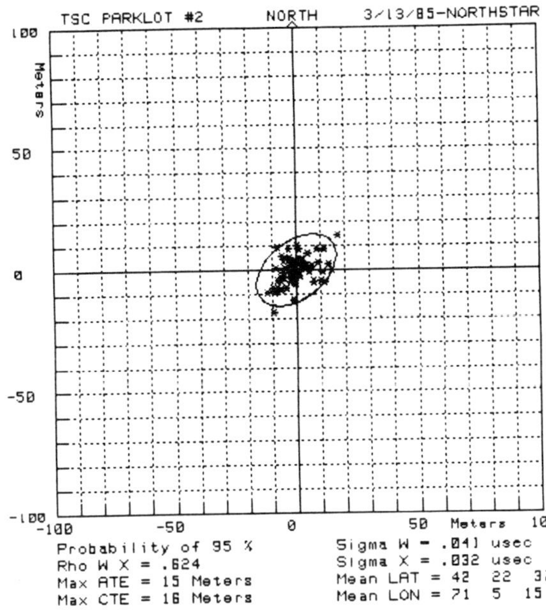


**NORTHSTAR**

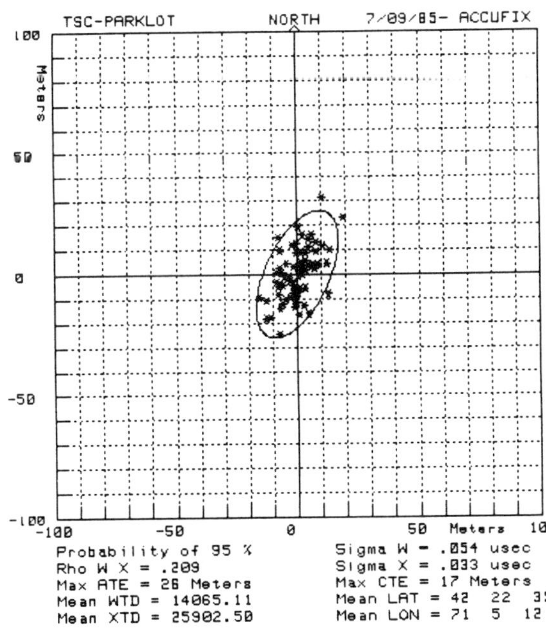


*Fig. 3—Representative LORAN-C Scatter Plots at TSC*

entered. ASF is the TD correction necessary to compensate for propagation variation caused by terrain. If ASF is constant in an area, position outputs will be accurate once the ASF is determined.

Predicted TDs were calculated for each position using the standard USCG EEE-10A program which was provided by the Radio Navigation Division of USCG Headquarters (G-NRN-3). ASF numbers were calculated through differencing observed versus predicted TDs, then an average ASF value was produced for the area. ASF values for the static positions are shown in Table 3.

In Boston, the WX triad produces the best accuracy, with the WY triad producing slightly less accurate fixes. Resultant position variations are shown in Figure 4. The WX triad produced offsets of less than 150 meters, while the WY triad was within 300 meters. The vector diagrams suggest that the grid was warped at the RAYNAV and Lynn Marsh sites. The observed ASF variations were sufficiently large that correct identification of individual streets in urban situations was not possible.

## *UNDERWAY NAVIGATION TESTS*

Thirty controlled navigation transits were performed during the January through March 1985 tests and eight additional transits were performed in August 1985. During these runs navigation data was collected at 1 second intervals for the LTN-700, 10 second intervals for the LORAN-C sets and 2–5 second intervals for the T-Set. Twenty-five of these transits were done on urban streets, 10 were done in rural/suburban areas and 3 were done in a suburban, downtown setting. The total data set for these observations contained more than 500,000 items, with over 10,000 positions. Analysis was primarily accomplished through position plots. When positions strayed from routes, other data, such as signal to noise information, the number of SVs tracked, HDOP, etc., were examined.

### *Highway Transit*

Both GPS and LORAN-C navigators provided accurate navigation information. When the MTF passed under overpasses and bridges, all receivers noted signal attenuation with lower SNR numbers. However, the units continued to track unless the MTF was stopped under the structure. Figure 5 shows representative examples for GPS and LORAN-C sets.

Table 3—ASF Values at Survey Points

| SITE | Position Reference | | ASF (usec) | | |
|---|---|---|---|---|---|
| | LATITUDE | LONGITUDE | W | X | Y |
| ARTERY #10 | N 42°21′59.53″ | W 71°03′49.92″ | 0.67 | 2.07 | 0.39 |
| ARTERY #12 | N 42°22′0.55″ | W 71°03′56.06″ | 0.83 | 1.92 | 0.27 |
| ARTERY #13 | N 42°22′1.13″ | W 71°04′7.30″ | 0.94 | 2.34 | 0.43 |
| TSC | N 42°21′51.00″ | W 71°05′9.00″ | 1.00 | 2.47 | 0.40 |
| BEDFORD COMMON | N 42°29′29.87″ | W 71°16′44.81″ | 0.85 | 1.47 | 0.50 |
| RAYNAV 1957 | N 42°28′36.13″ | W 71°17′15.26″ | 0.02 | 1.83 | 0.15 |
| LYNN MARSHES | N 42°25′51.23″ | W 70°59′49.75″ | 1.11 | 2.94 | 1.45 |
| AVERAGE | | | 0.77 | 2.13 | 0.51 |

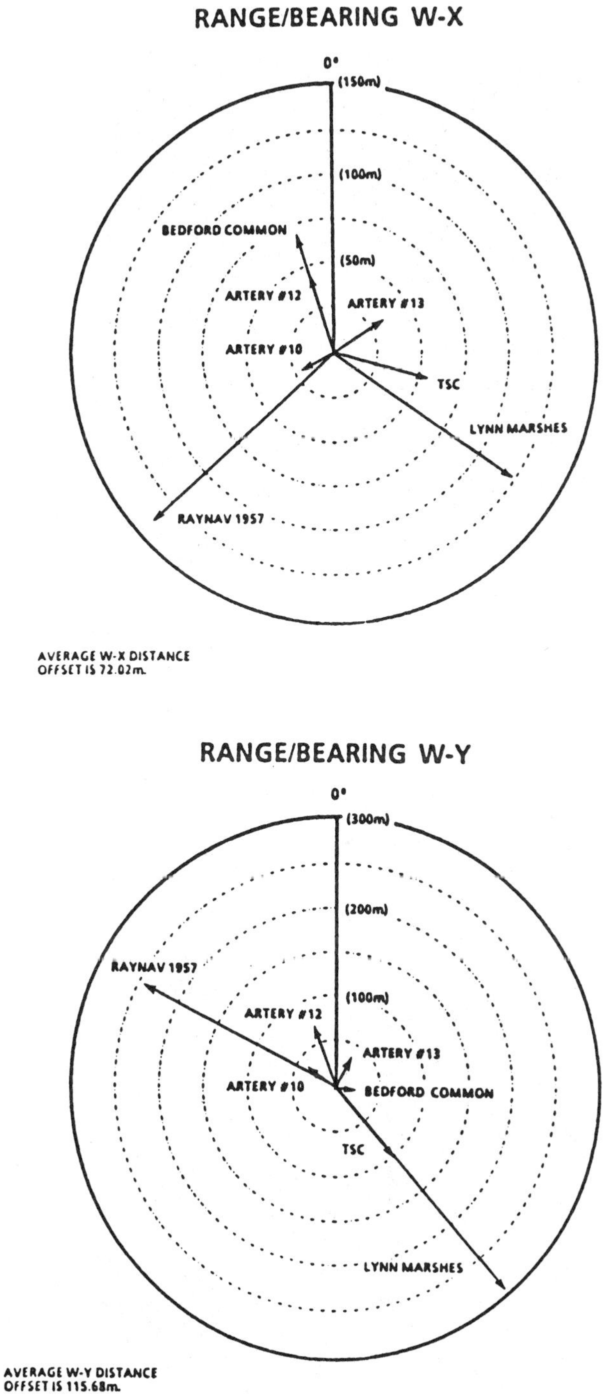


*Fig. 4—Position Variations for Survey Points*

**LORAN-C**

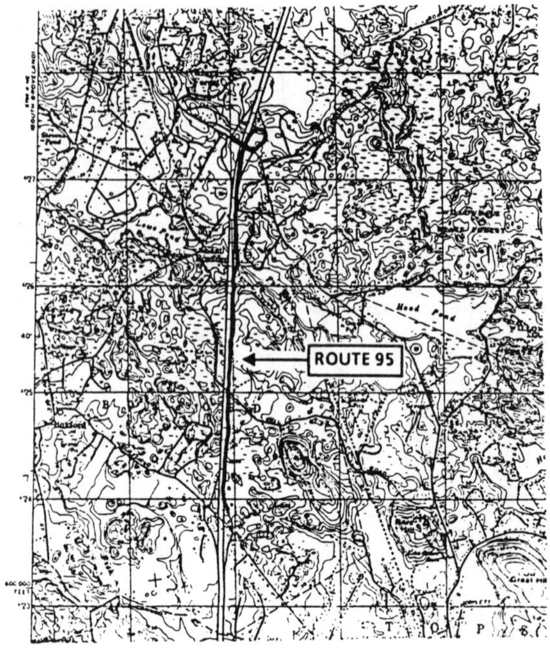


**GPS**

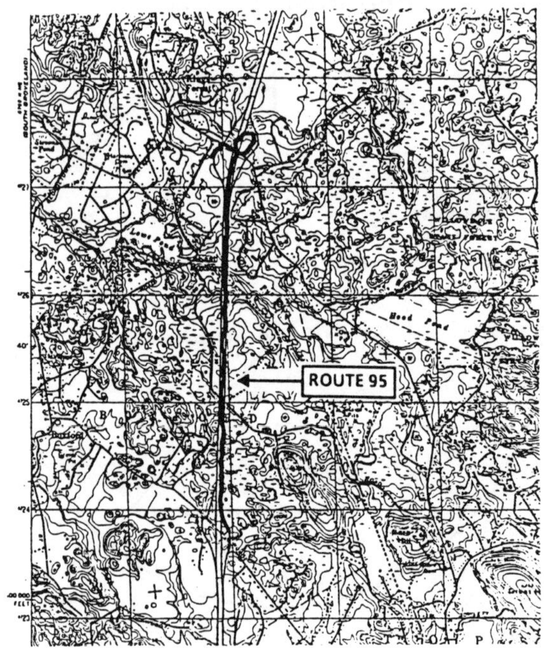


*Fig. 5—Representative Highway Transit Position Plots*

### *Suburban/Rural Road Transits*

Transits were made along Topsfield/Ipswich road. It is a secondary, macadam road, flanked by utility service lines. Both systems provided reasonably accurate track information. For the LORAN-C units, this was a surprise because troubles have been previously encountered due to power lines on this route.

### *Small City (urban)*

Transits were made in downtown Salem, a small city with a well defined business district. The streets featured no power lines and were flanked by 3–4 story buildings. About one third of the transit was over a railroad tunnel. The route included several turns, representative of normal urban transit. Neither system provided a satisfactory navigation track. The LORAN-C receivers maintained lock, but the position reference had errors. The GPS receiver tracked four satellites less than 20 percent of the time, and three satellites, the minimum necessary for a fixed altitude navigation solution, less than 50 percent of the time.

### *Large City (urban)*

The majority of urban transits were made in Boston, where it was possible to get a mix of wide and narrow streets, flanked by structures of varying heights. Data were taken on a number of routes, then two were selected for repeat runs. The first route involved transit of Cambridge Street, continuation on Tremont Street, turning onto Boylston Street and ending adjacent to the Boston Common, a public park. Structures started at 3–4 stories, abruptly changing to high rise, 25 stories, in the Government Center area, 6–15 stories along Tremont Street and were non-existent at the Boston Common. The street width varied from 100 to 150 feet.

A second route involved a loop around Boston Garden, transiting Commonwealth Avenue to Massachusetts Avenue and return. Structures were non-existent at the start, transitioning to 4–6 stories along Commonwealth Avenue and then disappearing upon return to the Boston Gardens. The Ritz Carlton Hotel, a high rise structure, exists at the Southeastern intersection of Commonwealth Avenue and Arlington Street. Commonwealth Avenue is about 200 feet wide, with a center island covered with trees. The routes are shown in Figure 6.

### *GPS*

The GPS sets were not able to maintain an accurate navigation position when driven through areas where the streets were flanked by high rise buildings (greater than 10 stories). Mixed results were obtained in less constricted areas. While six satellites were typically above a 7.5 degree elevation angle, it was apparent that buildings often blocked signals. On Tremont Street, the high rise route, four satellites were tracked less than 50 percent of the time, and three less than 70 percent. By restricting the LTN-700 to an altitude hold fix (3 satellites), it was possible to make reasonably accurate runs on Commonwealth Avenue (3 satellites available 75–95 percent of the time). The Ritz Carlton Hotel blocked signals on Arlington Street and the GPS track skipped to the

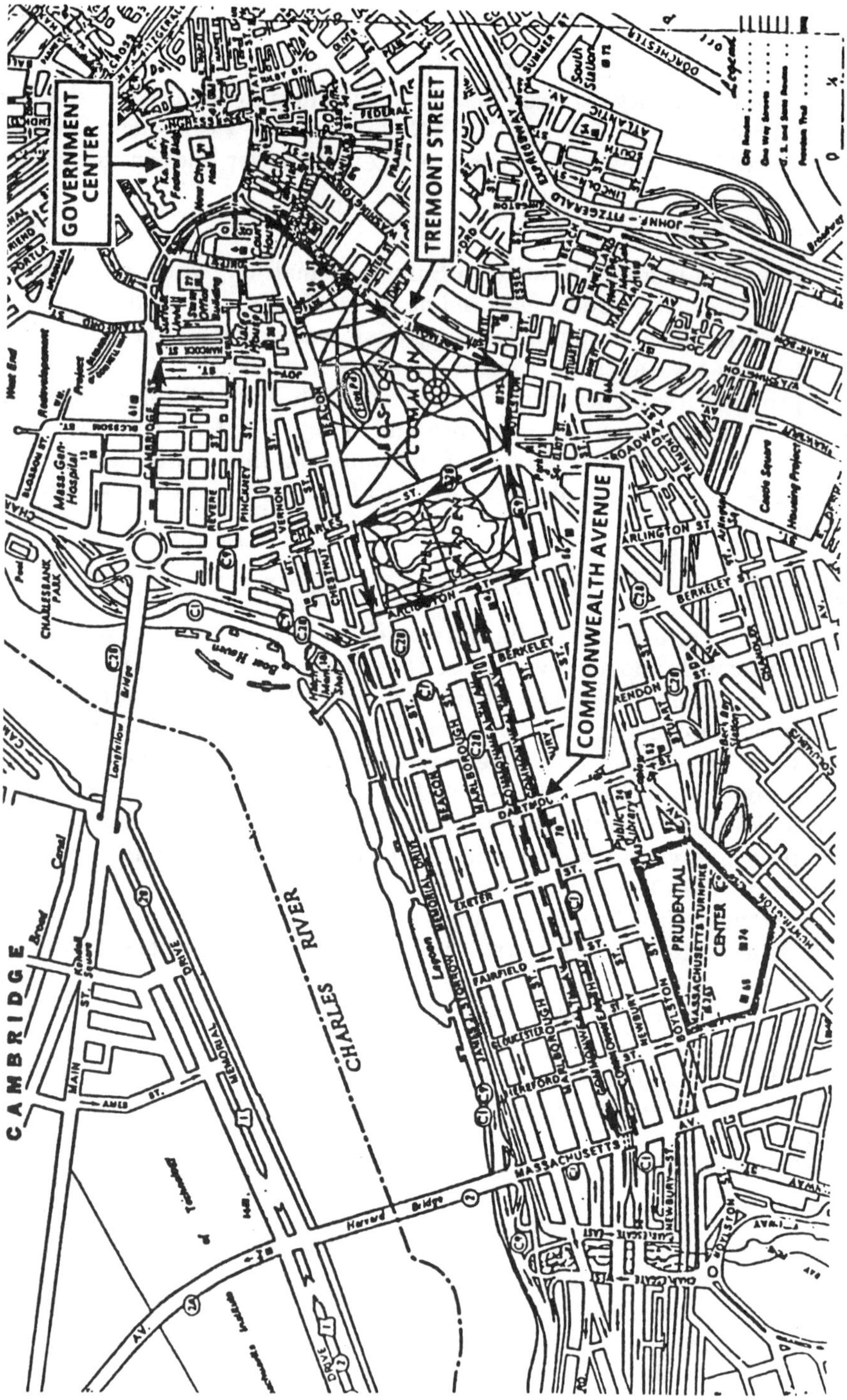


*Fig. 6—Navigation Transit Test Routes in Boston*

other side of the Boston Garden. Representative position plots are shown in Figure 7.

### *LORAN-C*

The LORAN-C sets were able to maintain lock and provide a reasonably accurate navigation reference except in the Government Center area. This was part of the Tremont Street run, characterized by high rise buildings and underground subways. Both LORAN-C receivers indicated a drop in SNR of over 20 dB in the Government Center area. Representative position plots are shown in Figure 8. A transit from one of the survey markers on Nashua street is included with the Tremont Street run. The Ritz Carlton did not significantly shift the LORAN-C receiver positions.

## INTERFERENCE TESTS

A series of controlled observations were made to detect receiver sensitivity to emissions in the vicinity of power substations, UHF/VHF television transmitting sites and at modular telephone transmitter sites (approximately 800 mHz). The LORAN-C receivers lost lock and indicated low SNR when within 400 feet of the power station, but recovered as the MTF left the area. The GPS receivers indicated no problem at any of the sites.

## FOLIAGE TESTS

No foliage of significance was present for the January through March tests. The T-Set was tested for 1 day at a number of sites along Lockwood Lane in Topsfield, MA. Foliage, from a mixture of oak, pine and maple trees, varied from none to a leafed arch, 50 feet off the road. Signal attenuation was readily apparent, with SNR levels dropping 7 to 10 dB when satellite elevation was less than 40 degrees. Rising satellites were not seen until above 20 degrees, with the signal strength at initial acquisition approximately 10 dB lower than baseline, unobstructed measurements. These tests and results are considered preliminary because of the limited scope of the tests. Additional tests will be conducted at another time.

## DISCUSSION OF OBSERVATIONS AND RESULTS

### *GENERAL*

Because of differing ellipsoid reference systems, the task of melding an area radionavigation system like LORAN-C or GPS with local maps is formidable. These tests were specifically designed to minimize translation effects. The evaluation stressed establishing the capabilities of the radionavigation systems to provide accurate fixing in a terrestrial environment.

Problems with the navigation solutions for both systems were noted in the Government Center area. Receivers detected a decrease in SNR (LORAN-C) or signal blockage (GPS). In view of the great difference in system characteristics, the coincidence of difficulty is noted, but no attempt to establish a common cause was made. Additional investigation is indicated if either system is seriously considered for application in the Boston area.

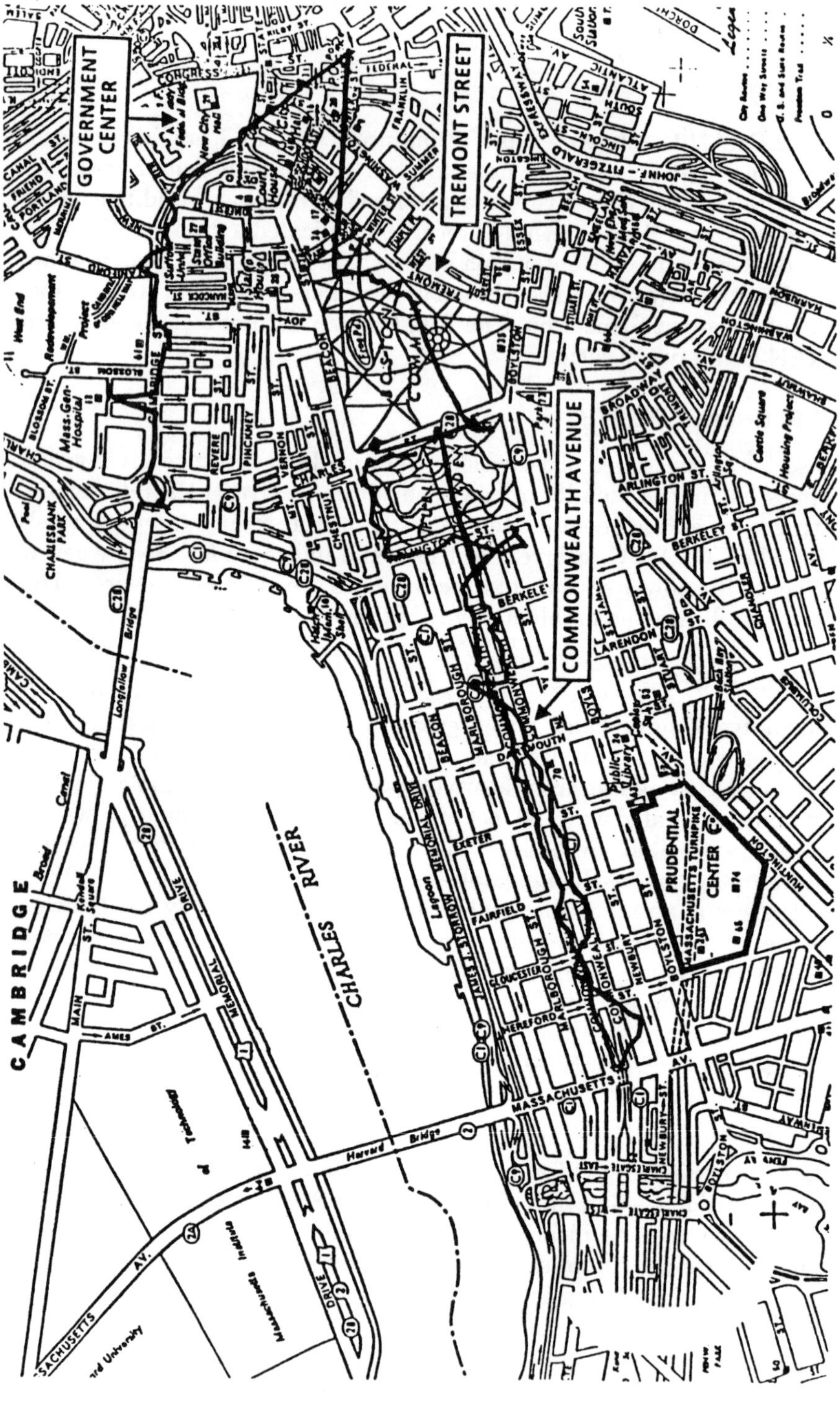


*Fig. 7—Representative GPS Position Plots in Boston*

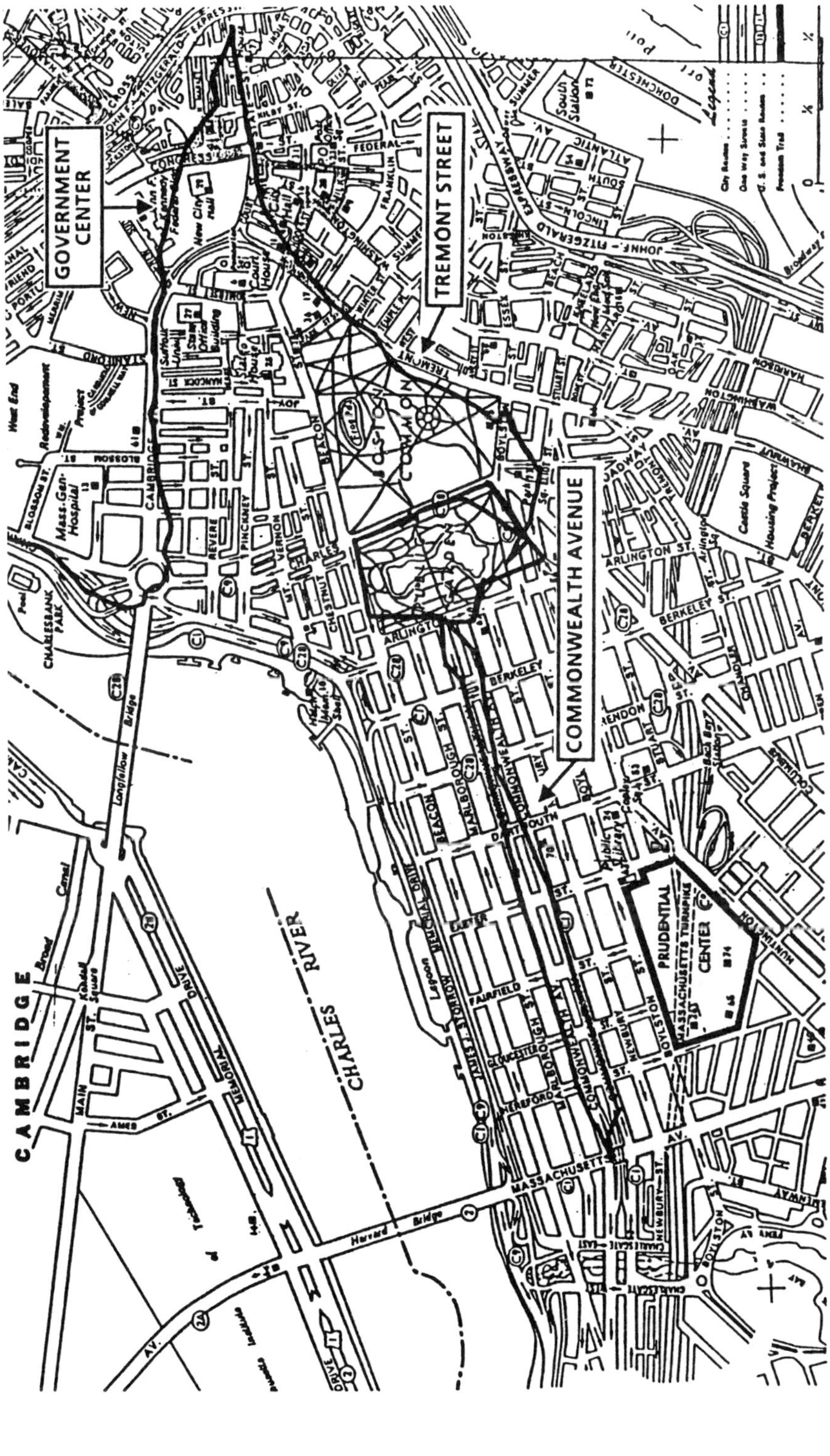


Fig. 8—Representative LORAN-C Position Plots in Boston.

### GPS

The navigation test runs on highways and rural roads were excellent, indicating that the system may provide an excellent capability. The tests also showed that a satellite visibility problem exists in the Boston and Salem urban environments with the current satellite constellation. A question which arises when conducting tests with the Block 1 satellites is the applicability of results to the operational system. Subsequent to the data analysis, satellite visibility profiles of the 18 and 21 GPS satellite configuration for the Boston area were developed. These are shown in Figures 9 and 10. Calculations were made at 30 minute intervals. In order to "see" four satellites, an unobstructed horizon to the 20 degree level is required. Some improvement is realized if the three spare satellites are operational and in the correct orbits. Three or more satellites are available if the receiver encounters an unobstructed horizon of 30 degrees.

The signal blockages with corresponding loss of position noted during these urban tests are, therefore, representative of what will be encountered during operations with the final system. It appears that sensor inputs for velocity, direction and altitude will be necessary for an urban navigator which uses the GPS as its principal navigation reference.

The signal attenuation observed during foliage tests also may be significant. Additional foliage tests will be conducted in 1986.

### LORAN-C

Tests of the LORAN-C system continue to show that use of the system for terrestrial navigation is feasible when coverage is adequate. Two results will be noted.

### ASF

The ASF variations, hence position variations, determined in these tests, fall well within the maritime accuracy goal of 1/4 nautical mile for coastal navigation, but are too large to permit identification of individual urban streets in Boston and Salem. Establishment of multiple calibration points throughout a terrestrial service area might keep the accuracy at acceptable levels.

### Power Station Interference

These tests reaffirmed the fact that LORAN-C signals are masked by noise when in the vicinity of utility stations. No effort was made to investigate sensitivity of the receivers to power lines along roadways. Power lines were buried where urban transits were made during these tests.

## ACKNOWLEDGEMENTS

Throughout the test period, Mr. David Pietrazewski, of the Coast Guard Research and Development Center, provided daily status reports on the GPS system. This information was used to ensure that any problems noted by the receivers were caused by local phenomenon rather than the GPS system. LCDR Robert Gazlay, of the Radionavigation Information Branch, USCG Headquarters (G-NRN-3) provided the computer programs which were used to predict

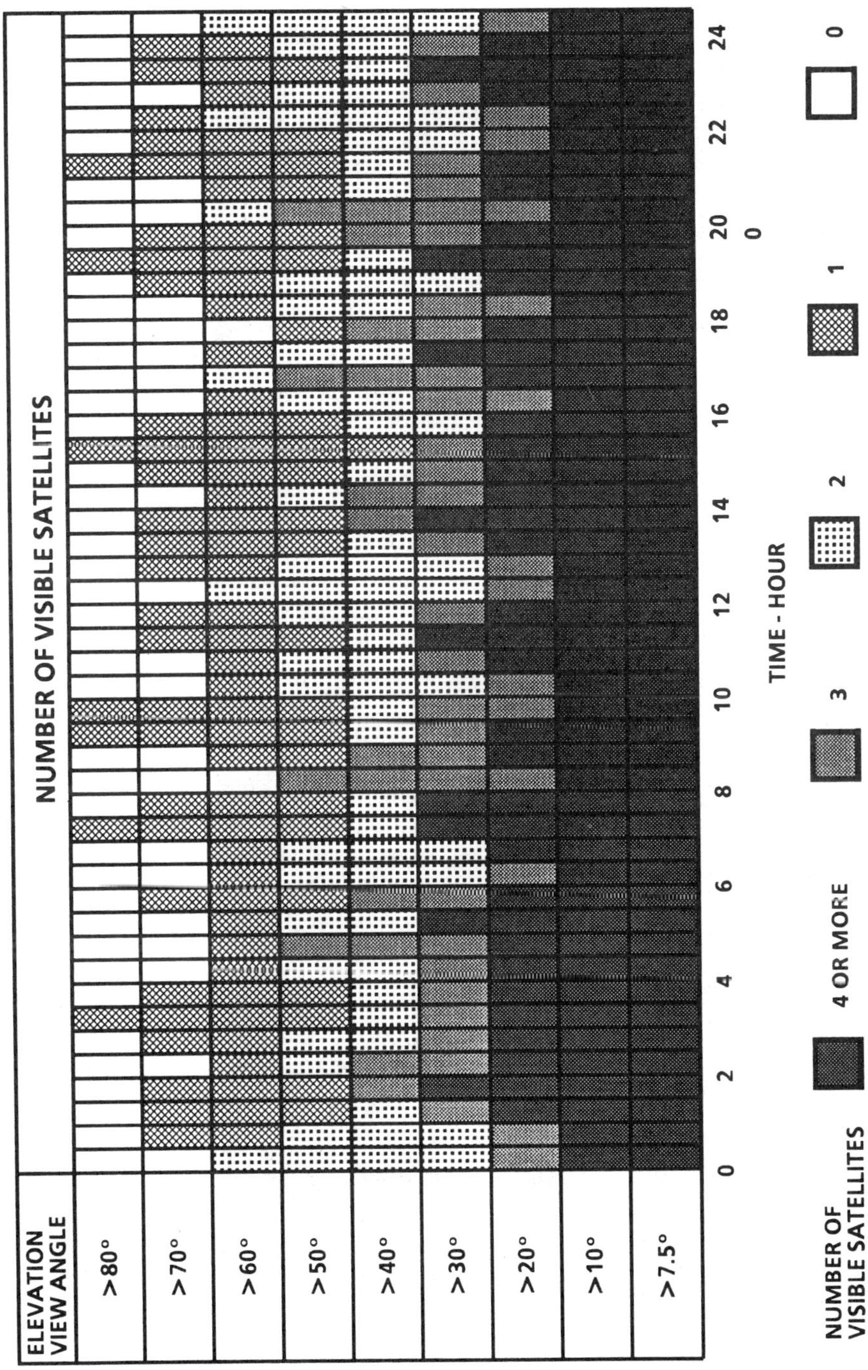


*Fig. 9—GPS Satellite Visibility in Boston (18 Satellite Configuration)*

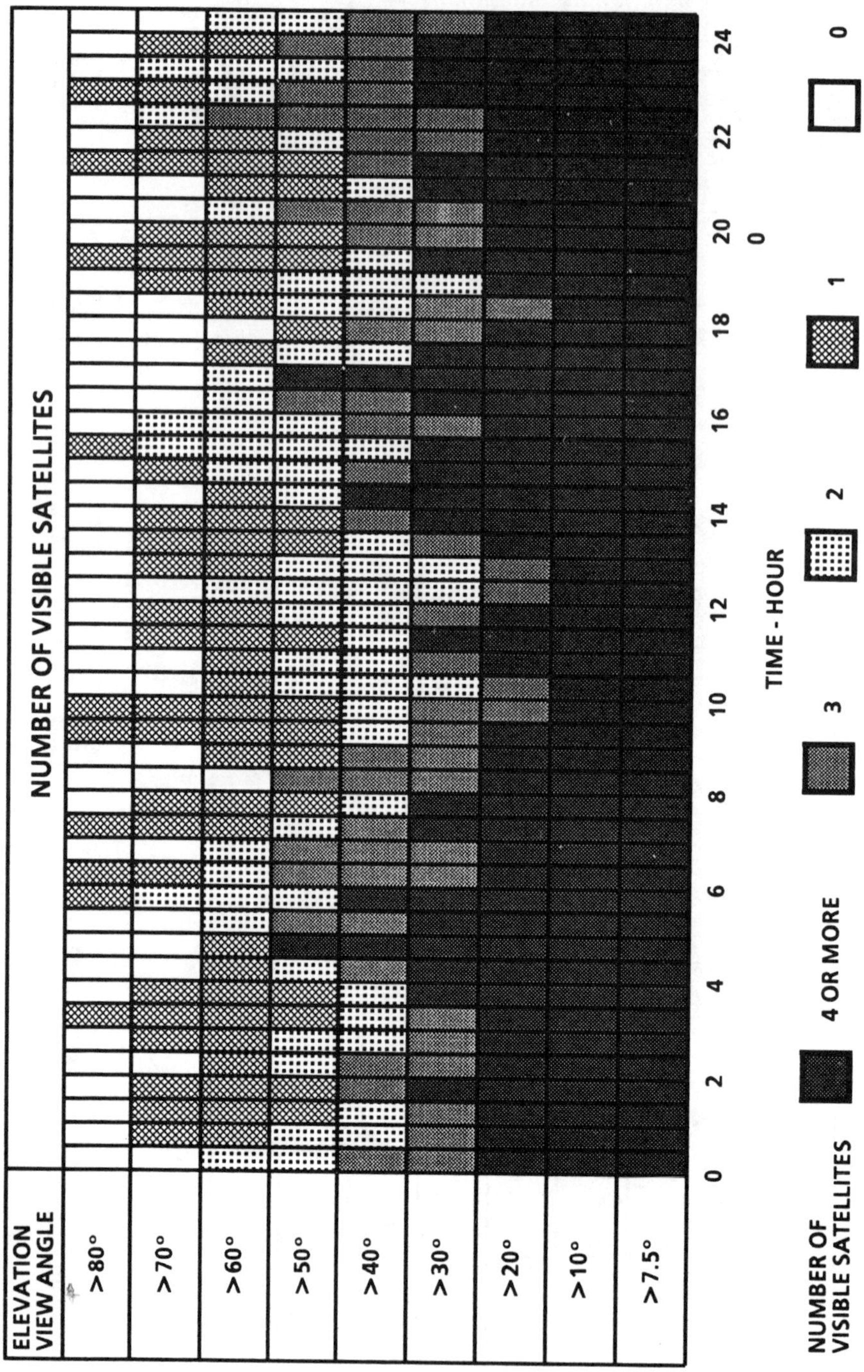


*Fig. 10—GPS Satellite Visibility in Boston (21 Satellite Configuration)*

LORAN-C time differences as well as those necessary to transform positions from one coordinate system to another. CWO Daniel Slagle, of the Coast Guard Research and Development Center, provided the computer programs, normally used for the LORAN-C Harbor Monitor Program, which were used to analyze static data. The assistance of these people facilitated the conduct of these tests and their efforts are gratefully acknowledged.